Auditory and Phonetic Processes in Speech Perception

Auditory and Phonetic Processes in Speech Perception

Editors

Richard Wright
Benjamin V. Tucker

MDPI • Basel • Beijing • Wuhan • Barcelona • Belgrade • Manchester • Tokyo • Cluj • Tianjin

Editors

Richard Wright
Department of Linguistics
University of Washington
Seattle
United States

Benjamin V. Tucker
Department of Communication
Sciences & Disorders
Northern Arizona University
Flagstaff
United States

Editorial Office
MDPI
St. Alban-Anlage 66
4052 Basel, Switzerland

This is a reprint of articles from the Special Issue published online in the open access journal *Brain Sciences* (ISSN 2076-3425) (available at: www.mdpi.com/journal/brainsci/special_issues/phonetic_speech).

For citation purposes, cite each article independently as indicated on the article page online and as indicated below:

LastName, A.A.; LastName, B.B.; LastName, C.C. Article Title. *Journal Name* **Year**, *Volume Number*, Page Range.

ISBN 978-3-0365-7413-4 (Hbk)
ISBN 978-3-0365-7412-7 (PDF)

Contents

brain sciences

MDPI

Article

Differences between Monolinguals and Bilinguals in Phonetic and Phonological Learning and the Connection with Auditory Sensory Memory

Laura Spinu [1,2,*], Jiwon Hwang [3] and Mariana Vasilita [2]

1 CUNY-Kingsborough Community College, Brooklyn, NY 11235, USA
2 CUNY-The Graduate Center, New York, NY 10016, USA
3 Stony Brook University, Stony Brook, NY 11794, USA
* Correspondence: laura.spinu@kbcc.cuny.edu; Tel.: +1-718-368-5296

Abstract: Bilingualism has been linked with improved function regarding certain aspects of linguistic processing, e.g., novel word acquisition and learning unfamiliar sound patterns. Two non mutually-exclusive approaches might explain these results. One is related to executive function, speculating that more effective learning is achieved through actively choosing relevant information while inhibiting potentially interfering information. While still controversial, executive function enhancements attributed to bilingual experience have been reported for decades. The other approach, understudied to date, emphasizes the role of sensory mechanisms, specifically auditory sensory memory. Bilinguals outperformed monolinguals in tasks involving auditory processing and episodic memory recall, but the questions whether (1) bilinguals' auditory sensory memory skills are also enhanced, and (2) phonetic skill and auditory sensory memory are correlated, remain open, however. Our study is innovative in investigating phonetic learning skills and auditory sensory memory in the same speakers from two groups: monolinguals and early bilinguals. The participants were trained and tested on an artificial accent of English and their auditory sensory memory was assessed based on a digit span task. The results demonstrated that, compared to monolinguals, bilinguals exhibit enhanced auditory sensory memory and phonetic and phonological learning skill, and a correlation exists between them.

Keywords: bilingualism; auditory sensory memory; phonetic and phonological learning

Citation: Spinu, L.; Hwang, J.; Vasilita, M. Differences between Monolinguals and Bilinguals in Phonetic and Phonological Learning and the Connection with Auditory Sensory Memory. *Brain Sci.* **2023**, *13*, 488. https://doi.org/10.3390/brainsci13030488

Academic Editors: Naseem Choudhury and Guillaume Thierry

Received: 11 January 2023
Revised: 6 February 2023
Accepted: 8 March 2023
Published: 14 March 2023

1. Introduction

For decades, the psycholinguistic literature has reported the existence of a bilingual cognitive advantage [1,2] whereby bilingual language experience is thought to enhance cognitive functions and ultimately contribute to cognitive reserve. However, the bilingual advantage has polarized academics as a controversial, difficult to replicate phenomenon, earning the nickname of "Loch Ness monster" [3,4]. In a different meta-analysis based on 46 original research studies, Van den Noort et al. [5] report that a majority of the articles on the topic (54.3%) found beneficial effects of bilingualism on cognitive control tasks; however, 28.3% found mixed results and 17.4% found evidence against its existence. Following DeLuca et al. [6], we take the position that bilingual effects on cognition exist, but they are conditional. It is no coincidence that the 2021 meeting of the world's largest conference on bilingualism, the International Symposium on Bilingualism, hosted two theme sessions entitled *Biases in research: Who counts as 'authentic' bilingual speaker—and how can we tell?* and *Language proficiency measures—what exactly are we measuring?* Several reasons, both methodological and conceptual in nature, have been invoked as potentially underlying the conflicting bilingual advantage findings [7,8]. These include individual differences such as talent [9], language-pair factors [10], the fact that the bilingual advantage may be most prominent during early and late stages of life, but less noticeable during adulthood [11], and experimental task complexity across studies [2]. Among these factors, the fact that all speakers have access to non-linguistic ways of improving cognitive function, the lack of a well-defined operational description of bilingualism, and the omission of lower-level, sensorimotor functions in considering the relationship between language and cognition have received heightened attention in recent literature. It is the third aspect we address in more detail in the current paper.

As mentioned above, numerous studies on bilingual cognition have explored the potential advantages associated with bilingualism on executive function. As Poarch and Krott [12] explain, the view that bilingualism has cognitive benefits is based on the theoretical assumption that bilingual individuals experience constant cross-linguistic activation and interaction during language processing [13,14]. To enable the use of the correct language in a given context, the need arises for a cognitive control mechanism permitting speakers to resolve the conflict between languages that are actively competing with each other. Such a cognitive control mechanism already exists for non-verbal processing, specifically executive function(s) [7,15]—also referred to generically as cognitive control. Executive function refers to a set of processes considered necessary for the cognitive control of behavior, including (in most models) attentional control, inhibitory control, working memory, and cognitive flexibility or shifting. Because frequent switching between languages is speculated to employ this mechanism, the expectation arises that the more this happens, the greater the enhancement of cognitive function [16]. Miyake et al. (2000) investigated the separability of shifting, updating, and inhibition, reporting that these three executive functions have differential contributions to performance on complex frontal lobe tasks [17]. Given that the frontal lobes are involved in language processing [18,19] and brain adaptations have been observed in the frontal regions in bilinguals [20], executive functions are likely to be involved in multiple aspects of language learning [6], though other mechanisms are likely to be involved as well.

Over a decade ago, Simmonds et al. posed the question why previous bilingualism research had largely ignored sensorimotor aspects of learning [21]. Indeed, an understudied area of research pertaining to bi- and multilingualism is their connection with cognitive aspects outside of the frequently explored set of executive functions. Because, as shown in Figure 1, language experience involves extensive use of sensorimotor mechanisms [21,22], such as motor (articulatory) control, somatic memory, and auditory sensory memory (iconic memory in the case of signed languages), the question arises whether these lower-level functions are also enhanced by bilingual experience outside of one's native language. Furthermore, if that is the case, the contribution of sensorimotor functions to cognitive function and whether a connection exists between sensorimotor and executive functions also needs to be clarified. Lindenberger (1994) and Lindenberger et al. (2000) posited a connection between the two in the cognitive permeation hypothesis [23,24], noting that sensorimotor aspects of behavior are more attention-demanding in older adults than in young adults, which leads to increased competition between sensorimotor and cognitive tasks for scarce attentional resources. Reviewing the research on the coupling between sensorimotor and cognitive aging, Schäfer et al. (2006) conclude that they are causally related and functionally interdependent and that age-associated increments in cognitive resource demands of sensorimotor functioning are malleable by experience [25]. Their recommendation is for future studies to attempt to shed further light on functional and etiological links between sensorimotor and cognitive aging and their interaction.

Exploring the connection between sensorimotor and cognitive functions also has the potential to shed more light on a phenomenon that has received heightened attention recently, specifically phonetic and phonological learning. Experimental research has shown that bilingual individuals (of various backgrounds) tend to outperform monolinguals in tasks requiring them to produce or perceive novel sounds or accents of a known language. For instance, ref. [2] trained monolinguals and bilinguals on vocabularies differentiating words that contained foreign phonetic contrasts. Their findings suggested a bilingual advantage in phonetic learning, which is influenced by the level of difficulty of the specific phonetic contrast being learned and by the similarity between the learners' native language and the target language (a similar conclusion was drawn by [26] in their study that investigated the acquisition of rhotics longitudinally). In a study focusing on non-native contrasts, ref. [27] report enhanced speech perception abilities in multilinguals and bilinguals compared to monolinguals, whose ability to discriminate a non-native contrast did not differ from that of the bilingual and multilingual group before training).

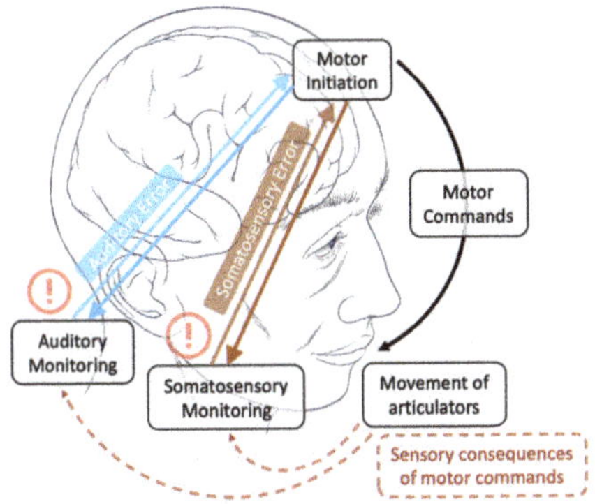

Figure 1. Sensorimotor systems involved in speech (adapted here from [21]).

Using a more naturalistic approach focusing on the global learning of a different accent and thus expanding on the production or perceptual discrimination studies employing sounds in isolation, ref. [28] compared Canadian monolinguals and bilinguals in an experiment that involved two tasks: imitating and spontaneously reproducing a novel foreign accent spoken in Sussex, England. The target sound (i.e., the glottal stop), which was already present in the speakers' production, was mapped differently to surface forms in the novel accent (i.e., as the only allophonic realization of word-final coronal stops). The results suggest more effective learning in bilinguals. Although the two groups performed very similarly during the training when they were asked to imitate what they heard immediately, bilinguals produced their glottal stop significantly more frequently than monolinguals during the post-training session. A follow-up study [29] employed a novel accent that was artificially created to have four phonetic features that differed from standard American English. The decision to use an artificially constructed accent instead of a natural one was made to allow better control over the measurements of the input and the output in the experiment. Early bilinguals of various language backgrounds consistently outperformed monolinguals . These findings are in line with a bilingual advantage found in phonetic and phonological learning that is robust enough to override the various issues speculated to cause conflicting results in the executive function studies discussed previously. Departing from the more widespread executive function work, we address the question whether a link exists between phonetic and phonological learning and auditory sensory memory. Given the complexity of phonetic and phonological learning, we expect it to be underlain by multiple mechanisms, including executive function, but in the current paper we narrow down the investigation to auditory sensory memory precisely because this connection has been understudied to date.

Turning to the work on auditory sensory memory, the digit span task (with a suffix) is a paradigm commonly employed to investigate this type of memory in behavioral studies. The suffix effect, as described by previous studies [30,31], refers to the difficulty in recalling a spoken sequence caused by the addition of an irrelevant speech item at the end. Typically, participants are presented with sequences of digits or letters that are arranged in a random order, followed by either a silent interval [32] or a suffix of equivalent duration (e.g., the word "go"). When compared to the items followed by a silence, the items closest to the suffix display an increase in errors, with the final item showing the largest increase in errors. This is in contrast with near-perfect performance in the control condition. Replicated consistently in a variety of studies, the suffix effect is thought to reflect an automatic type of processing that is characteristic of the functioning of auditory sensory memory [33–35].

It should be added that other types of memory, such as working memory, are likely active in digit span recall [36]. It is believed that information about the stimulus heard most recently can be accessed simultaneously by both auditory sensory memory and working memory. As a result, it can be challenging to differentiate the effects of auditory sensory memory and those of working memory processes, such as rehearsal, long-term retrieval, or chunking [37]. However, empirical studies have been able to distinguish the separate effects of working memory rehearsal and auditory sensory memory to digit span recall, along with their accompanying theoretical interpretations [38,39]. The general view has been that auditory serial recall tasks enable the separation of performance effects resulting from working memory rehearsal, which affects the first items in a longer list (i.e., primacy effects), from performance effects resulting from auditory recency, which applies to the last items in a list (i.e., recency effects). Based on this, we consider performance on the terminal items of a list mainly to reflect the working of auditory sensory memory, while not excluding the possibility of interference from additional mechanisms interacting with it, such as working memory where they need to hold incoming L2 information while decoding it. One should note, however, that recent work by Sofologi et al. [40] showed no differences in working memory between monolingual and bilingual students of the same age, while at the same time finding a bilingual advantage in inhibitory control and cognitive change. The authors conclude that when learning a (first or second) language, working memory does not correlate to all executive functions but forms a separate cognitive function. These findings are supported by Yang's 2017 study [41], which concluded that knowing two languages does not guarantee bilingual working memory advantages over monolinguals, but the advantage might be linked to bilinguals' unique L2 use environment. On the other hand, the relationship remains unclear: Morales et al. [42] found an advantage for bilingual children in working memory that was especially evident when the task contained additional executive function demands.

While research has shown a bilingual advantage in tasks involving auditory processing [43] and episodic memory recall [44], very few studies have investigated auditory sensory memory in the context of bilingualism. Philipp-Muller et al. [45] administered a digit recall task and used an algorithm to analyze the digit recall data and examine the mechanism underpinning the differences in memory performance in bilingual and monolingual participants. The Rational Transpositional Error Algorithm (RTEAlgorithm) showed that bilinguals made significantly fewer transpositional errors than monolinguals in the recall task. This study, however, did not specifically investigate performance on the terminal items of digit sequences and therefore its findings are not conclusive with respect to auditory sensory memory. More recently, ref. [46] administered a suffixed adaptive digit span task to bilinguals and monolinguals from the undergraduate population of the University of Toronto, and compared them in overall accuracy, accuracy by serial position, maximum number of digits recalled, and the percentage of participants who reached the longest digit span. The results showed that bilinguals have longer digit spans and higher accuracy than monolinguals across all serial positions within every list length. This suggests an advantage for bilinguals not only in terms of recently heard items, which are attributable to auditory sensory mechanisms (known as recency effect), but also for the items heard at the beginning of longer list lengths, which are owed to working memory (known as primacy effect). While [46] concluded that bilingual experience results in enhanced auditory sensory memory, further studies are needed to consolidate this finding and to explore the connection between this type of memory and phonetic and phonological learning, especially as the former has been suggested to have a significant role in the latter [18].

In sum, based on the research on phonetic and phonological learning and auditory sensory memory, which were both found to be enhanced in bilinguals, it is plausible to assume a link between the two, and further speculate that the mechanism supporting phonetic and phonological learning is partially supported by the work of auditory sensory memory. The experiment described in the following sections addresses the possible existence of a correlation between phonetic and phonological learning and auditory sensory memory.

2. Experiment: Materials and Methods

The aim of the current study was to address the prediction put forth in the previous section, which postulates a link between phonetic and phonological learning and auditory sensory memory, an experiment was designed to include a novel accent learning task, following [28,29] and a digit span task with a suffix [46]. The experiment was conducted with monolingual and bilingual speakers in person in a quiet room, inside a sound-attenuated booth, on the CUNY Kingsborough Community College campus and comprised the following parts: a language background questionnaire, a translation task for bilingual participants (from English into their other language)—not discussed here, a novel accent learning task that included three blocks (i.e., baseline, training, and testing), and a digit span task with a suffix. Preliminary findings of this study (covering a subset of the participants and only the results obtained for the phonetic and phonological learning task) were reported in [29].

2.1. Hypotheses

Our predictions are primarily based on previous findings suggesting that there is an advantage for bilinguals in phonetic and phonological learning [2,27–29] and in serial memory tasks [45], including those specifically focusing on auditory sensory memory [46].

Hypothesis 1. *Bilinguals will outperform monolinguals on the phonetic and phonological learning tasks.*

Hypothesis 2. *Bilinguals will display enhanced auditory sensory memory compared to monolinguals.*

Hypothesis 3. *A significant correlation exists between auditory sensory memory and phonetic and phonological learning.*

2.2. Language Background Questionnaire

Participants were individually administered an abbreviated version of the LEAP-Q questionnaire [47]. Following [46], participants who were included in the bilingual group met two primary criteria: (1) self-reported native or near-native proficiency level in both languages, and (2) exposure to both languages prior to school age (i.e., 6–7 years). Monolinguals were defined as individuals who reported speaking English natively and, in some cases, a second language at a level of conversational or beginner proficiency.

2.3. Testing Phonetic and Phonological Learning: The Novel Accent Learning Task

2.3.1. Stimuli

An artificial accent of English (henceforth Model Speech), was created such that it differed in four distinct ways from standard North American English (Figure 2):

1. **Tapping:** intervocalic /l/ → [ɾ] e.g., 'color' →[kʌɾɚ]
2. **Diphthongization:** the vowel /ɛ/ → [jɛ] after an onset consonant, e.g., 'bed' → [bjɛd]
3. **Vowel epenthesis:** voiceless clusters of the form sC̥ → səC̥ e.g., 'spy' → [səp^{h}aj]
4. **Intonation change:** tag questions were realized with a novel Mid-Low-High (MLH) pattern. Tag questions (e.g., *isn't it?*) are typically produced with either rising or falling intonation in standard American English.

The stimuli consisted of short sentences containing either one single feature e.g., *You make a good spy,* where *spy* was realized as [səp^{h}aj] (epenthesis), two features combined e.g., *She put a spell on him,* where [spɛl] was realized as[səp^{h}jɛl] (epenthesis and diphthongization), or all four of them (e.g., *You set the speed alone, didn't you?* where the vowel in the word *set* was diphthongized, epenthesis occured in the word *speed*, tapping affected the [l] in *alone*, and the tag question *didn't you?* was realized with a MLH contour).

The features were distributed as follows: 20 tapped /l/, 20 diphthongized vowels, 20 epenthesized vowels and 10 tag questions. The reason we included a lower number of tag questions compared to the other novel features was that they were found impressionistically to be highly salient and their presence in higher numbers was deemed to have a distracting effect on the listeners.

The total list of stimuli comprised 40 sentences (of which 20 contained single features, 15 contained combinations of two features, and 5 contained all four features). A highly trained monolingual female phonetician recorded the full list of stimuli using the Model Speech and also in her natural Northeastern US accent (for comparison). The consistent presence of all novel features in the artificial accent was verified acoustically (see Figure 2).

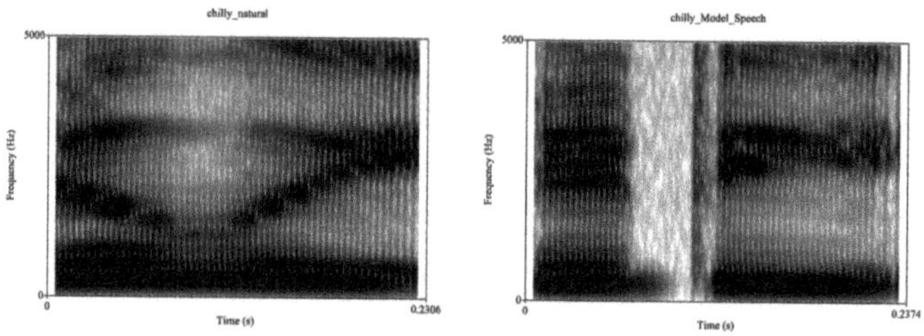

Tapping: [ɪli] (baseline) and [ɪɾi] (Model Speech)

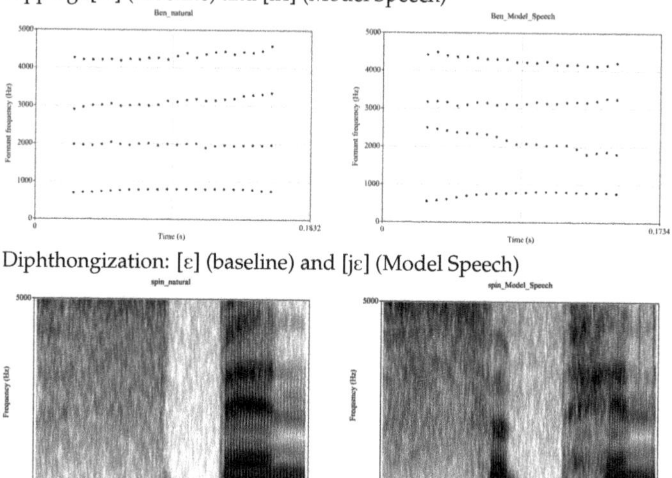

Diphthongization: [ɛ] (baseline) and [jɛ] (Model Speech)

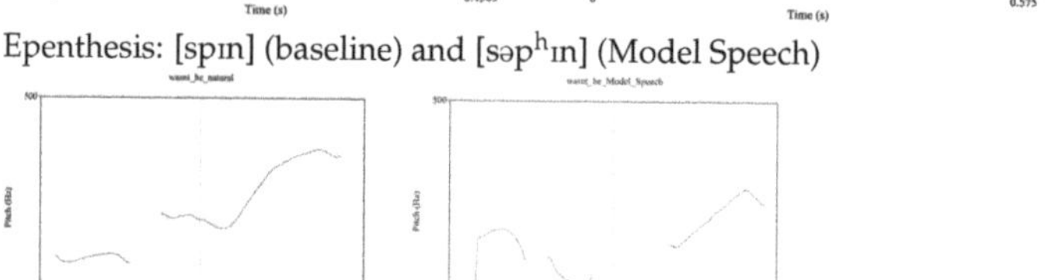

Epenthesis: [spɪn] (baseline) and [səpʰɪn] (Model Speech)

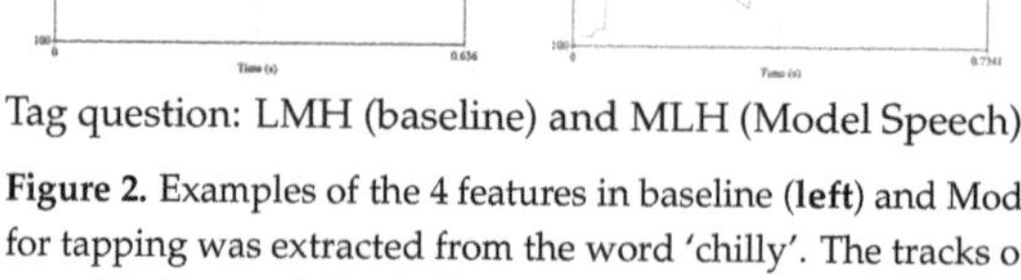

Tag question: LMH (baseline) and MLH (Model Speech)

Figure 2. Examples of the 4 features in baseline (**left**) and Model Speech (**right**). The VCV sequence for tapping was extracted from the word 'chilly'. The tracks of formants 1–4 are obtained from the vowel in the word 'Ben'. The spectrograms for epenthesis are obtained from the word 'spinning'. Pitch tracks for the sequence 'wasn't he?' illustrate the intonation change feature.

2.3.2. Procedure

The experimental procedure for this task started with the recording of 40 baseline sentences containing all structures of interest, followed by a two-part training phase. In the first part, participants listened to 40 sentences spoken in the Model Accent continuously, in the absence of orthographic input. In the second part, they listened to each of the same 40 sentences and were asked to immediately imitate it in the novel accent (see [48] for the role of imitation in phonetic and phonological learning), while also being able to see its orthographic transcription on a computer screen. In the testing phase, they read the baseline sentences again, this time aiming to reproduce the novel accent without any audio prompts. This task was administered using PsychoPy [49].

2.3.3. Data Processing and Analysis

Data processing consisted of categorical judgments provided by the same trained monolingual phonetician who recorded the Model Speech sentences (Note: the rater is not an author and had obtained her PhD prior to her collaboration on this study). The absence or presence of each target feature was scored with a 0 or 1, respectively, resulting in a mean accent score for each participant and for each block, as well as an overall score per participant averaging over the three blocks (baseline, training/imitation, and testing). While the scoring process was not blind, with the rater having access to language background information for the participants, the judgments were based on spectrographic evidence (as shown in Figure 2) and not on impressionistic data. While we anticipate conducting a number of acoustic analyses to be reported in a future study, including measurements of continuous parameters such as duration, pitch and formant values, as well as other pertinent measures for each of the four features employed, the current study is based on the categorical ratings only. The statistical analyses we conducted for the current study include a series of ANOVAs that compared various aspects of the two groups' performance across the different features, blocks, and sentence types (that is, containing 1, 2, or 4 features together), detailed in the following sections. See Appendix A for a detailed description of the variables employed.

2.4. *Testing Auditory Sensory Memory: The Digit Span Task with Suffix*

2.4.1. Stimuli

The stimuli consisted of sequences of digits varying in length (from a minimum of 2 digits to a maximum of 9). After each digit sequence, the word "recall" was presented, which served as a suffix. Both the digits (1 through 9) and the suffix (i.e., "recall") were generated using a natural-sounding synthetic male voice. The task was adaptive, presenting digit sequences of a specific length in blocks of five trials each. For example, a listener was first presented with 5 trials of 2-digit sequences, then 5 trials of 3-digit sequences, and so on and so forth, until they were no longer able to correctly recall at least 3 out of the 5 trials within a block. At that point, the task was terminated. Thus, the task could end earlier for some listeners compared to others, depending on their performance.

2.4.2. Procedure

The default template of PsyScope [50] digit span was modified to construct this task. The task was designed to be adaptive, beginning with a practice block of two digits and progressing to longer sequences if the participant accurately recalled at least three of the five trials at each sequence length. As a result of the adaptive nature of the task, the highest sequence length achieved varied among participants, resulting in a different number of blocks presented depending on their individual memory capacity.

2.4.3. Data Processing and Analysis

The software (PsyScope) automatically generated scores for each sequence, including the number of correct and incorrect responses and the maximum digit sequence length reached by each participant. Overall accuracy for each serial position for the longer digit sequences was subsequently obtained. A MATLAB script [51], specifically developed to compare the digits presented at each serial position with the participants' response and determine accuracy based on whether a match was found was also used. The algorithm searched for insertions or deletions by aligning a participant response string and the input string presented and counting the number of digits in each string to see if there was a discrepancy. If the number of digits in the response string was equal to the number of digits in the input string, then the answer was included in the analysis, but if the number of digits was not equal between the response and input strings, the answer was excluded. The algorithm evaluated the responses that were included digit-by-digit, employing a graded scoring method that assigned weighted scores based on transpositional distance. The goal of this graded scoring system was to award a higher score to transposed response digits that were closer to their original position in the participant response.

The z-scores were used to compare the proportion of participants from each group who were able to reach the longest digit sequence (i.e., nine digits). In a series of ANOVAs, group (monolingual/bilingual) and sequence length (2 through 9) were included as the independent factors and digit span (i.e., a single score per subject consisting of the highest list length reached), accuracy, and the algorithm score as the dependent variables.

Lastly, correlation analyses were performed to identify any potential relationships between the accent scores obtained (both on separate blocks—training and testing—and overall) and digit accuracy, maximum digit length reached, and both the raw and algorithm-based scores. All variables of interest are described in Appendix A.

2.5. Participants

The participants were 62 undergraduate students, 31 monolingual (mean age = 23.6, SD = 6.08, 8 male, 23 female) and 31 early bilingual (mean age = 22.33, SD = 4.6, 9 male, 22 female). As previously described in Section 2.2, early bilingual participants were characterized by a native or near-native level of proficiency in both languages and early exposure to them, defined as prior to school age (i.e., 6–7 years). Bilinguals' other languages included Arabic, Cantonese, Hebrew, Russian, Spanish, Urdu, Thai, and (Haitian/Jamaican/St. Lucian) Creole. Both age of acquisition and proficiency level were self-reported. Monolinguals were defined as individuals who reported speaking English natively and, in some cases, an additional language at a conversational or beginner level. Two of the participants (one from each group) were excluded from the analyses related to phonetic and phonological learning (and consequently the correlation analyses) due to technical issues leading to the loss of their voice recordings for the novel accent learning task, but their data were included in the analyses associated with auditory sensory memory.

2.6. Results

The results of the study are presented in three separate subsections, the first two reporting the findings for each of the two experimental tasks, and the third presenting the correlations between phonetic and phonological learning and auditory sensory memory.

2.6.1. Phonetic and Phonological Learning: The Novel Accent Learning Task

Figure 3 shows the average scores for monolinguals and bilinguals grouped by the number of novel accent features (1, 2, or 4) in the three different conditions (i.e., Baseline, Training, Testing). Bilinguals outperformed monolinguals across the board, in both the Training (imitation) and Testing conditions, with a more pronounced decrease in performance for monolinguals in Testing as the number of features present per sentence increased.

Figure 4 shows the average scores for monolinguals and bilinguals for all four novel features in the three different conditions (i.e., Baseline, Training, Testing). Bilinguals outperformed monolinguals across the board, in both the Training (imitation) condition (except for the diphthongization feature) and the Testing condition, but the differences in Training were more pronounced with tapping and tag questions. In Testing, monolinguals performed best with tag questions, followed by epenthesis, and performed most poorly on the tapping feature. An ANOVA with *Accent Score* as the dependent variable and *Group (monolingual/bilingual), Block (baseline/training/testing), Feature (dipthongization/tapping/epenthesis/tag question)* and *Number of features per sentence (1/2/4)* as independent variables revealed significant main effects of all independent variables (Group: $F(1, 12587) = 148.98$, $p < 0.001$, Block: $F(2, 12587) = 1768.32$, $p < 0.001$, Feature: $F(3, 12587) = 50.12$, $p < 0.001$, and Number of features per sentence $F(2, 12587) = 89.81$, $p < 0.001$), and also of the interactions between Group × Block, Group × Feature, Block × Feature, Block × Number of features per sentence, Feature × Number of features per sentence, Group × Block × Feature and Block × Feature × Number of features per sentence. Post hoc tests using the Bonferroni correction revealed that each block differed significantly from the other two, and tapping differed significantly from all other features. The three configurations for number of features per sentence (1, 2, or 4) also differed significantly from each other.

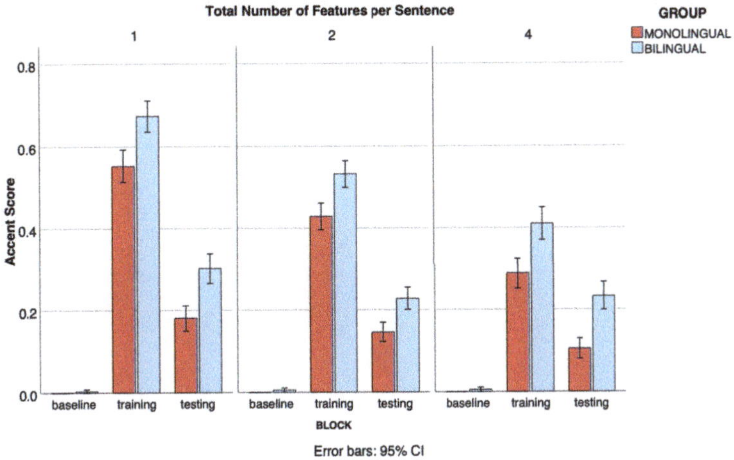

Figure 3. Mean of accent scores grouped by the number of novel accent features (1, 2, or 4) per sentence obtained by monolinguals and bilinguals in Baseline, Training and Testing.

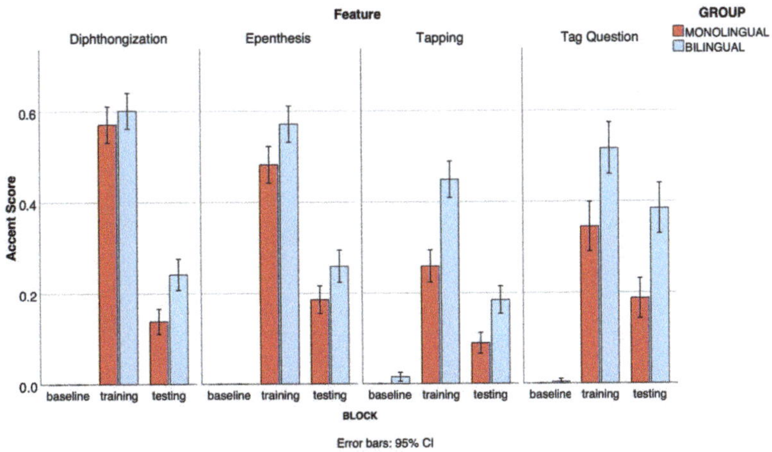

Figure 4. Mean of accent scores for each novel accent feature obtained by monolinguals and bilinguals in Baseline, Training and Testing.

2.6.2. Auditory Sensory Memory: The Digit Span Task with Suffix

Figure 5 presents the proportion of participants (monolingual or bilingual) who were able to advance to each sequence length. Participants from both groups began to "drop out" at sequence length = 6, but those from the monolingual group dropped out in greater proportions than the bilinguals—a slight difference at first, with 93.5% of bilinguals and 90.3% of monolinguals reaching sequence length = 6, which becomes larger as the sequence length increases, with 54.8% of bilinguals and 48.4% of monolinguals reaching sequence length = 7. Only 25.8% of bilinguals and 9.7% of monolinguals were able to complete successfully the 7-digit block and move to the 8-digit block. Only participants from the bilingual group moved on to the 9-digit block. However, none of these consistently recalled these sequences, which means that the 8-digit sequence was the longest sequence recalled reliably by participants in this experiment. Based on the use of a z-score to evaluate the proportions of the two populations still present at the 8-digit sequence ($z = 1.6622$, one-tailed), we conclude that significantly more bilinguals reached this list length compared to monolinguals ($p < 0.05$).

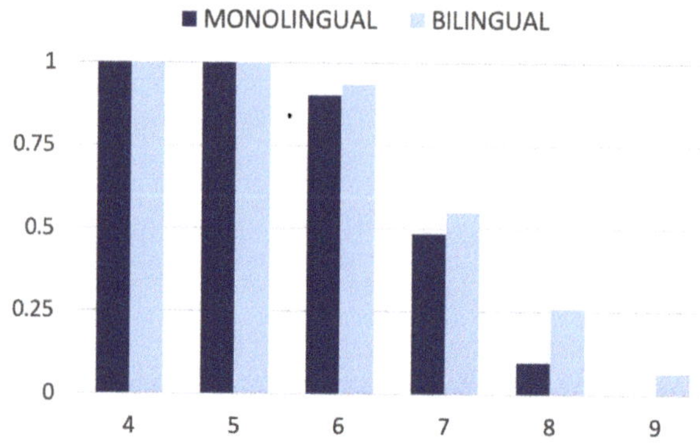

Figure 5. Digit span task: proportion of group who reached each list length.

The two groups did not differ significantly with respect to the average maximum digit length reached, which was 6.8 for bilinguals and 6.5 for monolinguals. For a finer-grained perspective, Figure 6 displays the two groups' accuracy broken down by sequence length. As previously described, there were five trials in each block for a given list length, and a participant needed to answer at least 3 (out of the 5 trials) correctly in order to advance to the next (higher) sequence length. This means that even when a sequence length has been successfully completed by all of the participants, overall accuracy for that sequence length is not necessarily 100% (for example, sequence length = 4). From sequence length = 4 onwards, bilinguals had higher accuracy than monolinguals across the board (except for sequence length = 5, for which the accuracy of both groups was 85%). As the sequence length and consequently difficulty level of a block increased, the group differences became larger. The mean accuracy for monolinguals for the 6-, 7-, and 8- digit sequences was 52.8%, 31%, and 6%. For the same sequence lengths (in increasing order), the bilingual group's overall accuracy was 62.7%, 49.4%, and 42.1%. A one-way analysis of variance showed that Accuracy was significantly affected by Group, $F(1, 1731) = 23.67$, $p < 0.01$ and Sequence Length, $F(7, 1731) = 121.94$, $p < 0.01$, and by these two factors' interaction, $F(6, 1731) = 4.15$, $p < 0.01$. In a series of post hoc comparisons (with the Bonferroni correction) accuracy for list lengths 5 and 6 was significantly different from all of the other sequence lengths, while no significant differences were found in overall accuracy for sequence lengths 2, 3, and 4, and accuracy for sequence lengths 7 and 8 were significantly different from all other list lengths except for each other. A series of one-way ANOVAs performed separately for each list length showed significant effects of Group on Accuracy at sequence length 7, $F(1, 157) = 5.62$, $p = 0.019$ and sequence length 8, $F(1, 51) = 6.75$, $p = 0.012$. The results for

Accuracy are very similar with those we obtained for the algorithm-based scores, so in the interest of space we will not be discussing the latter here, but only in the following section focusing on the correlations between variables.

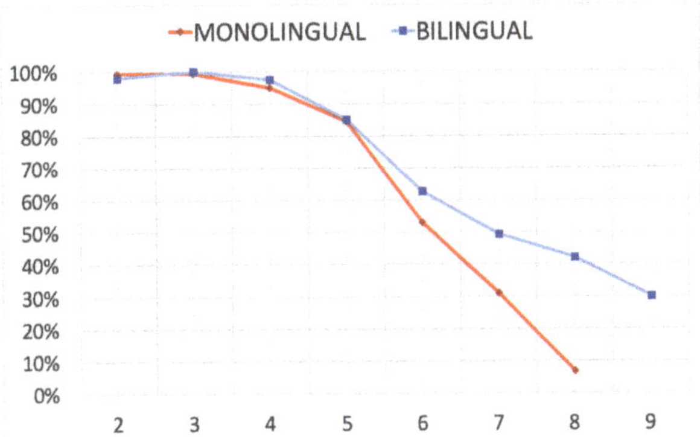

Figure 6. Memory task: mean accuracy for monolinguals and bilinguals for each sequence length.

Figure 7 takes a closer look at the 7- and 8-digit sequences by showing the two groups' mean accuracy at each serial position. A small number of the participants' responses had to be excluded in order to create these plots in cases where the length of the response differed from the length of the input (for example, shorter responses such as "46,382" when the input had been "95,164,832").

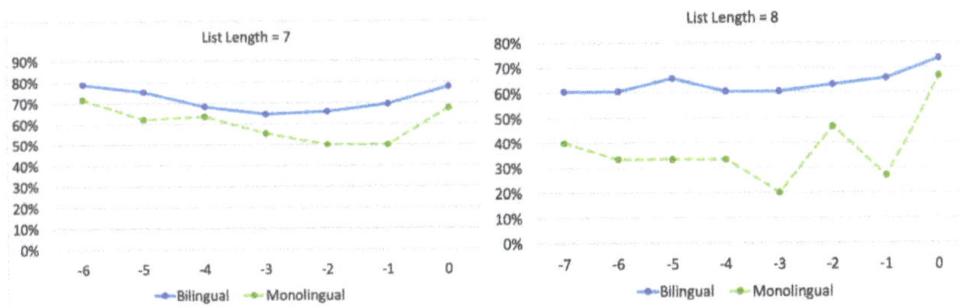

Figure 7. Digit span task: mean accuracy at each serial position for 7-digit (**right**) and 8-digit sequences (**left**). The terminal item is labeled with a 0, and each item preceding it is labeled in terms of its distance from the terminal item (e.g., −1 for the penultimate item, −2 for the antepenultimate item, etc.).

For both sequence lengths, we observe small primacy as well as recency effects, as for both of the groups the accuracy for initial and final items tended to be higher than that of items from the middle of the sequence, except for the initial items in the 8-digit sequence for the bilinguals. For the sequences containing 7 digits, bilinguals display higher accuracy than monolinguals at all serial positions, a difference that is smaller for the initial items but gradually becomes larger for each position that follows them through the preterminal position. In final position, probably because of a recency effect, the two groups perform more similarly than in the penultimate position.

Moving on to the sequences comprising 8 digits, we observe a similar pattern to that described above for 7-digit sequences. Bilinguals have higher accuracy than monolinguals at all serial positions and recency effects are noted for both groups, while only monolinguals display a slight primacy effect. Other patterns that can be observed include the overall lower accuracy for both groups (as might be expected due to the increased difficulty of having to recall the longer sequence), and the larger difference in group performance at all serial positions. Monolinguals show a spike in accuracy for the antepenultimate position, not noted with the 7-digit sequence length. This may have to do with individual factors, considering that only about 10% of the monolingual participants were able to reach this sequence length.

2.6.3. Correlations between Phonetic and Phonological Learning and Auditory Sensory Memory

Lastly, we consider the potential link between phonetic and phonological learning performance and auditory sensory memory. Correlation analyses were performed using the following variables:

- **Phonetic and phonological learning:** Accent Score (both Overall, collapsing performance on the Training and Testing blocks, and separately for each of these two blocks)
- **Auditory sensory memory:** Maximum Sequence Length reached, Overall Digit Accuracy obtained in the memory task, and Algorithm-based Score, that is, the digit recall score obtained by taking into account permutation errors, with bigger penalties for items displaced at longer distances.

Table 1 shows the Pearson correlations that were significant in a two-tailed analysis when monolinguals and bilinguals (n = 60) were considered together, while Tables 2 and 3 show the Pearson correlations that were significant in a one-tailed analysis when monolinguals and bilinguals were considered separately (n = 30 for each group).

Table 1. Significant correlations between variables associated with phonetic and phonological learning (arranged vertically) and variables associated with auditory sensory memory (arranged horizontally) when all participants were considered together. Gray shading indicates the strength of the correlation (light gray = weak correlation, medium gray = moderate correlation, dark gray = strong correlation); n.s. = not significant.

	Max Sequence Length	Overall Accuracy	Algorithm-Based Score
Accent Score (Overall)	r(58) = 0.497 $p < 0.001$	n.s.	r(58) = 0.504 $p < 0.001$
Accent Score (Testing)	r(58) = 0.479 $p < 0.001$	r(58) = 0.297 $p < 0.05$	r(58) = 0.469 $p < 0.001$
Accent Score (Training)	r(58) = 0.445 $p < 0.001$	r(58) = 0.312 $p < 0.05$	r(58) = 0.459 $p < 0.001$

Table 2. Significant correlations between variables associated with phonetic and phonological learning (arranged vertically) and variables associated with auditory sensory memory (arranged horizontally) for the MONOLINGUAL group. Gray shading indicates the strength of the correlation (light gray = weak correlation, medium gray = moderate correlation, dark gray = strong correlation); n.s. = not significant.

	Max Sequence Length	Overall Accuracy	Algorithm-Based Score
Accent Score (Overall)	r(28) = 0.379 $p < 0.05$	n.s.	n.s.
Accent Score (Testing)	r(28) = 0.370 $p < 0.05$	n.s.	n.s.
Accent Score (Training)	r(28) = 0.336 $p < 0.05$	n.s.	n.s.

To summarize the above, several significant correlations were found between variables associated with phonetic and phonological learning and auditory sensory memory (and more generally serial memory, given the difficulty of excluding the effects of working memory). Specifically, the accent scores obtained by participants in both the training (imitation) condition and in testing, as well as (in some cases) the compounded overall scores, correlated with the maximum digit span, overall digit accuracy, and corrected algorithm scores obtained by the same participants. These correlations, however, were stronger and more numerous in the bilingual group. When considered separately, only three positive correlations were significant for the monolingual group, all of which were weak correlations. For comparison, 9 correlations were significant based on the data from bilingual speakers, of which 7 were strong correlations and 2 were of moderate strength. Figure 8 provides visual representations for some of these correlations.

Table 3. Significant correlations between variables associated with phonetic and phonological learning (arranged vertically) and variables associated with auditory sensory memory (arranged horizontally) for the BILINGUAL group. Gray shading indicates the strength of the correlation (light gray = weak correlation, medium gray = moderate correlation, dark gray = strong correlation); n.s. = not significant.

	Max Sequence Length	Overall Accuracy	Algorithm-Based Score
Accent Score (Overall)	r(28) = 0.572 $p < 0.001$	r(28) = 0.525 $p = 0.001$	r(58) = 0.617 $p < 0.001$
Accent Score (Testing)	r(28) = 0.539 $p = 0.001$	r(28) = 0.518 $p < 0.05$	r(28) = 0.581 $p < 0.001$
Accent Score (Training)	r(28) = 0.499 $p < 0.05$	r(28) = 0.451 $p < 0.05$	r(28) = 0.527 $p < 0.001$

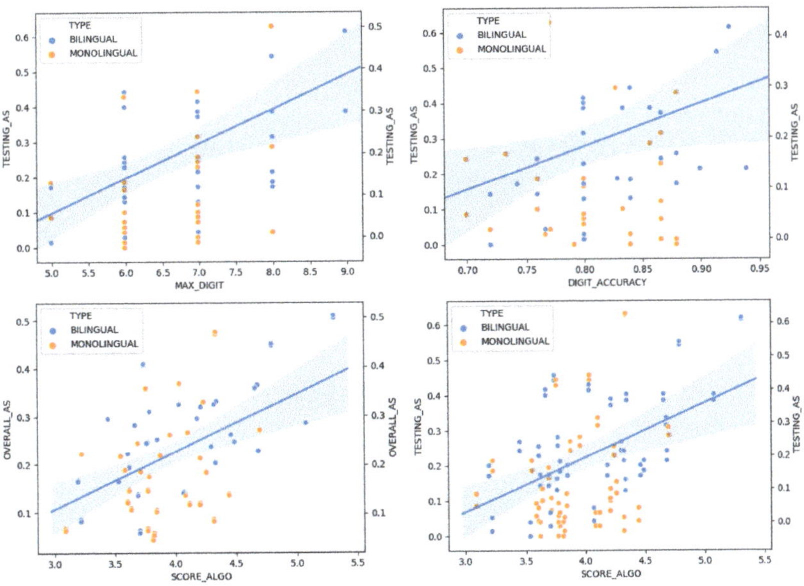

Figure 8. Regression plots for pairs of variables reflecting phonetic and phonological learning (Testing AS, Overall AS) and variables associated with auditory sensory memory (Max Digit, Digit Accuracy, and Score Algo). AS = accent score. Max Digit = maximum sequence length reached. Digit Accuracy = overall accuracy obtained in the digit span task, Score Algo = the corrected algorithm-based score obtained by taking into account permutation errors, with bigger penalties for items displaced at longer distances.

3. Discussion

Our study supports the hypotheses we formulated, replicating earlier results (Hypothesis 1, [2,27–29] and Hypothesis 2, [46]) and also reporting a novel finding (Hypothesis 3). Specifically, bilinguals obtained higher performance scores on the novel accent learning task for all four features tested (Hypothesis 1). This was most apparent in the testing phase, but bilinguals also outperformed monolinguals in the training (imitation) block for three of the four features. More generally, bilinguals also obtained higher scores than monolinguals on sentences containing 1, 2, or 4 different features, with monolinguals showing a return to baseline in the more complex case of sentences requiring all four features to be expressed, reflecting an inability to manifest the newly learned patterns even though they were able to produce them in isolation. Both bilinguals and monolinguals performed better on the novel intonation pattern in tag questions than on the other three patterns, possibly due to the fact that it was a more global (suprasegmental) phenomenon of longer duration and higher salience compared to the other features.

Hypothesis 2 was supported by the finding that there is a bilingual advantage in auditory sensory memory (manifested as better performance on the items that preceded the suffix, see Figure 7), which became more pronounced as the task's complexity (i.e., the length of the sequence to be recalled) increased. This suggests bilinguals have a longer auditory sensory memory span than monolinguals. This assumption is also supported by the percentage of participants in each group who reached the longest digit sequences (that is, 8 and 9 digits), and the two groups' performance in terms of accuracy both when we considered the sequences as a whole and when we broke them down by serial position. Notably, the increase from 7- to 8-digit sequences caused a substantial drop in accuracy in the monolingual group (from 31% to 6%), while the decrease in accuracy was more gradual in bilinguals (from 49.4% to 42.1%). Additionally, the higher accuracy exhibited by bilinguals with the items positioned at the start of longer sequences suggests potential enhancement of their working memory as well, in line with earlier behavioral [45,52–54], and electrophysiological findings [55], but in contrast with studies which found no differences in working memory between bilinguals and monolinguals [56,57]. Lastly, Hypothesis 3 was also supported, as significant positive correlations were found between variables reflecting phonetic and phonological learning and variables associated with auditory sensory memory. We found this relationship to be much stronger in bilinguals compared to monolinguals.

One of the immediately arising questions in light of our results is whether the fact that the link between phonetic and phonological learning and auditory sensory memory was much stronger in bilinguals supports the idea that auditory sensory memory plays a crucial part in this type of learning. While we believe this to be the case, based on arguments we discuss in what follows, we would like to clarify that our study has not investigated the existence of a causal relationship between the two, but simply established that a relationship exists. The possibility remains that bilingual experience leads to the independent enhancement of both phonetic and phonological learning on the one hand and auditory sensory memory on the other hand, without the former being supported by the latter. Future studies are needed to elucidate this question.

In support of the involvement of auditory sensory memory in phonetic and phonological learning, Calabrese [18] discusses a mechanism involving two distinct modes of speech perception, the phonemic and the phonetic mode [58]. Listeners are posited to engage in the top-down, "phonemic" mode of perception when they perceive stimuli containing native-language phonological categories. This mode enables rapid unfolding of speech perception because it is able to ignore non-contrastive aspects of perceptual representations. But if perception were exclusively phonemic, that would mean that listeners are unable to to perceive allophonic variation, which would make languages unlearnable. It is the "phonetic" (or bottom-up) perception that enables access to allophonic details. "Phonetically-relevant perception" is thus crucial in order to learn allophonic variation and access sound contrasts in both native and non-native languages, as well as for acquiring foreign sounds. To achieve this, the perceptual system is assumed to contain a memory component for preserving acoustically accurate representations of the received signal making it possible for novel representations to be stored in order to (eventually) construct a new phonological system. While Calabrese uses the term echoic memory for this specific type of memory, it has more recently been referred to as auditory sensory memory [36]. A part of the bottom-up perceptual component, auditory sensory memory is posited to play a part in language learning (and more specifically in phonetic and phonological learning). From this perspective, the concept of phonological "deafening" for adults (to non-native sounds) does not describe an inability to hear or access the acoustic signal. Instead, it refers to their inability to translate the new cue pattern characterizing the non-native sound into a permissible phonological representation. Crucially, auditory sensory memory makes it possible for these novel acoustic patterns to be heard and preserved. Following sufficient articulatory training, acoustic patterns captured by auditory sensory memory can eventually be adapted into admissible phonological representations, at which point a learner has become able to acquire the non-native sound.

Other studies have acknowledged the role played by sensorimotor systems in language learning. Earlier findings [59,60] point to the existence of a specific left lateralized auditory mirror neuron system engaged in auditorily triggered speech imitation which [19] found be more active in "poor" speech imitators.Simmonds et al. (2011) also discuss how learning to speak a second language also has effects on auditory and somatosensory feedback systems, and emphasize the motor and sensory complexities involved in learning to speak a second language as an adult [21]. Their suggestion is that adult second language learners might benefit from a mute period of intense auditory exposure to a second language before attempting to produce the sounds. This mute period could prove to be "beneficial in enabling the learner to hear (and thus produce) subtly different phonetic features, new phoneme distinctions and unfamiliar sequences of stress patterns". Future neurolinguistic findings may shed more light in this respect, also taking into account the involvement of the insula region, which was identified as a key component of accent processing, possibly playing a role in sensory-perceptual processing [61], and supporting conscious awareness and regulation of accent features [62]. This may help in understanding the observed differences between monolinguals and bilinguals in phonetic and phonological learning because these differences may partially also be due to the two groups' recruiting different cognitive resources to achieve learning, with more conscious and effortful processing in the case of monolinguals.

Other than the relatively reduced number of participants, our study is subject to the methodological limitations we have pointed out in the introduction, such as not being able to obtain homogenous groups of bilingual speakers with respect to their experience with each of the languages they speak [6]. Given the lack of a unitary definition of "bilingual", two people with very similar linguistic backgrounds and abilities might readily place themselves in different groups [63]. Among many possible scenarios, speakers might not feel confident enough to report bilingual knowledge if the second language is mostly practiced passively (e.g., their parents speak it at home but they only speak it occasionally), if they do not have the same competencies as their native-speaking relatives (e.g., a heritage speaker of Chinese or Arabic in the United States might not consider themselves bilingual because they cannot read or write in this language), or if the second language they speak is in some ways similar to another one, to the point where they feel they speak a somehow inferior version of that language (e.g., Haitian Creole speakers reporting they speak "broken French"). Other problems with self-reports include the fact that speakers may not accurately record the age of first exposure to a given language or how often and in what ways they were exposed to it (e.g., they may not be aware of extended trips abroad in their early childhood) and thus under-report their experience. While all of our bilingual participants reported (near-) native competence in both languages, high variability emerged in their performance on the short translation task administered at the beginning of the experiment (the analysis of which we have not yet completed). This inability to control for bilingual experience has been acknowledged as a major challenge in the study of bilingual cognition, thus future studies may benefit from the use of standardized language tests in order to evaluate a speaker's proficiency. If such tests enabling finer-grained assessment of bilingual abilities are incorporated to experimental procedures, this may result in higher replicability, rendering more comparable the results of different studies [64]). Very recent work in neurolinguistics also supports this position as it indicates that proficiency (even more than age of acquisition)—is a critical factor differentiating the functional organization of bilingual language processing, a finding which has also been "underlined by structural neuroimaging investigations" [65].

Despite the mean group differences, in the current study we saw a number of monolinguals performing as well as the top bilinguals, for instance the top 10 performers on the novel accent learning task included 3 monolinguals, as did the top 10 participants with the highest digit span reached. In terms of overall accuracy on the digit span, 4 monolinguals were among the top 10 performers. This highlights another methodological complication: other than the use of multiple languages, several factors have been found to modulate the development of cognitive functioning, including socio-economic status [66], physical activity [67], circadian rhythm and sleep [68], dietary intake [69], and musical expertise [70]. Language learning constitutes one out of several possible ways of engaging in cognitive training, and cognitive training itself is only one of the lifestyle factors also known to affect cognitive function [8]. Studying any one of these aspects in isolation might obscure other meaningful relationships or be subject to confounds preventing us from observing significant effects, and contributing to replication failure. According to [71], we may expect future work to uncover that distinct effects of language on cognitive operations arise from interdependent functions. In consequence, research studies exploring directly how multiple levels of processing interact with one another have the potential to offer a more far-reaching view of how exactly language shapes our mind.

4. Conclusions

Our study focused on the sensorimotor bases of language—and more specifically phonetic and phonological—learning in bilinguals. We replicated the bilingual advantage previously observed in phonetic and phonological learning [2,27–29] and auditory sensory memory [46], though the differential roles of working memory and auditory sensory memory have yet to be determined more precisely. Our study also showed a significant correlation between phonetic and phonological learning and auditory sensory memory, which was stronger in bilinguals in comparison to monolinguals. Whereas higher-level cognitive functions are likely to be at play in the execution of complex tasks such as phonetic and phonological learning, it is important not to underestimate the role played by lower-level, sensorimotor functions as well, so a full picture of the mechanism supporting this type of learning may be obtained. These findings thus raise questions about the role of sensorimotor mechanisms in language learning and suggest that incorporating a sensorimotor perspective in future studies on bilingual cognition may be a fruitful research direction.

Author Contributions: Conceptualization, L.S. and J.H.; methodology, L.S. and J.H.; software, L.S. and M.V.; validation, L.S., J.H. and M.V.; formal analysis, L.S. and J.H.; investigation, L.S. and J.H.; resources, L.S., J.H. and M.V.; data curation, L.S., J.H. and M.V.; writing—original draft preparation, L.S.; writing—review and editing, L.S.; visualization, L.S., J.H. and M.V.; supervision, L.S.; project administration, L.S. and M.V.; funding acquisition, L.S. All authors have read and agreed to the published version of the manuscript.

Funding: This research was funded by PSC-CUNY grant number # 61667-00 49.

Institutional Review Board Statement: The study was conducted in accordance with the Declaration of Helsinki, and approved by the Institutional Review Board (or Ethics Committee) of the City University of New York (protocol code # 2018-0824, date of approval 07/17/2018).

Informed Consent Statement: Informed consent was obtained from all subjects involved in the study.

Data Availability Statement: Not applicable.

Conflicts of Interest: The authors declare no conflict of interest.

Appendix A

VARIABLES MEASURING PHONETIC AND PHONOLOGICAL LEARNING

Accent Score (Overall) The absence or presence of the four novel features associated with the Model Speech (in the appropriate environment) was scored with a 0 or 1, respectively. This was achieved for all three blocks completed by the participants (Baseline, Training, and Testing). Subsequently, an overall score was obtained for each participant, indicating the percentage of novel features produced (out of their total utterances). Note that the features were not expected to appear in the Baseline block which reflected participants' natural accents.

Accent Score (Training) The absence or presence of the four novel features associated with the Model Speech (in the appropriate environment) was scored with a 0 or 1, respectively. This was achieved separately for the Training block, which consisted of participants' imitation of sentences uttered in the Model Speech accent immediately after hearing each one of them. Subsequently, an overall score was obtained for each participant, indicating the percentage of novel features produced during Training (out of their total utterances).

Accent Score (Testing) The absence or presence of the four novel features associated with the Model Speech (in the appropriate environment) was scored with a 0 or 1, respectively. This was achieved separately for the Testing block, which consisted of participants' re-reading of the sentences presented during the Baseline block, being prompted to now utter them in the Model Speech accent they had received training on, to the best of their ability. Subsequently, an overall score was obtained for each participant, indicating the percentage of novel features produced during Testing (out of their total utterances).

VARIABLES MEASURING AUDITORY SENSORY MEMORY

Max Sequence Length This variable measures the longest digit sequence reached by a participant. The first digit sequence presented contained two digits only (and also served as a practice block) following which, if a participant correctly recalled at least 3 out of a total of 5 trials per block, the next digit sequence would be presented (one digit longer than the one that had just been completed). The task was adaptive therefore this part of the experiment would end whenever a participant failed to successfully recall at least 3 trials for a given sequence.

Overall Digit Accuracy A participant's overall accuracy in the digit span task. Since each sequence length included 5 trials, errors were possible even when participants were able to advance successfully to the next sequence length. Participants from both groups started making errors from sequence length = 4 and higher.

Algorithm-based Accuracy Score A 'corrected' score that took into account the similarity between a digit sequence input and a participant's response. Thus, a response string that was very similar to the input (for instance by the transposition of 2 digits) received a higher score than a response in none of the digits matched the input (the Overall Digit Accuracy would have assigned both such sequences an identical score of 0). The algorithm searched for insertions or deletions by aligning a participant response string and the input string presented and counting the number of digits in each string to see if there was a discrepancy. If the number of digits in the response string was equal to the number of digits in the input string, then the answer was included in the analysis, but if the number of digits was not equal between the response and input strings, the answer was excluded. The algorithm evaluated the responses that were included digit-by-digit, employing a graded scoring method that assigned weighted scores based on transpositional distance. The goal of this graded scoring system was to award a higher score to transposed response digits that were closer to their original position in the participant response.

References

1. Bialystok, E.; Craik, F.I.M.; Binns, M.A.; Ossher, L.; Freedman, M. Effects of bilingualism on the age of onset and progression of MCI and AD: Evidence from executive function tests. *Neuropsychology* **2014**, *28*, 290–304. [CrossRef]
2. Antoniou, M.; Liang, E.; Ettlinger, M.; Wong, P.C.M. The bilingual advantage in phonetic learning. *Biling. Lang. Cogn.* **2015**, *18*, 683–695. [CrossRef]
3. Paradis, M. The Loch Ness monster approach to bilingual language lateralization: A response to Berquier and Ashton. *Brain Lang.* **1992**, *43*, 534. [CrossRef] [PubMed]
4. Marzecová, A. Bilingual advantages in executive control—A Loch Ness Monster case or an instance of neural plasticity. *Cortex* **2015**, *73*, 364–366. [CrossRef] [PubMed]
5. Van den Noort, M.; Struys, E.; Bosch, P.; Jaswetz, L.; Perriard, B.; Yeo, S.; Barisch, P.; Vermeire, K.; Lee, S.H.; Lim, S. Does the bilingual advantage in cognitive control exist and if so, what are its modulating factors? A systematic review. *Behav. Sci.* **2019**, *9*, 27. [CrossRef]
6. DeLuca, V.; Rothman, J.; Bialystok, E.; Pliatsikas, C. Redefining bilingualism as a spectrum of experiences that differentially affects brain structure and function. *Proc. Natl. Acad. Sci. USA* **2019**, *116*, 7565–7574. [CrossRef]
7. Bialystok, E. Bilingualism and executive function: What's the connection?In *Bilingual Cognition and Language: The State of the Science across Its Subfields*; Miller, D., Bauram, F., Rothman, J., Eds.; John Benjamins: Amsterdam, The Netherlands, 2018 ; pp. 283–305.
8. Valian, V. Bilingualism and cognition. *Biling. Lang. Cogn.* **2015**, *18*, 3–24. [CrossRef]
9. Obler, L.K.; Fein, D.E. *The Exceptional Brain: Neuropsychology of Talent and Special Abilities*; Guilford Press: New York, NY, USA, 1988.
10. Higby, E.; Kim, J.; Obler, L.K. Multilingualism and the brain. *Annu. Rev. Appl. Linguist.* **2013**, *33*, 68–101. [CrossRef]
11. Bialystok, E.; Fergus, I.M.C.; Luk, G. Bilingualism: Consequences for Mind and Brain. *Trends Cogn. Sci.* **2012**, *16*, 240–250. [CrossRef]
12. Poarch, G.J.; Krott, A. A bilingual advantage? An appeal for a change in perspective and recommendations for future research. *Behav. Sci.* **2019**, *9*, 95. [CrossRef]
13. Marian, V.; Spivey, M. Competing activation in bilingual language processing: Within- and between-language competition. *Biling. Lang. Cogn.* **2003**, *6*, 97–115. [CrossRef]
14. Thierry, G.; Wu, Y.J. Brain potentials reveal unconscious translation during foreign-language comprehension. *Proc. Natl. Acad. Sci. USA* **2007**, *104*, 12530–12535. [CrossRef] [PubMed]
15. Luk, G.; Green, D.W.; Abutalebi, J.; Grady, C. Cognitive control for language switching in bilinguals: A quantitative meta-analysis of functional neuroimaging studies. *Lang. Cogn. Process.* **2011**, *27*, 1479–1488. [CrossRef] [PubMed]
16. Green, D.W.; Abutalebi, J. Language control in bilinguals: The adaptive control hypothesis. *J. Cogn. Psychol.* **2013**, *25*, 515–530. [CrossRef]
17. Miyake, A.; Friedman, N.P.; Emerson, M.J.; Witzki, A.H.; Howerter, A.; Wager, T.D. The unity and diversity of executive functions and their contributions to complex "Frontal Lobe" tasks: A latent variable analysis. *Cogn. Psychol.* **2000**, *41*, 49–100. [CrossRef] [PubMed]
18. Calabrese, A. Auditory representations and phonological illusions: A linguists perspective on the neuropsychological bases of speech perception. *J. Neurolinguist.* **2012**, *25*, 355–381. [CrossRef]
19. Reiterer, S.M.; Hu, X.; Erb, M.; Rota, G.; Nardo, D.; Grodd, W.; Winkler, S.; Ackermann, H. Individual differences in audio-vocal speech imitation aptitude in late bilinguals: Functional neuroimaging and brain morphology. *Front. Psychol.* **2011**, *2*, 271. [CrossRef] [PubMed]
20. Pliatsikas, C.; DeLuca, V.; Voits, T. The many shades of bilingualism: Language experiences modulate adaptations in brain structure. *Lang. Learn.* **2020**, *70*, 133–149. [CrossRef]
21. Simmonds, A.J.; Wise, R.J.; Leech, R. Two tongues, one brain: Imaging bilingual speech production. *Front. Psychol.* **2011**, *2*, 166. [CrossRef]
22. Kröger, B.J.; Birkholz, P.; Neuschaefer-Rube, C. Towards an articulation-based developmental robotics approach for word processing in face-to-face communication. *Paladyn* **2011**, *2*, 82–93. [CrossRef]
23. Lindenberger, U.; Baltes, P.B. Sensory functioning and intelligence in old age: A strong connection. *Psychol. Aging* **1994**, *9*, 339–355. [CrossRef] [PubMed]
24. Lindenberger, U.; Marsiske, M.; Baltes, P.B. Memorizing while walking: Increase in dual-task costs from young adulthood to old age. *Psychol. Aging* **2000**, *15*, 417–436. [CrossRef] [PubMed]
25. Schäfer, S.; Huxhold, O.; Lindenberger, U. Healthy mind in healthy body? A review of sensorimotor-cognitive interdependencies in old age. *Eur. Rev. Aging Phys. Act.* **2006**, *3*, 45–54. [CrossRef]
26. Kopeckova, R. The bilingual advantage in L3 learning: A developmental study of rhotic sounds. *Int. J. Multiling.* **2016**, *13*, 410–425. [CrossRef]
27. Tremblay, M.-C.; Sabourin, L. Comparing behavioral discrimination and learning abilities in mono- linguals, bilinguals and multilinguals. *J. Acoust. Soc. Am.* **2012**, *132*, 3465–3474. [CrossRef] [PubMed]
28. Spinu, L.E.; Hwang, J.; Lohmann, R. Is there a bilingual advantage in phonetic and phonological acquisition? The initial learning of word-final coronal stop realization in a novel accent of English. *Int. J. Biling.* **2018**, *22*, 350–370. [CrossRef]
29. Spinu, L.; Hwang, J.; Pincus, N.; Vasilita, M. Exploring the use of an artificial accent of English to assess phonetic learning in monolingual and bilingual speakers. *Proc. Interspeech* **2020**, 2377–2381.

30. Bloom, L.C. Two-component theory of the suffix effect: Contrary evidence. *Mem. Cogn.* **2006**, *34*, 648–667. [CrossRef]
31. Crowder, R.G. Echoic memory and the study of aging memory systems. In *New Directions in Memory and Aging: Proceedings of the G. A. Talland Memorial Conference*; Poon, L., Fozard, J., Cermak, L., Arenberg, D., Thompson, L., Eds.; Originally Published 1980; Routledge—Taylor & Francis Group, Psychology Press: London, UK, 2014; pp. 181–204.
32. Pilotti, M.; Beyer, T.; Yasunami, M. Top-down processing and the suffix effect in young and older adults. *Mem. Cogn.* **2002**, *30*, 89–96. [CrossRef]
33. Cowan, N. On short and long auditory stores. *Psychol. Bull. 1984*, 96, 341–370. [CrossRef]
34. Crowder, R.G. Mechanisms of auditory backward masking in the stimulus suffix effect.*Psychol.* **1978**, *85*, 502–524. [CrossRef]
35. Greene, R.L.; Crowder, R.G. Modality and suffix effects in the absence of auditory stimulation. *J. Verbal Learn. Verbal Behav.* **1984**, *23*, 371–382. [CrossRef]
36. Nees, M.A. Have we forgotten auditory sensory memory? Retention intervals in studies of nonverbal auditory working memory. *Front. Psychol.* **2016**, *7*, 1892. [CrossRef] [PubMed]
37. Cowan, N. What are the differences between long-term, short-term, and working memory? *Prog. Brain Res.* **2008**, *169*, 323–338.
38. Jones, D.M.; Hughes, R.W.; Macken, W.J. Perceptual organization masquerading as phonological storage: Further support for a perceptual-gestural view of short-term memory. *J. Mem. Lang.* **2006**, *54*, 265–281. [CrossRef]
39. Jones, D.M.; Macken, W.J.; Nicholls, A.P. The phonological store of working memory: Is it phonological and is it a store? *J. Exp. Psychol. Learn. Mem. Cogn.* **2004**, *30*, 656–674. [CrossRef] [PubMed]
40. Sofologi, M.; Zafiri, M.; Pliogou, V. Investigating the relationship of working memory and inhibitory control: Bilingual education and pedagogical implications in elementary school. *Int. J. Learn. Teach. Educ. Res.* **2020**, *19*, 163–183. [CrossRef]
41. Yang, E. Bilinguals' Working Memory (WM) Advantage and Their Dual Language Practices. *Brain Sci.* **2017**, *7*, 86. [CrossRef]
42. Morales, J.; Calvo, A.; Bialystok, E. Working memory development in monolingual and bilingual children. *J. Exp. Child Psychol.* **2013**, *114*, 187–202. [CrossRef]
43. Krizman, J.; Marian, V.; Shook, A.; Skoe, E.; Kraus, N. Subcortical encoding of sound is enhanced in bilinguals and relates to executive function advantages. *Proc. Natl. Acad. Sci. USA* **2012**, *109*, 7877–7881. [CrossRef]
44. Ljungberg, J.K.; Hansson, P.; Andrés, P.; Josefsson, M.; Nilsson, L.G. A longitudinal study of memory advantages in bilinguals. *PLoS ONE* **2013**, *8*, e73029. [CrossRef] [PubMed]
45. Philipp-Müller, N.; Spinu, L.; Rafat, Y.; Rand, J. Serial memory error patterns in bilinguals and monolinguals. *Proc. Mtgs. Acoust.* **2018**, *35*, 060007. [CrossRef]
46. Spinu, L. Serial memory mechanisms in monolingual and bilingual speakers. *Int. J. Biling.* **2022**, 13670069211070977. [CrossRef]
47. Marian, V.; Blumenfeld, H.K.; Kaushanskaya, M. The Language Experience and Proficiency Questionnaire (LEAP-Q): Assessing language profiles in bilinguals and multilinguals. *J. Speech Lang. Hear. Res.* **2007**, *50*, 940–967. [CrossRef]
48. Adank, P.; Hagoort, P.; Bekkering, H. Imitation improves language comprehension. *Psychol. Sci.* **2010**, *21*, 1903–1909. [CrossRef]
49. Peirce, J.W.; Gray, J.R.; Simpson, S.; MacAskill, M.R.; Höchenberger, R.; Sogo, H.; Kastman, E.; Lindeløv, J. PsychoPy2: Experiments in behavior made easy. *Behav. Res. Methods* **2019**, *51*, 195–203. [CrossRef] [PubMed]
50. Cohen, J.; MacWhinney, B.; Flatt, M.; Provost, J. PsyScope: A new graphic interactive environment for designing psychology experiments. *Behav. Res. Methods Instrum. Comput.* **1993**, *25*, 257–271. [CrossRef]
51. *MATLAB*, Version: 9.13.0 (R2022b); The MathWorks Inc.: Natick, MA, USA, 2022. Available online: https://www.mathworks.com (accessed on 5 February 2023).
52. Comishen, K.J.; Bialystok, E. Increases in attentional demands are associated with language group differences in working memory performance. *Brain Cogn.* **2021**, *147*, 105658. [CrossRef]
53. Ma, X.; Ma, X.; Li, P.; Liu, Y. Differences in working memory with emotional distraction between proficient and non-proficient bilinguals. *Front. Psychol.* **2020**, *11*, 1414. [CrossRef]
54. Signorelli, T.; Obler, L.K. Working memory in simultaneous interpreters. In *Memory, Language, and Bilingualism: Theoretical and Applied Approaches*; Altarriba, J., Isurin, L., Eds.; Cambridge University Press: New York, NY, USA, 2013; pp. 95–125.
55. Morrison, C.; Kamal, F.; Taler, V. The influence of bilingualism on working memory event-related potentials. *Biling. Lang. Cogn.* **2019**, *22*, 191–199. [CrossRef]
56. Bialystok, E.; Craik, F.I.M.; Luk, G. Lexical access in bilinguals: Effects of vocabulary size and executive control. *J. Neurolinguist.* **2008**, *21*, 522–538. [CrossRef]
57. Engel de Abreu, P.M. Working memory in multilingual children: Is there a bilingual effect? *Memory* **2011**, *19*, 529–537. [CrossRef]
58. Werker, J.F.; Logan, J. Cross-language evidence for three factors in speech perception. *Percept. Psychophys.* **1985**, *37*, 35–44. [CrossRef]
59. Aziz-Zadeh, L.; Ivry, R.B. The human mirror neuron system and embodied representations. *Adv. Exp. Med. Biol.* **2009**, *629*, 355–376. [PubMed]
60. DD'Ausilio, A.; Pulvermüller, F.; Salmas, P.; Bufalari, I.; Begliomini, C.; Fadiga, L. The motor somatotopy of speech perception. *Curr. Biol.* **2009**, *19*, 381–385. [CrossRef]
61. Chee, M.W.; Soon, C.S.; Lee, H.L.; Pallier, C. Left insula activation: A marker for language attainment in bilinguals. *Proc. Natl. Acad. Sci. USA* **2004**, *101*, 15265–15270. [CrossRef] [PubMed]
62. Ghazi-Saidi, L.; Dash, T.; Ansaldo, A.I. How native-like can you possibly get: FMRI evidence for processing accent. *Front. Hum. Neurosci.* **2015**, *9*, 587. [CrossRef] [PubMed]

63. de Bruin, A.; Dick, A.S.; Carreiras, M. Clear theories are needed to interpret differences: Perspectives on the bilingual advantage debate. *Neurobiol. Lang.* **2021**, *2*, 1–46. [CrossRef]
64. Marian, V.; Hayakawa, S. Measuring bilingualism: The quest for a "bilingualism quotient". *Appl. Psycholinguist.* **2021**, *42*, 527–548. [CrossRef]
65. Del Maschio, N.; Abutalebi, J. Neurobiology of bilingualism. In *Bilingual Cognition and Language: The State of the Science across Its Subfields;* Miller, D., Bauram, F., Rothman, J., Eds.; John Benjamins: Amsterdam, The Netherlands , 2018; pp. 325–346.
66. Noble, K.G.; McCandliss, B.D.; Farah, M.J. Socioeconomic gradients predict individual differences in neurocognitive abilities. *Dev. Sci.* **2007**, *10*, 464–480. [CrossRef]
67. Best, J.R. Effects of physical activity on children's executive function: Contributions of experimental research on aerobic exercise. *Dev. Sci.* **2010**, *30*, 331–551. [CrossRef] [PubMed]
68. Kuula, L.; Pesonen, A.-K.; Martikainen, S.; Kajantie, E.; Lahti, J.; Strandberg, T.; Tuovinen, S.; Heinonen, K.; Pyhälä, R.; Lahti, M.; et al. Poor sleep and neurocognitive function in early adolescence. *Sleep Med.* **2015**, *16*, 1207–1212. [CrossRef] [PubMed]
69. Kim, J.Y.; Wang, S.W. Relationships between dietary intake and cognitive function in healthy Korean children and adolescents. *J. Lifestyle Med.* **2017**, *7*, 10–17. [CrossRef]
70. Zuk, J.; Benjamin, C.; Kenyon, A.; Gaab, N. Behavioral and neural correlates of executive functioning in musicians and non-musicians. *PLoS ONE* 2014, 9, e99868. [CrossRef] [PubMed]
71. Hayakawa, S.; Marian, V. Consequences of multilingualism for neural architecture. *Behav. Brain Funct.* **2019**, *15*, 1–24. [CrossRef] [PubMed]

brain sciences

MDPI

Article

Increased Pre-Boundary Lengthening Does Not Enhance Implicit Intonational Phrase Perception in European Portuguese: An EEG Study

Ana Rita Batista [1], Dinis Catronas [1], Vasiliki Folia [2] and Susana Silva [1,*]

1 Center for Psychology at University of Porto, Faculty of Psychology and Educational Sciences, Psychology Department, University of Porto, Rua Alfredo Allen, s/n, 4200-135 Porto, Portugal
2 Lab of Cognitive Neuroscience, School of Psychology, Aristotle University of Thessaloniki, University Campus, 54124 Thessaloniki, Greece
* Correspondence: susanamsilva@fpce.up.pt

Abstract: Prosodic phrasing is the segmentation of utterances into prosodic words, phonological phrases (smaller units) and intonational phrases (larger units) based on acoustic cues—pauses, pitch changes and pre-boundary lengthening. The perception of prosodic boundaries is characterized by a positive event-related potential (ERP) component, temporally aligned with phrase boundaries—the Closure Positive Shift (CPS). The role of pre-boundary lengthening in boundary perception is still a matter of debate: while studies on phonological phrase boundaries indicate that all three cues contribute equally, approaches to intonational phrase boundaries highlight the pause as the most powerful cue. Moreover, all studies used explicit boundary recognition tasks, and it is unknown how pre-boundary lengthening works in implicit prosodic processing tasks, characteristic of real-life contexts. In this study, we examined the effects of pre-boundary lengthening (original, short, and long) on the EEG responses to intonational phrase boundaries (CPS effect) in European Portuguese, using an implicit task. Both original and short versions showed equivalent CPS effects, while the long set did not elicit the effect. This suggests that pre-boundary lengthening does not contribute to improved perception of boundaries in intonational phrases (longer units), possibly due to memory and attention-related constraints.

Keywords: prosodic phrasing; prosodic boundaries; closure positive shift; boundary perception; pre-boundary lengthening; implicit boundary recognition task

Citation: Batista, A.R.; Catronas, D.; Folia, V.; Silva, S. Increased Pre-Boundary Lengthening Does Not Enhance Implicit Intonational Phrase Perception in European Portuguese: An EEG Study. *Brain Sci.* **2023**, *13*, 441. https://doi.org/10.3390/brainsci13030441

Academic Editors: Antoine Shahin and Heather Bortfeld

Received: 27 December 2022
Revised: 20 February 2023
Accepted: 2 March 2023
Published: 4 March 2023

1. Introduction

Prosody refers to the melodic (pitch movements), timbral (voice quality) and rhythmic events (pauses, changes in the velocity of speech like pre-boundary lengthening) that coexist with speech sounds (vowels and consonants) in natural speech [1]. Prosodic processing has been long since acknowledged as an important component of speech perception [2]. Prosody may convey affective (i.e., expression of emotions) and/or linguistic information. Prosody-related linguistic information enables the segmentation of utterances into speech units with acoustically defined boundaries [3–7], which is commonly referred to as prosodic phrasing. Prosodic words, phonological phrases (relatable to clause components like noun or verb phrases) and intonational phrases (relatable to clauses) form a hierarchy of prosodically defined speech units [1,4,8]. The role of prosodic phrasing in speech processing is twofold: it allows the segmentation of larger speech units into smaller ones for chunking purposes, and it provides a context to solve syntactic or semantic ambiguities in speech, as in garden-path sentences [1,9–12]. Prosody is not the sole contributor to speech segmentation—syntax and discourse context also play a role—but it is a fundamental one [13].

The perception of prosodic units has been a topic of interest to linguistics, psycholinguistics and, more recently, to cognitive neuroscience. In 1999, Steinhauer and colleagues [12] found a neural correlate of phrase boundary perception in Event-Related Potential (ERP), opening a new strand of research on speech processing. In this first study [12], the authors found that sentences with two intonational phrase boundaries (B versions) elicited two positive deflections, while sentences with one boundary (A versions of the same sentences) elicited only one. Since these positive deflections were temporally aligned with intonational phrase boundaries, the ERP component was taken to reflect a phrase closure mechanism and was termed Closure Positive Shift (CPS). The discovery was replicated in other languages such as English (e.g., [14]), German (e.g., [15]) and Dutch (e.g., [16]). CPS was also observed in silent reading, i.e., without acoustic input [17]. Critically, Pannekamp et al. [18] showed that CPS is also elicited in the absence of syntactic, semantic, and lexical content in the sentences, a finding that directly links CPS with the prosodic aspects of phrase boundary processing.

The literature on the perception of Intonational Phrase (IPH) boundaries emphasizes three prosodic boundary cues: pause (silent interval after the last word in the phrase boundary), pitch change, and pre-boundary lengthening (extension of the final word or syllable in the phrase) [5,14,19]. Phrase boundaries containing these cues are more easily identified and considered more salient when compared to sentences lacking prosodic boundary markers [7]. However, the extent to which each acoustic cue or cue combinations contribute to IPH boundary perception remains unclear.

Available findings regarding pre-boundary lengthening cues suggest that these modulate phrase boundary perception, at least when coupled with boundary-related pitch information. Most studies have focused on phrasal units at the phonological phrase level (clause constituents), and all point to a relevant contribution from pre-boundary lengthening, equivalent to that of other cues (pitch, pause). Scott [20] manipulated pre-boundary lengthening in natural speech phonological phrases containing pitch information (e.g., (Kate) or Pat and Tony will come vs. (Kate or Pat) and Tony will come) and found that it modulates boundary recognition. Aasland and Baum [8], found that increasing pre-boundary lengthening in natural speech modulates behavioral recognition of phonological phrases (e.g., (Pink and black) and green vs. (Pink) and black and green) in neurotypical participants as efficiently as pauses and pitch markers. Holzgrefe-Lang et al. [19] used a combination of ERP and behavioral measures to investigate cue weighting in the recognition of phonological phrases (e.g., (Mona) or Lena and Lola vs. (Mona or Lena) and Lola). They manipulated pitch and pre-boundary lengthening independently and asked participants to perform an explicit boundary recognition task while the EEG was recorded. They found that a combination of pitch change and pre-boundary lengthening is necessary to elicit CPS. In contrast, at least one study using larger units—intonational phrases (e.g., (If you want to keep ahead,) it is very necessary to take time to do exercises)—found that the pause was a more powerful perceptual cue than both final lengthening and pitch cues, with the latter two cues showing up as perceptually equivalent [7].

Available findings suggest, thus, that the role of pre-boundary lengthening may differ across phonological phrases and intonational phrases, being weaker in the latter. This would be in line with findings of slightly different EEG responses to phonological phrases vs. intonational phrases [4], suggesting that these two units may not be equivalent when it comes to boundary recognition mechanisms. On the other hand, research on acoustic cues has consistently used explicit tasks (recognize boundary structure), but not implicit ones—as in the original CPS studies [12,18], where listeners were asked to perform a prosody-unrelated task while listening to sentences. Therefore, it is yet unknown if pre-boundary lengthening remains a relevant cue for prosodic phrasing in intonational phrase units and implicit (inattentive) prosodic processing, which is likely the most realistic context of speech perception, since listeners in natural settings tend to focus on lexico-syntactic information.

In the present study, we investigated the role of pre-boundary lengthening in the implicit detection of intonational phrase boundaries as measured by the CPS component. To that end, we used the original CPS paradigm [12,18], in which natural speech sentence pairs with one (version A) vs. two IPH boundaries (version B), reflecting clause-like constituents, that are presented to participants while they perform a lexical recognition task. The principle embedded in this paradigm is that responses to A vs. B versions should not differ in the first IPH boundary (IPH1, common to both), but only in the second IPH boundary (IPH2, available in B but not in A). From this viewpoint, a CPS effect means that the B–A difference is positive at the IPH2 boundary and larger than the B–A difference in the IPH1 boundary. As a first necessary step to proceed with our ultimate goal, we investigated whether the CPS effect was present in our original set of natural speech stimuli. Consequently, in order to determine the effect of pre-boundary lengthening (ultimate goal), we manipulated the original set of sentence pairs twice: first, by reducing the amount of pre-boundary lengthening (short set), secondly, by increasing it (long set). In both manipulations, we kept the pitch- and pause-related cue values of the original set. According to our hypothesis, if increased pre-boundary lengthening enhances boundary detection, we should see increased CPS effects in long compared to original, and in original compared to short. If the opposite holds true (longer prosodic units like IPHs and/or implicit tasks diminish the impact of pre-boundary lengthening) other patterns may be expected.

2. Materials and Methods

2.1. Participants

According to a priori power analysis, we would need at least 28 participants to capture a medium effect size with 80% power and a critical alpha of 0.05. Fifty-four native speakers of European Portuguese enrolled in this study. Thirteen were excluded due to excessive EEG artifacts (more than 30% of contaminated trials, n = 6) or outlier voltage values (n = 7). The final sample consisted of 41 participants (31 female, 10 male), aged 18-45 (M = 21.8; SD = 5.88), with a mean of 13.7 years of formal education (SD = 2.27; range 11–20). None reported hearing problems or epilepsy. All participants gave informed consent according to the declaration of Helsinki. The project was approved by the ethics committee of the Faculty of Psychology and Educational Sciences of the University of Porto (Ref. 2022/01-10).

2.2. Stimulus Materials

Following the CPS paradigm, we created a set of 48 European Portuguese sentence pairs, which either had the potential to generate one phrase boundary (two clauses, A version) or two (three clauses, B version). These were syntactically simple declarative sentences composed of high-frequency words (frequency data taken from the Porlex database [21]), verbs, and other syntactic constituents such as "and" or "but". The two sentences in each pair had similar lexical content, and they were matched for the number of words and syllables (for A versions, mean number of syllables = 25.0; SD = 2.4; for B versions, mean number of syllables = 25.3; SD = 2.1). In the A versions, sentences contained one potential phrase boundary at an early position in the sentence, and in the B version an additional one at a later position. These sentence pairs were read by a native Portuguese speaker (female) in a sound booth and digitally recorded at 24 bit a sampling rate of 48 kHz. All files were normalized to +70 dB rms. An example of each version (A vs. B) is provided below ((1); # denotes phrase boundaries and IPH1/2 the identity of the preceding IPH; the complete list of sentences is presented in Appendix A).

(1):

A: (O João comprou carne) #IPH1 (o Jorge e a Luísa trouxeram saladas e bebidas). João bought meat # Jorge and Luísa brought salads and drinks.

B: (O João comprou carne) #IPH1 (O Jorge trouxe saladas) #IPH2 (e a Luísa trouxe bebidas). João bought meat # Jorge brought salads # and Luísa brought drinks.

Since a one-to-one mapping of syntactic units onto prosodic ones is not mandatory [22], we made perceptual and acoustic validations of prosodic structure. Prior to running the experiment, the initial pool of 48 pairs (AB) of spoken sentences was rated for the clarity in number of IPHs by four independent annotators (judges), among whom there was a foreign listener (naive to the Portuguese language). Annotators were asked to state whether A versions had clearly two IPHs, and whether B versions had three by answering Yes, No or Not sure (Appendix B). In cases where more than one annotator answered No or Not sure to one or both versions of a sentence, the pair was rejected. The final selection of 30 AB pairs was made. The set was acoustically validated for differences between the IPH boundaries of A vs. B versions regarding pause length and pitch change (expected to be equivalent at IPH1 but not IPH2). Pauses at IPH2 (version B) had an average length of 361 ms (SD = 118 ms), while in version A they were undetectable at the corresponding locations. Pause length at the end of IPH1 was similar across the two versions (A: M = 364; SD = 110; B: 384 ms, SD = 124 ms). Pitch cues were measured by computing the change in fundamental frequency in the last 200 ms of the IPH. As expected, the analysis showed rising pitch trends for stimuli at IPH2 in B versions (M = 78.81 Hz) that were not seen in the A version at the same position (M = 1.24 Hz).

The critical validation concerned pre-boundary lengthening at the end of IPH1 (version A and B) and IPH2 (version B only, Figure 1). Pre-boundary lengthening was defined as a larger-than-one ratio between the last stressed syllable of the IPH (in which pre-boundary lengthening is expected to begin) and the first stressed syllable of the sentence. Note that, in line with Oyedeji et al. [23], we assume that boundary cues (pitch and lengthening) start to appear in the last stressed syllable of the IPH (Figure 1), at least in European Portuguese (EP). In EP, the last stressed syllable of IPHs is not always the last syllable of the word. Instead, stress tends to occur at the penultimate syllable (trochaic pattern, though other situations exist). Since EP has, at least, partly, a stress-based rhythm, post tonic syllables (i.e., the last syllable, or the two final syllables) are usually marked by vowel reduction (conversion to schwa, e.g., salada becomes salade) or deletion (elimination of vowel, e.g., carn for carne), except in highly specific statements such as greeting calls [24]. Therefore, applying length changes in the last syllable would make the word sound unnatural, as if only the consonant had been lengthened.

Example (2) indicates the position of the first syllable of the sentence (upper case) and that of the last stressed syllable in the IPH (underlined)

(2):

A: (O JoÃO comprou <u>car</u>ne) # (o Jorge e a Luísa trouxeram saladas e be<u>bi</u>das). João bought meat # Jorge and Luísa brought salads and drinks.

B: (O JoÃO comprou <u>car</u>ne) # (O Jorge trouxe sa<u>la</u>das) # (e a Luísa trouxe be<u>bi</u>das). João bought meat # Jorge brought salads # and Luísa brought drinks.

The original set was edited twice, generating two additional sets: short and long. In the short set, pre-boundary lengthening in both A and B versions was set to half (in most cases eliminating pre-boundary lengthening). In the long set, pre-boundary lengthening was doubled. As a result of the transformations made in IPH1, IPH2 boundaries (B versions) were time-shifted in short (earlier) and long sentences (later, Figure 2). These manipulations were made using the software Praat (https://www.fon.hum.uva.nl/praat/ Amsterdam, The Netherlands; Version 6.51.52, accessed on 10 October 2021), specifically by creating a duration tier wherein we marked the time limits of the last stressed syllable and multiplied (long) or divided (short) the original duration by a factor of 2. Stimuli were, then, resynthesized, generating additional audio files with the desired transformations.

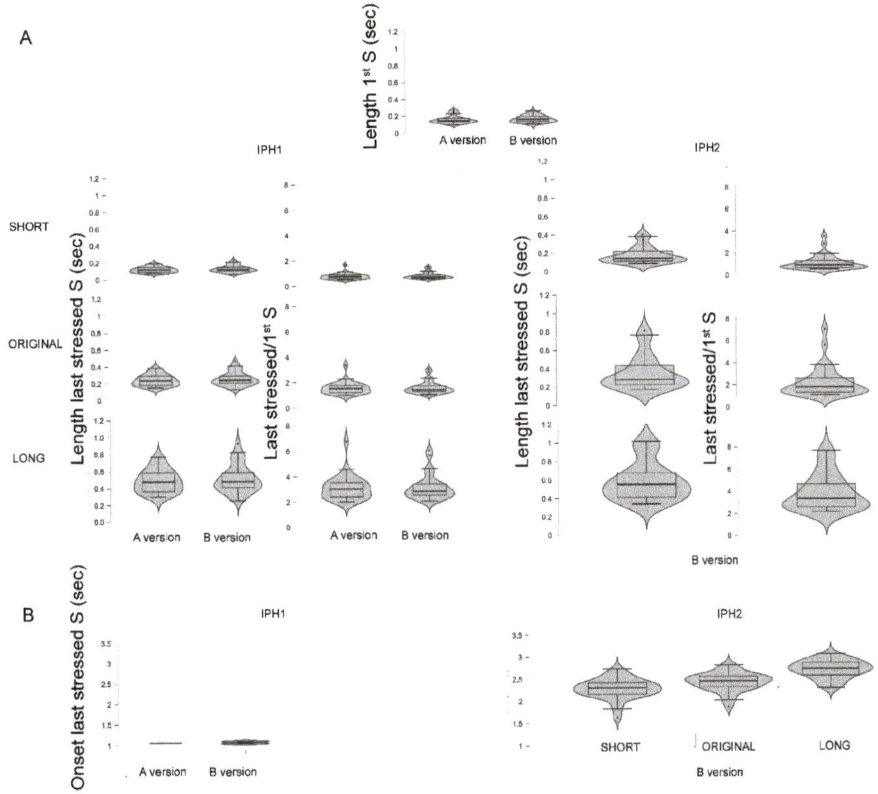

Figure 1. Properties of stimulus materials: (**A**) Length of first syllable (up), length of last stressed syllable and pre-final lengthening (length of last stressed/length of first) for IPH1 (down, left) and IPH 2 (down, right); (**B**) Onset time of last stressed syllable in IPH1 and IPH2. Note: IPH = intonational phrase; though IPH2 was present only in B versions, EEG analyses used B versions onset times in A for comparison between absent (**A**) and present boundary (**B**).

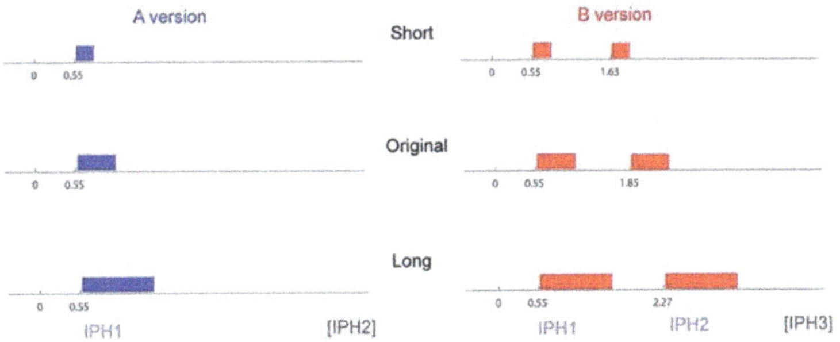

Figure 2. IPH boundaries (rectangles represent last stressed syllable of each IPH) at A and B versions, for short vs. original vs. long stimulus sets. For EEG analysis, IPH2 boundaries in A versions were copied from B versions.

2.3. Procedure

Participants were instructed to listen to the sentences while performing a vigilance task, which was unrelated to the experimental manipulation, i.e., they did not perform explicit judgments about prosodic phrase boundaries. Participants listened to several sentences heard from the speakers connected to the stimulation computer. At the end of each sentence, a word appeared on the screen. Participants were then asked to judge if each of the words was part of the previously heard sentence or not, pressing two different keys on the computer keyboard (YES or NO). After receiving the instructions, participants performed practice trials and possible doubts were clarified. Counterbalancing across participants was performed by switching the label of the key numbers on the computer keyboard, creating two versions of the task, 1 (YES/NO) and 2 (NO/YES). For each of these two versions, we created two variants by pseudo randomizing the order of trials. The goal was to avoid consecutive presentations of the A and B versions of the same nuclear sentence or two length-related conditions of the same sentence.

Each trial was structured as follows: a fixation cross signaled the onset of the auditory presentation of the sentence; after the offset of the sentence, a blank screen was presented for 200 ms; the probe word appeared on the screen until a response was provided; and the response was followed by an interstimulus interval of 2000 ms, during which the screen was blank.

The experiment was run in an acoustically shielded room and lasted around 40 min, head preparation included.

2.4. EEG Recording and Preprocessing

Participants were seated in a comfortable chair in front of the stimulation computer. We then placed on their scalp an electrode cap with 64 active channels positioned according to the 10-20 system (FP1, FPz, FP2, AF7, AF3, AFz, AF4, AF8, F7, F5, F3, F1, Fz, F2, F4, F6, F8, FT7, FC5, FC3, FC1, FCz, FC2, FC4, FC6, FT8, T7, C5, C3, C1, Cz, C2, C4, C6, T8, TP7, CP5, CP3, CP1, CPz, CP2, CP4, CP6, TP8, P9, P7, P5, P3, P1, Pz, P2, P4, P6, P8, P10, PO7, PO3, POz, PO4, PO8, O1, Oz, O2, Iz). Two external electrodes placed at the mastoids were added to allow re-referencing during preprocessing. An additional electrode was placed below the left eye to record vertical eye movements (VEOG).

EEG data was collected using a Biosemi ActiveTwo system (https://www.biosemi.com/ Amsterdam, The Netherlands; Accessed on 1 July 2021) with 512 Hz sampling rate. Before the experiment started, signal quality was checked and kept under the system-recommended thresholds. Participants were asked to move as little as possible and try to blink only between trials.

We preprocessed EEG data with the Fieldtrip toolbox [25] for MATLAB (https://www.mathworks.com/ Massachusetts, United States of America; Accessed on 15 August 2022). After trial definition based on sentence-onset triggers, trials with vertical and horizontal eye movement artifacts were marked based on visual analysis. Trials with other types of artifacts detectable by variance inspection were also marked, as well as defective channels. Contaminated trials were rejected, and bad channels were interpolated using nearest neighbor averaging. Clean trials were baseline corrected (200 ms pre-trigger), detrended, re-referenced to the mastoid electrodes, and band-pass filtered between 0.01 and 30 Hz. Finally, trials of each condition were averaged per subject, and then grand averaged.

Triggers were placed at the onset of each sentence, in line with previous approaches to CPS [14]. However, since we noted early divergences between A and B versions in this scenario, we adopted new, IPH-related baselines, by applying baseline correction to each trial at the 200 ms period preceding the two relevant events (see Results: the minimum boundary onset time for IPH1 (common to short, original and long), and the same for IPH2 (different time points across length conditions). We then extracted the time window between 150 ms and 650 ms post-boundary onset for both IPH1 and IPH2 and ran the statistical analysis. The IPH2 trigger point was based on B versions, where the boundary was present (Figure 2), and it was applied to the A versions. Thus, at this point, we were comparing presence (B) vs. absence (A) of a boundary.

2.5. Statistical Analysis

Time-averaged voltage values per subject and region of interest (nine regions: anterior, central, posterior x left, mid right) were extracted for the time windows between 150 and 650 ms post boundary onset (post onset of last stressed syllable).

First, we analyzed the CPS effect per length and topography, considering the B–A difference at IPH2 (expected to be higher than the one at IPH1 in case of significant effect) minus the B–A difference at IPH1 as dependent variable. The expected CPS effect (the previous value, expressing the interaction IPH x length) would, thus, consist of a positive value. Once observing length effects, we compared the length conditions two at a time to locate significant length-related differences. Finally, we moved into the analysis of IPH (1 vs. 2) x version (A x B) interactions in each length condition (short vs. original vs. long) to verify whether and where values were significant.

In all analyses, the critical alpha value was set to 0.05. Greenhouse Geiser corrections were made for sphericity violations. Complementary Bayesian analyses were run in case of marginal results. We calculated Bayes Factors (BF) with default priors to further investigate the alternative hypothesis over the null one (BF_{10}), using the JASP software (https://jasp-stats.org/ Amsterdam, The Netherlands; Version 0.16.0, Accessed on 15 September 2022) [26]. Unlike traditional null-hypothesis-significance-testing, which relies on dichotomous information (significant vs. non-significant results), Bayes factors quantify the relative predictive performance of two alternative hypotheses (alternative vs. null, or null vs. alternative), measuring the strength of evidence in favor of one over the other [27,28]. Bayes factors are particularly relevant to strengthen claims of null effects and clarify marginal results, and this was how we have mostly used them in the present study. Following the heuristics provided in van Doorn et al. [28], we considered BFs between 1 and 3, 3 and 10, 10 and 30 and above 30 as weak, moderate, strong and very strong evidence in favor of the alternative hypothesis. While BFs above 1 support the alternative hypothesis, BFs below 1 indicate evidence in favor of the null hypothesis, and evidence here becomes stronger as values decrease: BFs between 1 and 0.33 provided weak evidence, between 0.33 and 0.10 moderate, between 0.10 and 0.03 strong, and below 0.03 very strong.

3. Results

3.1. CPS Effect

The repeated measures ANOVA with length, caudality and laterality as factors, and the CPS effect ((B–A at IPH2)—(B–A at IPH1)) as dependent variable (Figure 3) showed a significant main effect of length, $F(1,40) = 3.40$, $p < 0.038$, $\eta_p{}^2 = 0.078$, without further interactions with topographical factors. Positive values for short and original indicated the expected effect (B–A at IPH2 greater than B–A at IPH1), while the negative values for long suggest that the effect is, at least, absent.

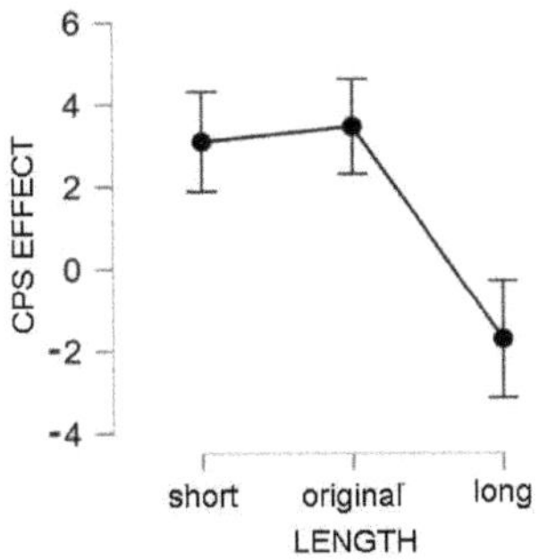

Figure 3. CPS effect across the three length (pre-boundary lengthening) conditions. Vertical bars represent 95% confidence intervals; The CPS effect refers to whole-scalp-averaged voltage values resulting from B–A at IPH 2 minus B–A at IPH 1 and is expected to be positive.

3.2. Pairwise Comparisons across Length Conditions

Comparisons between short and original showed no significant differences $F(1,40) = 0.041$, $p = 0.84$, $\eta_p{}^2 = 0.001$. Long differed significantly from original $F(1,40) = 4.66$, $p = 0.037$, $\eta_p{}^2 = 0.10$ and marginally from short $F(1,40) = 4.66$, $p = 0.052$, $\eta_p{}^2 = 0.091$, both comparisons showing medium effect sizes. Bayes factors revealed strong evidence of differences between short and long ($BF_{10} > 30$) and no evidence for the comparison short–original ($BF_{10} = 0.098$), suggesting that both short and original elicit relevant differences from long.

3.3. Interaction IPH x Version Per Length

For the original stimulus set, the interaction was significant, $F(1,40) = 5.87$, $p = 0.020$, $\eta_p{}^2 = 0.13$, and we found the expected CPS effect: B showed higher voltages than A at IPH2, while the reverse happened at IPH1 (Figure 4).

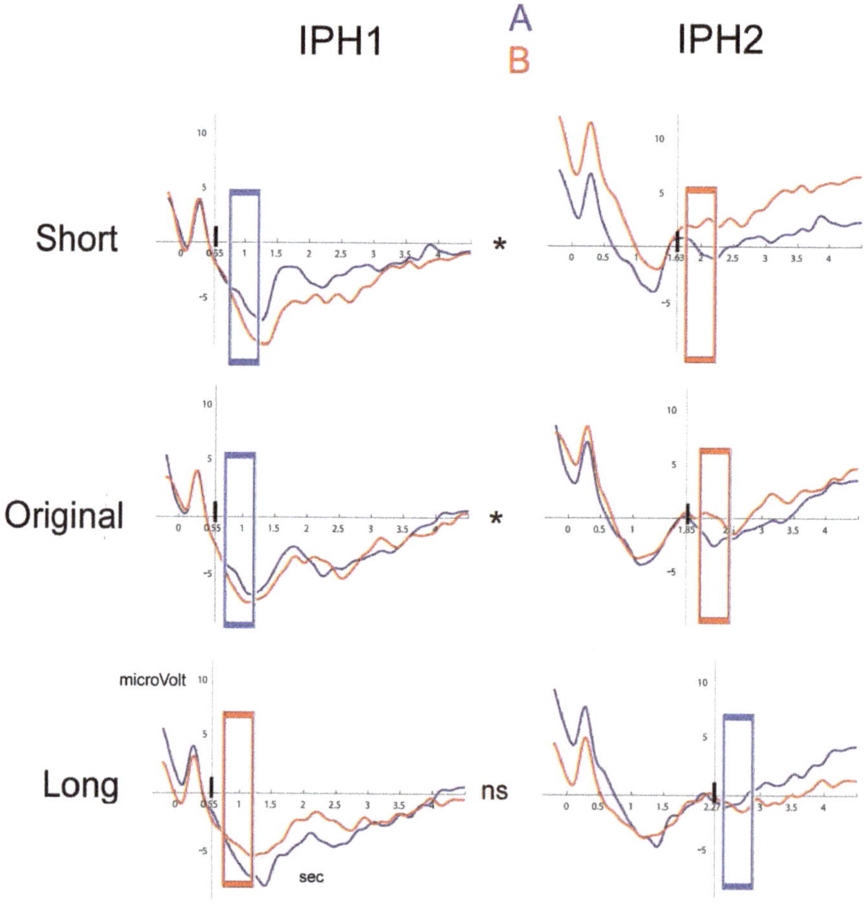

Figure 4. ERPs to IPH1 and IPH2 boundaries (onset of last stressed syllable in the IPH) across the three levels of pre-boundary lengthening. Baseline corrections were applied to the time points marking the onset of the last stressed syllable of both IPH1 and IPH2. The waveforms represent average voltage for the right-central region. Asterisks and ns (non-significant) refer to the significance of the IPH x version interaction (CPS effect). Rectangles indicate the 500 ms time window used for analysis: red rectangles indicate B > A and blue ones A > B.

For the short stimulus set, we found a similar scenario, though the interaction was marginal, $F(1,40) = 3.65$, $p = 0.064$, $\eta_p^2 = 0.083$. BFs, however, provided strong evidence for interaction ($BF_{10} > 30$).

For the long stimulus set, a different pattern emerged, with the interaction IPH x version losing significance, $F(1,40) = 1.59$, $p = 0.22$, $\eta_p^2 = 0.038$. Note that, despite the lack of significance, the expected direction of the CPS effect was even reversed, with A showing higher voltages than B at IPH2, while the reverse happened at IPH1 (Figure 4).

In summary, while longer-than-original pre-boundary length values appear to attenuate the CPS effect, shorter-than-original values do not seem to make a difference.

4. Discussion

In the current study, we wanted to determine whether increased pre-boundary lengthening leads to enhanced implicit intonational phrase boundary detection as measured with the CPS ERP component in European Portuguese (EP). To that end, we manipulated a set of natural speech sentence pairs both by reducing pre-boundary lengthening (short set) and enlarging it (long set). We found that pre-boundary lengthening seems to affect phrase boundary processing, but not in the expected way: while both short and original elicited similar CPS responses, responses to long did not show the CPS effect, differing from both short and original. Therefore, variations in pre-boundary lengthening did not have the expected effect on phrase boundary perception, and this may be accounted for by several reasons.

One reason (1) could be that, since other prosodic-boundary cues were available (at least pitch and pauses), listeners simply ignored the variations in lengthening when processing boundaries. This would explain the lack of difference between short and original. To explain the difference between original (CPS effect) and long (no CPS effect), we would have to hypothesize an additional mechanism wherein the enlarged lengthening that was applied to long versions was excessively unnatural, and these sounded like speech aberrations.

Besides the fact that listeners had other cues, why would listeners ignore pre-boundary lengthening in particular? Based on the literature (see introduction), our hypothesis was that dealing with large units (IPHs instead of phonological phrases, as in previous research) and/or deviating listeners' attention to non-prosodic information (implicit instead of explicit task) could diminish the weight of pre-boundary lengthening in IPH boundary perception when compared to phonological phrases (increased weight), in line with [7] vs. [8,19,20]. Why would these differences matter? Concerning the use of IPHs (clause-like) instead of phonological phrases (clause-component-like), we may hypothesize that longer units (IPHs) make it harder to maintain a reference syllable length in memory for comparison with the last stressed syllable, where pre-boundary lengthening takes place: if we admit that the first syllable is indeed a reference, then IPHs would require a much larger time window for memory maintenance than simple phonological phrases. As a result, listeners would focus more on short-time-window cues (pitch change) or even absolute cues (pause) for boundary processing. The reason for implicit tasks having a hypothetical negative effect on the use of pre-boundary lengthening may be related to the previous reasoning: faced with larger and more complex amounts of information to process (IPHs), listeners would be more available for short-term or absolute (less memory-demanding) cues in an implicit task than in an explicit one—where they were prompted to focus on prosodic patterns and, thus, have more chances to rely on all available cues. One way to test this possibility would be to carry out the experiment with synthetic speech, such that all boundary cues except lengthening were removed.

Regarding the failure in obtaining a CPS effect in long sentences (unlike what happened in short and original), it may have occurred because artificial lengthening, made without any pitch-related compensation, made the pitch contour unnatural, and this caused the atypical response we saw. A way to test this could be comparing long sets with vs. without pitch corrections for length. Moreover, even though we followed the procedures described in a previous study on EP, we agree that the possibility of extreme manipulations cannot be ruled out. For example, the ratios obtained for lengthening may have been bloated due to the frequent sentence-initial speed-up, and thus the manipulations based on those ratios were too extreme. A counter argument for this possibility is that, if manipulations were too extreme, they would affect the difference in short-original too (besides original-long), in the sense that the difference would be either shocking or noticeable. For instance, with a lengthening of 2, the difference between original and short would be 2–1, 1 point of difference for lengthening ratios; if lengthening was 1.5 for original, it would be 0.75 for short −0.75 difference). We saw no differences in the ERPs for short vs. original sets, suggesting that listeners did not feel unnatural length reduction in short versions and, furthermore, they did not discriminate between short from original. Future studies should address this question and consider other ways of calculating the degree of lengthening, for example, by using average syllable durations as a reference point instead of IP-initial syllables.

Another reason (2) for the lack of differences between short and original sets may be that listeners—instead of ignoring the length cue—tolerated the difference between short and long, something that makes sense in light of sociolinguistic explanations. In European Portuguese, clear differences are found between north and south dialects concerning pre-boundary lengthening, with northern variants showing increased values [29]. Since Portugal's capital (and most urban) city is in the center-south, upper-class dialects in the north tend to adapt to south features, including shorter pre-boundary lengthening. On the other hand, our experiment was run in the north, where most participants were, thus, exposed to both north (due to location) and south (due to dialect adaptation) variants. This may have placed short and original sentences at similar levels of familiarity, thus preventing listeners from perceiving the short sentences as unnatural. One way of testing this hypothesis would be running the experiment with southern participants, listening to our northern speaker. In case the short stimuli showed advantage over the original (northern) one, this would indicate that familiarity plays a role in how pre-boundary lengthening is used in phrasing.

Besides the suggestions for future studies presented above, other ideas arise from the limitations of this study. Perhaps the biggest limitation is that we did not compare phonological phrases with intonational phrases, nor implicit with explicit tasks. This precludes us proposing more solid interpretations of our findings and make such comparisons a priority for the near future. As also mentioned in the methods, we saw early divergences in the responses to A and B versions, but we should only see these divergences at the boundary of IPH2. The most likely cause for this is the fact that we used natural speech in both versions and, hence, it was difficult to achieve perfect acoustic equivalence between the two versions, especially when speech units were relatively long. Perhaps future studies can apply post-recording prosodic manipulations to make versions A and B identical up to the onset of the IPH2 boundary. Moreover, looking at the ERP waveforms, one may question whether the time windows used for analysis were too short. It is indeed possible that CPS responses were prolonged beyond this time window of interest. However, the waveforms also clearly indicate that the pattern found in the time window analysis remains till the end of the sentence. Finally, regarding our motivated choice to manipulate the last stressed syllable of the IPH, future studies could investigate what happens when the last syllable is manipulated instead.

Despite its limitations, this study is, to our knowledge, the first to determine the role of pre-boundary lengthening in the implicit recognition of intonational phrase boundaries, raising new questions to address in the future. Our findings suggest that prosodic units at different scales may recruit different types of acoustic cues when it comes to perceptual segmentation, namely that pre-boundary lengthening is not recruited in the perception of IPHs when other cues are available. Finally, it is also possible that sociolinguistic factors have a strong influence in the process of prosodic boundary recognition.

Author Contributions: Conceptualization, A.R.B., S.S. and V.F.; methodology, D.C., S.S. and V.F.; software, S.S.; formal analysis, A.R.B. and S.S.; investigation, A.R.B., D.C. and S.S.; data curation, A.R.B., D.C. and S.S.; writing—original draft preparation, A.R.B., D.C., S.S. and V.F.; writing—review and editing, A.R.B., D.C., S.S. and V.F.; visualization, A.R.B. and S.S.; supervision, S.S.; project administration, S.S.; funding acquisition, S.S. All authors have read and agreed to the published version of the manuscript.

Funding: This research was funded by the Portuguese Foundation for Science and Technology (FCT), grant numbers CPUP UIDB/00050/2020; and PTDC/PSI-GER/5845/2020.

Institutional Review Board Statement: Ethical statement: The study was conducted in accordance with the Declaration of Helsinki and approved by the Ethics Committee of Faculty of Psychology and Educational Sciences at University of Porto (protocol code 2022/01-10, date of approval 17 January 2022).

Informed Consent Statement: Informed consent was obtained from all subjects involved in the study.

Data Availability Statement: Data availability: stimulus materials (database and audio), as well as statistical analyses are available at osf link https://osf.io/gs3ep/?view_only=ae7fec4571b741ecbaca7f13a8e7df80.

Acknowledgments: We are grateful to José Sousa and Ana Mesquita for logistic support. We thank all our participants. Our gratitude also goes to R.G. who played an important role in preparing the experiment, collecting data and sketching the analysis, but was unable to follow and approve the current version of the manuscript due to professional constraints.

Conflicts of Interest: The authors declare no conflict of interest.

Appendix A

Sentence pairs

ID	A Version	B Version
1	O João comprou carne, o Jorge e a Luísa trouxeram saladas e bebidas.	O João comprou carnes, o Jorge trouxe salada, e a Luísa trouxe bebidas.
2	A carne está estragada, o marisco e a fruta aguentaram-se bastante bem.	A carne está estragada, o marisco ficou bem, e a fruta aguentou-se firme.
3	O meu vestido era preto, o da Ana e o da Luísa tinham tons de azul e laranja.	O meu vestido era preto, o da Ana era todo azul, e o da Luísa tinha laranja.
4	O Manuel apresentou, o Daniel e o Alexandre dançaram rumba e cha-cha-chá.	O Manuel apresentou, o Daniel dançou rumba, e o Alexandre cha-cha-chá.
5	Eu fiquei por lá sentada, a Cláudia e a Inês saíram logo para a sala de jantar.	Eu fiquei por lá sentada, a Cláudia foi para átrio, e a Inês saiu para a cozinha.
6	O avião é às três horas, o comboio da noite ou o barco já não interessam.	O avião é às três horas, o comboio sai às dez, e o barco já não interessa.
8	Perdi todo o subsídio, reclamei logo e contactei os serviços de finanças.	Perdi todo o subsídio, reclamei nas finanças, contactei a segurança social.
9	Eu saio de casa cedo, observo bem os ramos nus e os pássaros tão leves.	Eu saio de casa cedo, observo os ramos nus, e os pássaros tão leves.
10	Nós limpamos a casa, os pais e você tratam da roupa e de abrir a porta.	Nós limpamos a casa, os pais tratam da roupa, e vocês abrem a porta.
11	Tu contas-me tudo já, eu e o juiz fazemos o relatório completo do caso.	Tu contas-me tudo já, eu faço o relatório, e depois o juiz vai expor o caso.

ID	A Version	B Version
13	Os cães dormem fora, os gatos e os peixes ficam onde estão dentro de casa.	Os cães dormem fora, os gatos ficam em casa, e os peixes estão no aquário.
16	Ele até ensina bem, mas os testes e os trabalhos são difíceis e numerosos.	Ele até ensina bem, mas os testes são longos, e os trabalhos muito difíceis.
19	Eu gosto bastante dela, mas a paciência e a compreensão por vezes falham.	Eu gosto bastante dela, mas a paciência falha, e compreensão é bem difícil.
21	Tornou-se conhecido, e deixou de se interessar como antes pelo rigor.	Tornou-se conhecido, mas ao ficar célebre, descurou o trabalho e o rigor.
23	Aspirei bem os quartos, mas mesmo assim o pó continuou no ar abafado.	Aspirei bem os quartos, mas o pó não saiu todo, nem o ar ficou fresco.
27	A mesa já é velha, mas a madeira e a cor são muito bonitas e requintadas.	A mesa já é velha, mas a madeira é boa, e a cor parece-me requintada.
29	Embora esteja frio, já se sente um calorzinho do sol e um ar leve de verão.	Embora esteja frio, o sol já está brilhante, e sente-se um ar leve de verão.
30	Quando forem horas, tu e o secretário tratam desses papéis e dos telefonemas.	Quando forem horas, tu tratas desses papéis, e o secretário faz os telefonemas.
31	Segundo o que dizem, mãe e filha percebem muito de festas e receções.	Segundo o que têm dito, a mãe percebe de festas, e a filha sabe receber.
32	Se nos fores lá buscar, a Isabel e eu levamos as duas colunas e o amplificador.	Se nos fores lá buscar, levo as duas colunas, e a Isabel traz o amplificador.
35	Desde que ali entrou, deixou de se ouvir o barulho e a confusão de antes.	Desde que ali entrou, não se ouve barulho, nem houve mais confusão.
36	Quando logo saires, não te esqueças de levar a chave e os teus postais.	Quando logo saires, fecha bem à chave, e leva embora os teus postais.
38	Quando vocês chegarem, o João e eu vamos buscar-vos com a bagagem também.	Quando vocês chegarem, eu vou buscar-vos, e o João ajuda com a bagagem.
40	Contando que haja sol, a Helena e o Carlos trazem a prancha e a mota de água.	Contando que haja sol, a Helena traz a prancha, e o Carlos a mota de água.
42	Concordei com tudo, desde que pudesse ver e também experimentar por um dia.	Concordei com tudo, desde que pudesse ver, e depois experimentar por um dia.
43	Trabalhamos nesta sala, só se a Eva e o Pedro a pintarem e decorarem.	Trabalhamos nesta sala, se a Eva a pintar, e também se o Pedro a decorar.
44	Se for mesmo preciso, posso acabar ainda hoje as reportagens e as entrevistas.	Se for mesmo preciso, posso acabar isto hoje, e deixo para amanhã a entrevista.
45	Em situações destas, é melhor manter silêncio e também muita discrição.	Em situações destas, é melhor criar silêncio, e manter muita discrição.
46	Trabalhamos nesta sala, só se a Eva e o Pedro a pintarem e decorarem.	Trabalhamos nesta sala, se a Eva a pintar, e também se o Pedro a decorar.
48	Em situações destas, é melhor manter silêncio e também muita discrição.	Em situações destas, é melhor criar silêncio, e manter muita discrição.

Appendix B

Perceptual validation of number of IPHs in versions A (two IPHs) vs. B (three IPHs)

Sentences	Annotator 1—Foreigner	Annotator 2	Annotator 3	Annotator 4	Decision
1a	No	Yes	Yes	Yes	Accept
1b	No	Yes	Yes	Yes	Accept
2a	Yes	Yes	Yes	Yes	Accept
2b	No	Yes	Yes	Yes	Accept
3a	Yes	Not sure	Yes	Not sure	Reject
3b	No	Not sure	Yes	Not sure	Reject
4a	No	Yes	No	No	Reject
4b	No	Yes	Not sure	Not sure	Reject
5a	Yes	Not sure	Yes	Yes	Accept
5b	Yes	Not sure	Yes	Yes	Accept
6a	Yes	Not sure	Yes	Not sure	Reject
6b	No	Not sure	No	Yes	Reject
7a	No	Yes	No	No	Reject
7b	No	Yes	Yes	No	Reject
8a	Yes	Yes	No	Yes	Accept
8b	Yes	Yes	No	Yes	Accept
9a	Yes	Not sure	Not sure	Yes	Reject
9b	No	No	Yes	Yes	Reject
10a	Yes	No	Yes	Yes	Accept
10b	Not sure	Yes	Yes	Yes	Accept
11a	Yes	Not sure	Yes	Yes	Accept
11b	Yes	Not sure	Yes	Yes	Accept
12a	Yes	Yes	Yes	Yes	Accept
12b	Yes	Yes	Yes	Yes	Accept
13a	Yes	Yes	Yes	Yes	Accept
13b	Not sure	Yes	Yes	Yes	Accept
14a	Yes	Yes	Yes	Not sure	Reject
14b	No	Yes	No	Not sure	Reject
15a	Not sure	Yes	Yes	Yes	Accept
15b	Yes	Yes	Yes	Yes	Accept
16a	Yes	Yes	No	Not sure	Reject
16b	Yes	Yes	No	Not sure	Reject
17a	Yes	Yes	Yes	Yes	Accept
17b	Not sure	Yes	Yes	Yes	Accept
18a	Yes	Yes	Yes	Yes	Reject
18b	No	Not sure	Not sure	No	Reject
19a	Yes	Yes	Yes	Yes	Accept
19b	Not sure	Yes	Yes	Yes	Accept

Sentences	Annotator 1—Foreigner	Annotator 2	Annotator 3	Annotator 4	Decision
20a	Yes	Yes	Yes	Yes	Accept
20b	Yes	Yes	Yes	Yes	Accept
21a	Yes	Not sure	Yes	Yes	Accept
21b	Yes	Not sure	Yes	Yes	Accept
22a	Yes	Not sure	Yes	Yes	Accept
22b	Yes	Yes	Yes	Yes	Accept
23a	No	Yes	Not sure	Not sure	Reject
23b	Yes	Yes	Yes	No	Reject
24a	Yes	Yes	Yes	Yes	Accept
24b	Yes	Yes	Yes	Yes	Accept
25a	Yes	No	Not sure	Yes	Reject
25b	No	No	Not sure	Yes	Reject
26a	Yes	Yes	Not sure	Yes	Accept
26b	Yes	Yes	Not sure	Yes	Accept
27a	Yes	Yes	Yes	Yes	Accept
27b	Yes	Yes	Yes	Yes	Accept
28a	Yes	Yes	Not sure	Yes	Accept
28b	Yes	Yes	Not sure	Yes	Accept
29a	No	No	No	No	Reject
29b	Yes	Yes	Yes	Not sure	Reject
30a	Yes	Yes	Yes	Yes	Accept
30b	Yes	Yes	Yes	Yes	Accept
31a	Yes	Yes	Yes	No	Accept
31b	Yes	Yes	Yes	No	Accept
32a	Yes	No	Yes	Yes	Accept
32b	Yes	No	Yes	Yes	Accept
33a	No	No	No	No	Reject
33b	No	No	No	No	Reject
34a	No	Not sure	Not sure	Yes	Reject
34b	Yes	Not sure	Yes	No	Reject
35a	Yes	Yes	Yes	No	Accept
35b	Yes	Yes	Yes	Yes	Accept
36a	Yes	Yes	Yes	Yes	Accept
36b	Yes	Yes	Not sure	Yes	Accept
37a	No	No	Not sure	Yes	Reject
37b	Yes	No	Not sure	Yes	Reject
38a	Yes	Not sure	Not sure	Not sure	Reject
38b	No	Yes	Not sure	Not sure	Reject
39a	Yes	Yes	Yes	Yes	Accept
39b	Yes	Yes	Yes	Yes	Accept

Sentences	Annotator 1—Foreigner	Annotator 2	Annotator 3	Annotator 4	Decision
40a	Yes	Not sure	Yes	No	Reject
40b	Not sure	Not sure	Yes	No	Reject
41a	Yes	No	Yes	Yes	Accept
41b	Yes	Yes	Yes	Yes	Accept
42a	Yes	Yes	Yes	Yes	Accept
42b	Yes	Yes	No	Yes	Accept
43a	Yes	Yes	Yes	Yes	Accept
43b	Yes	Yes	No	Yes	Accept
44a	Yes	Yes	No	No	Reject
44b	No	Yes	No	No	Reject
45a	Not sure	Yes	Yes	Yes	Reject
45b	Not sure	No	Not sure	No	Reject
46a	Yes	No	Yes	Yes	Accept
46b	Yes	No	Yes	Yes	Accept
47a	Yes	Yes	Yes	Yes	Accept
47b	Yes	Yes	Yes	Yes	Accept
48a	Yes	Yes	Yes	Yes	Accept
48b	Yes	Yes	Yes	Yes	Accept

References

1. Hawthorne, K.; Gerken, L. From pauses to clauses: Prosody facilitates learning of syntactic constituency. *Cognition* **2014**, *133*, 420–428. [CrossRef] [PubMed]
2. Pisoni, D.B.; Sawusch, J.R. Some stages of processing in speech perception. In *Structure and Process in Speech Perception*; Cohen, A., Nooteboom, S.G., Eds.; Springer: Berlin/Heidelberg, Germany, 1975; Volume 11, pp. 16–35.
3. Anurova, I.; Vetchinnikova, S.; Dobrego, A.; Williams, N.; Mikusova, N.; Suni, A.; Mauranen, A.; Palva, S. Event-related responses reflect chunk boundaries in natural speech. *Neuroimage* **2022**, *255*, 119203. [CrossRef]
4. Li, W.; Yang, Y. Perception of prosodic hierarchical boundaries in Mandarin Chinese sentences. *Neuroscience* **2009**, *158*, 1416–1425. [CrossRef] [PubMed]
5. Selkirk, E. *Phonology and Syntax: The Relation between Sound and Structure*; The MIT Press: Cambridge, MA, USA, 1984.
6. Selkirk, E. Comments on intonational phrasing in English. In *Prosodies. With Special Reference to Iberian Languages*; Frota, S., Vigário, M., Freitas, M.J., Eds.; Mouton de Gruyter: Berlin, NY, USA, 2005; pp. 11–58.
7. Yang, X.; Shen, X.; Li, W.; Yang, Y. How Listeners Weight Acoustic Cues to Intonational Phrase Boundaries. *PLoS ONE* **2014**, *9*, e102166. [CrossRef] [PubMed]
8. Aasland, W.; Baum, S.R. Temporal parameters as cues to phrasal boundaries: A comparison of processing by left- and right-hemisphere brain-damaged individuals. *Brain Lang.* **2003**, *87*, 385–399. [CrossRef]
9. Kerkhofs, R.; Vonk, W.; Schriefers, H.; Chwilla, D.J. Discourse, Syntax, and Prosody: The Brain Reveals an Immediate Interaction. *J. Cogn. Neurosci.* **2007**, *19*, 1421–1434. [CrossRef]
10. Kjelgaard, M.M.; Speer, S.R. Prosodic Facilitation and Interference in the Resolution of Temporary Syntactic Closure Ambiguity. *J. Mem. Lang.* **1999**, *40*, 153–194. [CrossRef]
11. Steffman, J.; Katsuda, H. Intonational Structure Influences Perception of Contrastive Vowel Length: The Case of Phrase-Final Lengthening in Tokyo Japanese. *Lang. Speech* **2020**, *64*, 839–858. [CrossRef]
12. Steinhauer, K.; Alter, K.; Friederici, A.D. Brain potentials indicate immediate use of prosodic cues in natural speech processing. *Nat. Neurosci.* **1999**, *2*, 191–196. [CrossRef]
13. Frazier, L.; Carlson, K.; Cliftonjr, C. Prosodic phrasing is central to language comprehension. *Trends Cogn. Sci.* **2006**, *10*, 244–249. [CrossRef] [PubMed]
14. Peter, V.; McArthur, G.; Crain, S. Using event-related potentials to measure phrase boundary perception in English. *BMC Neurosci.* **2014**, *15*, 129. [CrossRef] [PubMed]
15. Holzgrefe, J.; Wellmann, C.; Petrone, C.; Truckenbrodt, H.; Hoehle, B.; Wartenburger, I. Brain response to prosodic boundary cues depends on boundary position. *Front. Psychol.* **2013**, *4*, 421. [CrossRef]

16. Bögels, S.; Schriefers, H.; Vonk, W.; Chwilla, D.J.; Kerkhofs, R. The Interplay between Prosody and Syntax in Sentence Processing: The Case of Subject- and Object-control Verbs. *J. Cogn. Neurosci.* **2010**, *22*, 1036–1053. [CrossRef]
17. Hwang, H.; Steinhauer, K. Phrase Length Matters: The Interplay between Implicit Prosody and Syntax in Korean "Garden Path" Sentences. *J. Cogn. Neurosci.* **2011**, *23*, 3555–3575. [CrossRef]
18. Pannekamp, A.; Toepel, U.; Alter, K.; Hahne, A.; Friederici, A.D. Prosody-driven Sentence Processing: An Event-related Brain Potential Study. *J. Cogn. Neurosci.* **2005**, *17*, 407–421. [CrossRef] [PubMed]
19. Holzgrefe-Lang, J.; Wellmann, C.; Petrone, C.; Räling, R.; Truckenbrodt, H.; Höhle, B.; Wartenburger, I. How pitch change and final lengthening cue boundary perception in German: Converging evidence from ERPs and prosodic judgements. *Lang. Cogn. Neurosci.* **2016**, *31*, 904–920. [CrossRef]
20. Scott, D.R. Duration as a cue to the perception of a phrase boundary. *J. Acoust. Soc. Am.* **1982**, *71*, 996–1007. [CrossRef]
21. Gomes, I.; Castro, S.L. Porlex: A lexical database in European Portuguese. *Psychologica* **2003**, *32*, 91–108.
22. Himmelmann, N.P.; Sandler, M.; Strunk, J.; Unterladstetter, V. On the universality of intonational phrases: A cross-linguistic interrater study. *Phonology* **2018**, *35*, 207–245. [CrossRef]
23. Oyedeji, M.; Annie, C.G.; Shari, R.B.; Miguel, O., Jr. Electrophysiological correlates of prosodic boundaries at different levels in Brazilian Portuguese. In Proceedings of the 19th International Congress of Phonetic Sciences, Melbourne, Australia, 5–9 August 2019.
24. Frota, S.; Cruz, M.; Fernandes-Svartman, F.; Collischonn, G.; Fonseca, A.; Serra, C.; Oliveira, P.; Vigário, M. Intonational variation in Portuguese: European and Brazilian varieties. In *Intonation in Romance*; Frota, S., Prieto, P., Eds.; Oxford University Press: Oxford, UK, 2015; pp. 235–283.
25. Oostenveld, R.; Fries, P.; Maris, E.; Schoffelen, J.-M. FieldTrip: Open Source Software for Advanced Analysis of MEG, EEG, and Invasive Electrophysiological Data. *Comput. Intell. Neurosci.* **2011**, *2011*, 156869. [CrossRef] [PubMed]
26. Love, J.; Selker, R.; Verhagen, J.; Marsman, M.; Grounau, Q.F.; Jamil, T.; Smira, M.; Epskamp, S.; Wild, A.; Ly, A.; et al. Software to Sharpen your stats. *Aps Obs.* **2015**, *28*, 27–29.
27. Biel, A.L.; Friedrich, E.V.C. Why You Should Report Bayes Factors in Your Transcranial Brain Stimulation Studies. *Front. Psychol.* **2018**, *9*, 1125. [CrossRef] [PubMed]
28. van Doorn, J.; Bergh, D.V.D.; Böhm, U.; Dablander, F.; Derks, K.; Draws, T.; Etz, A.; Evans, N.J.; Gronau, Q.F.; Haaf, J.M.; et al. The JASP guidelines for conducting and reporting a Bayesian analysis. *Psychon. Bull. Rev.* **2020**, *28*, 813–826. [CrossRef] [PubMed]
29. Vigário, M.; Frota, S. The intonation of Standard and Northern European Portuguese: A comparative intonational phonology approach. *J. Port. Linguist.* **2003**, *2*, 115. [CrossRef]

brain sciences

MDPI

Article

Explaining L2 Lexical Learning in Multiple Scenarios: Cross-Situational Word Learning in L1 Mandarin L2 English Speakers

Paola Escudero [1,2,*], Eline A. Smit [1,2,3] and Karen E. Mulak [1,2]

1 The MARCS Institute for Brain, Behaviour, and Development, Western Sydney University, Penrith, NSW 2751, Australia
2 Australian Research Council Centre of Excellence for the Dynamics of Language, Canberra, ACT 2601, Australia
3 Department of Linguistics, University of Konstanz, 78457 Konstanz, Germany
* Correspondence: paola.escudero@westernsydney.edu.au

Abstract: Adults commonly struggle with perceiving and recognizing the sounds and words of a second language (L2), especially when the L2 sounds do not have a counterpart in the learner's first language (L1). We examined how L1 Mandarin L2 English speakers learned pseudo English words within a cross-situational word learning (CSWL) task previously presented to monolingual English and bilingual Mandarin-English speakers. CSWL is ambiguous because participants are not provided with direct mappings of words and object referents. Rather, learners discern word-object correspondences through tracking multiple co-occurrences across learning trials. The monolinguals and bilinguals tested in previous studies showed lower performance for pseudo words that formed vowel minimal pairs (e.g., /dit/-/dɪt/) than pseudo word which formed consonant minimal pairs (e.g., /bɔn/-/pɔn/) or non-minimal pairs which differed in all segments (e.g., /bɔn/-/dit/). In contrast, L1 Mandarin L2 English listeners struggled to learn all word pairs. We explain this seemingly contradicting finding by considering the multiplicity of acoustic cues in the stimuli presented to all participant groups. Stimuli were produced in infant-directed-speech (IDS) in order to compare performance by children and adults and because previous research had shown that IDS enhances L1 and L2 acquisition. We propose that the suprasegmental pitch variation in the vowels typical of IDS stimuli might be perceived as lexical tone distinctions for tonal language speakers who cannot fully inhibit their L1 activation, resulting in high lexical competition and diminished learning during an ambiguous word learning task. Our results are in line with the Second Language Linguistic Perception (L2LP) model which proposes that fine-grained acoustic information from multiple sources and the ability to switch between language modes affects non-native phonetic and lexical development.

Keywords: cross-situational word learning; L1 mandarin L2 english; minimal and non-minimal word pairs; acoustic cues; language modes; L2LP model

Citation: Escudero, P.; Smit, E.A.; Mulak, K.E. Explaining L2 Lexical Learning in Multiple Scenarios: Cross-Situational Word Learning in L1 Mandarin L2 English Speakers. *Brain Sci.* **2022**, *12*, 1618. https://doi.org/10.3390/brainsci12121618

Academic Editors: Richard Wright and Benjamin V. Tucker

Received: 23 May 2022
Accepted: 15 November 2022
Published: 25 November 2022

1. Introduction

Learning a second language (L2) in adulthood is difficult. Adults typically require additional time and exposure to their target L2 compared to younger learners to reach native-like proficiency (e.g., [1,2]), and commonly exhibit prolonged difficulty with pronunciation [3–6], and lexical access [7,8]. Most researchers agree that these difficulties stem in part from differences between learners' L1 and L2 systems. In the realm of speech learning, whether an L2 phoneme contrast is easy or difficult for a learner to perceive highly depends on how L1 speech sounds compare to those in the target L2, as advocated by most cross-language and L2 speech learning models (e.g., [9–12]). For instance, the Second Language Linguistic Perception model (L2LP; [11–14]) describes possible L2 learning scenarios depending on the acoustic proximity and overlap between L1 and L2 phoneme categories.

Within the L2LP model, when two phonemes of an L2 contrast are acoustically closest to a single L1 category, learners face a NEW scenario, and have difficulty perceiving the difference between the "new" L2 contrast (e.g., English /i/-/ɪ/ for L1 Catalan, Japanese, Mandarin, Polish, Portuguese, or Russian speakers, cf. (for a review see [13]). Learners face a SIMILAR scenario when two L2 categories are close matches of an acoustically similar L1 contrast (e.g., English /æ/-/ɛ/ for L1 Spanish speakers or English /i/-/ɪ/ for L1 Japanese speakers, cf. [14]). In this case, learners can replicate their existing L1 categories in the L2 but adjust their boundary as needed so that they match the L2 contrast. The SIMILAR scenario is suggested to be less problematic for the L2 learner since, unlike the NEW scenario, it does not require creating new phonological categories [12,14,15].

Non-native vowel perception studies support the claim that differences in the acoustic realization of L1 and L2 phonemes influence L2 speech perception (e.g., [16,17]). For instance, Alispahic and colleagues [17] tested Australian English (AusE) and Peruvian Spanish native speakers' discrimination of Dutch vowel contrasts. They made predictions about which Dutch contrasts would be easy and difficult for AusE monolinguals to discriminate and which would be easy and difficult for Peruvian Spanish monolinguals based on the unique acoustic relationships between L1 and L2 categories for each language. Indeed, they found that performance matched these predictions for both groups of speakers, supporting the idea that acoustic properties impact listeners' perception of non-native sounds (see [18]) and that listeners use perceptual cues from their L1 when categorizing non-native contrasts.

L2LP additionally proposes that the difficulties and relative ease in perceiving certain L2 contrasts based on the L1–L2 acoustic relationship extend to word learning and recognition [11–13,19–23]. Specifically, L2 learning of word pairs differing in a single phonological category—also known as minimal pairs—is predicted and explained by L2 learners' perceptual difficulty, which in turn is based on the acoustic comparisons of L1 and L2 categories (e.g., [15,20,21]. For instance, [20] tested learning of Dutch minimal and non-minimal pairs in L1 Spanish speakers learning Dutch. Word pairs were separated into easy and difficult categories based on whether these contrasts exist in Spanish and are thus NEW for learners and predicted to be difficult, such as /i-ɪ/—Spanish only has /i/—or whether the L2 contrasts are similar to Spanish and predicted to be easy, such as /ɪ-a/. In contrast with Dutch native speakers who performed equally well for all minimal word pairs, the L1 Spanish speakers performed worse for the difficult minimal pairs compared to the easy minimal pairs, confirming the L2LP proposal that L2 perceptual difficulty influences L2 lexical representations and L2 word learning.

In the present study, we examined L1 Mandarin L2 English learners' ability to learn English words in an ambiguous cross-situational word learning paradigm (CSWL) containing English sounds that do not exist in their L1 and compared their performance to English monolinguals. Specifically, we tested whether these L2 learners (a) perform equally or worse than English monolinguals in ambiguous word learning situations and if so, whether (b) their performance on minimal pairs may be explained by the relationship between L1 and L2 vowels and consonants, as proposed by most L2 speech learning models, including the L2LP. CSWL paradigms resemble a common real-world word learning situation in which word-objects pairings are presented ambiguously, in the context of other candidate pairings. Across multiple encounters with the words and items, learners can derive the correct word-object pairing through bottom-up statistical tracking mechanisms (e.g., [24]) or top-down hypothesis testing mechanisms (e.g., [25]).

Previous studies have demonstrated that both simultaneous bilinguals and L2 learners (also known as sequential bilinguals) can learn the pseudo words we present in a statistical word learning task. We define simultaneous bilinguals as learners who were exposed to two languages from birth and sequential bilinguals as L2 learners with exposure to the L2 after acquiring a first language, with onset of L2 acquisition during childhood, adolescence or adulthood. In the case of our study, all learners were exposed to English as a foreign language at school and therefore are referred to as L2 learners or sequential bilinguals. While most simultaneous bilinguals acquire proficiency comparable to monolingual speakers of their two languages, L2 learning can yield different levels of proficiency, in the case of the present group and those tested in [26]. The participants in both groups tested here followed university education in English, thus their L2 proficiency was advanced enough to understand English at a tertiary education level. When learning English words in a CSWL task Singaporean English-Mandarin simultaneous bilinguals had higher word-learning accuracy than monolingual English speakers [27]. In contrast, no difference in CSWL between highly proficient L2 English speakers with heterogeneous L1 backgrounds and monolingual English listeners was found in [26]. In both CSWL studies, participants were tested on their ability to learn eight words, four of which differed from one another on their initial consonant, forming consonant minimal pairs (cMP; e.g., BON-TON), and four of which formed vowel minimal pairs (vMP; e.g., DIT-DUT). Pairing a word from one set with one from the other set formed a non-minimal pair (nonMP; e.g., BON-DIT). Even though the simultaneous bilinguals in [27] outperformed monolinguals in accuracy, their reaction time for vMPs was slower to that of monolinguals. This may have been due to difficulties distinguishing the words DIT (/dɪt/) and DEET /dit/, as the vowel /ɪ/ is not found in the Mandarin vowel inventory. Alternatively, a vowel bias in Mandarin could have impacted reaction time, as vowels appear to provide stronger lexical identity in Mandarin compared to consonants [28], unlike in English. This may result in delayed processing of words differing in a single vowel by English-Mandarin bilinguals. The contrasting findings between learner groups, namely bilinguals in [27] versus L2 learners in [26], may be due to their specific linguistic background or to the English variety of the stimuli (American versus Australian English).

We presented L1 Mandarin L2 English learners with the same CSWL task as in [26,27] including the same eight words (i.e., /dit/, /dɪt/, /dʊt/, /dut/, /bɔn/, /pɔn/, /tɔn/ and /dɔn/, and pairings (i.e., nonMPs, vMPs and cMPs), compared their performance to that of AusE monolinguals, and assessed whether the relationship between Mandarin and English vowels and consonants can explain L2 word learning. First, we expected our AusE monolinguals to perform similarly to those tested in [27], with higher performance for nonMPs and cMPs compared to vMPs. In contrast, we expected our L1 Mandarin L2 English learners to have more L1 interference and have less optimal L2 representations than simultaneous bilinguals and therefore predicted they would find vMPs more difficult than cMPs or nonMPs, with their high L2 proficiency leading to similar performance to AusE monolinguals in cMPs and nonMPs. This prediction is in line with the L2LP model and with many other cross-language and L2 speech learning models. Specifically, if L1 Mandarin L2 English learners continue to have L2 representations that are L1-like (as shown in [12]), English words containing the vowels /ɪ/, /ʌ/, /ʊ/ should be particularly difficult to master [29], as these vowels are not found in Mandarin and are acoustically very similar to Mandarin /i/, /a/, /u/, leading to a NEW scenario that has not been resolved despite advanced L2 proficiency. Although many previous studies have shown plasticity for L2 learners in the phonetic/phonology domain, L2 proficiency does not seem to have a clear correlation with mastering new contrasts [13]. Conversely, previous studies have shown English consonants appear to be easier to perceive for Mandarin speakers, suggesting that they constitute a SIMILAR scenario, leading to better L2 performance from the start [30,31].

Finally, we tested a developmental tenet of L2LP which poses that transition from naïve listening to high L2 proficiency results in L2 perceptual and lexical development, such as creating or shifting of phonological categories to better represent the L2 [11,12]. Previous results have been mixed, as no difference between naïve Spanish-speaking listeners and those who had been learning Dutch in an immersive environment has been found [13,20], while some L2 perception studies report a positive effect of L2 experience [32,33] and others find no effect [13,34,35]. L2 immersion in a city where the L2 is spoken is a further opportunity to learn and be surrounded by the L2 in daily life, as opposed to only in a classroom. Within the L2LP framework, language immersion is seen as richer and more impactful language exposure, which should lead to further learning and in turn to higher

L2 proficiency. Thus, if immersive experience with the specific target L2, namely AusE, influences performance, L2 word learning accuracy will be higher for L1 Mandarin L2 English learners who are immersed in the target L2 in Sydney, Australia than those who live in their home country (Shanghai, China). Our study therefore differs from previous studies in examining a homogeneous group of L2 learners who have Mandarin as their L1.

2. Materials and Methods

2.1. Participants

Sixty participants took part in the experiment. Thirty-one were AusE monolinguals from Sydney (M_{age} = 26, 21 females, 10 males), and 29 were L1 Mandarin L2 English participants who were divided in two groups according to their place of residence during testing: 11 lived in Sydney, Australia (MandSyd, M_{age} = 27.34, 9 females, 2 males) and 18 in Shanghai, China (MandShanghai, M_{age} =22, 10 females, 8 males). Participants tested in Sydney were undergraduate psychology students or people from the local community recruited through word-of-mouth, advertisements or the university's participant recruitment system. Mandarin speakers in Sydney were native in Mandarin and indicated to speak and understand (Australian) English at advanced to native level. Participants tested in Shanghai were recruited through word-of-mouth and were all native Mandarin speakers at East China Normal University. They had studied English for an average of 14 years. They used Mandarin-Chinese daily, and English occasionally at university. Specific data regarding the precise number of years of English experience per participant was not collected or is no longer available. Participants received course credit or $10 travel compensation for their participation. Written informed consent was obtained from all participants prior to the start of the experiment, and the study was approved by the Western Sydney University Human Research Ethics Committee.

2.2. Stimuli

2.2.1. Pseudo Spoken Words

Stimuli consisted of eight monosyllabic pseudo words recorded by a female speaker of AusE using infant-directed speech (IDS). The words originate from a prior CSWL study [36] and have been used in other word learning and CSWL studies [26,27,37–40] with no effect of item on word learning accuracy. For the present study, we chose to use the same IDS stimuli in order to directly compare our results to previous studies, which were aimed at testing word learning in infants versus adults. We also chose IDS stimuli because many previous studies have shown that IDS facilitates word learning in infants learning their native language [41,42] and adult second language learners [43–46].

Words followed a CVC structure following English phonotactics. Per word, two tokens were selected to match prosodic contours across all words, with one token having a rising prosodic contour whereas the second has a descending prosodic contour. Four words differing from one another on their initial consonant formed consonant minimal pairs and followed a /Cɔn/ structure (cMP; e.g., BON-TON). The other four words followed a /dVt/ structure, forming vowel minimal pairs (vMP) with one another (e.g., DIT-DUT). Pairing a word from one set with one from the other set formed a non-minimal pair (nonMP; e.g., BON-DIT).

2.2.2. Pseudo Visual Referents

Each word was paired with a visual referent, which consisted of colour pictures of pseudo items (see Figure 1). These pictures have been used in prior CSWL studies (e.g., [23,26,27,36,40,47]. Pictures were 210 × 206 pixels.

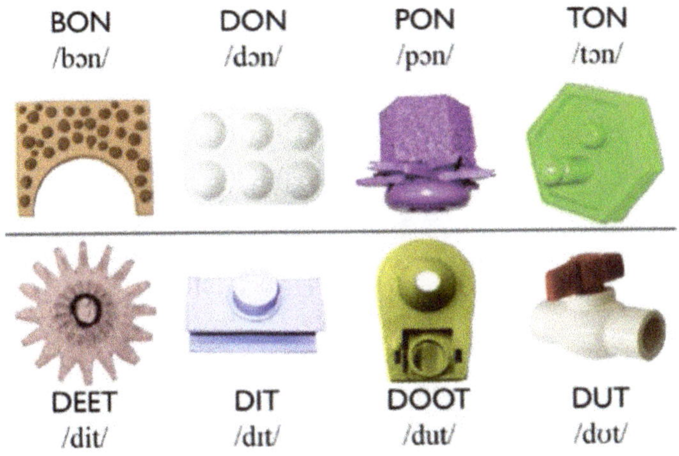

Figure 1. The eight pseudo words and their visual referents. The four words in the top row are minimally different in their initial consonant, whereas the words on the bottom are minimally different in their vowel. The vowel used for the consonant minimal pairs is /ɔ/ as in POT. Vowels used for the vowel minimal pairs are /i/ as in BEAT, /ɪ/ as in BIT, /u/ as in BOOT, and /ʊ/ as in PUT. Figure 1 originates from [27].

2.3. Procedure

After obtaining written consent, participants filled out a language background questionnaire. They then completed the CSWL task, comprising a learning phase followed by a test phase. For this, participants were seated in front of a laptop computer with a 17-inch monitor and were asked to wear headphones throughout the experiment. The experiment was run using the software package E-prime (version 2.0, Psychology Software Tools Inc., Sharpsburg, PA, USA).

The learning phase consisted of 36 trials (as in [26]), with each word-referent pairing presented nine times. As in the previous CSWL studies, participants were instructed to look at the images and listen to the sounds but were not informed that this was a word learning experiment. During each trial, two visual referents were presented on the screen on a white background, centered vertically. After the images had been on the screen for 500 ms (to keep the experimental design consistent with prior CSWL studies [26,27,38,40]), the auditory labels corresponding to each referent were played such that the referents were named left-to-right or right-to-left with 500 ms between tokens, without indication of which label belonged to which referent. Trials were randomized for each participant and were controlled to ensure that each image was presented simultaneously with every other image at least once and at most twice (for more specific details regarding the counterbalancing of the trials, see [26]). In total, there were 24 nonMPs pairs, 6 cMP pairs, and 6 vMPs. Trials lasted for 3.5 s leading to a total learning phase of approximately 3 min. Participants did not complete a familiarization test before the testing phase as this would defeat the purpose of the statistical learning paradigm.

Participants were tested directly after the learning phase and were told that they would view two images on the screen and would hear one word. They were instructed to press the left or right ALT key on the keyboard to indicate whether they thought the word corresponded to the left or right image, respectively. Trials in the testing phase used the same visual referent pairs as the learning phase, but the left and right designations of the images were randomized once (similar to [26]). In each trial participants heard four repetitions of the label corresponding to one image (the target word). The first token began 500 ms after presentation of the images, with 500 ms between tokens. Every word appeared as target word four or five times. As in the training phase, there were 36 trials in total with

24 nonMP trials, 6 cMP trials and 6 vMP trials. Trials were separated into three blocks of 12 trials, with block order counterbalanced between participants, and trial order within blocks randomized for each participant. Every trial lasted 6.5 s leading to a total test phase of approximately 4 min. An example of a learning and test trial is presented in Figure 2.

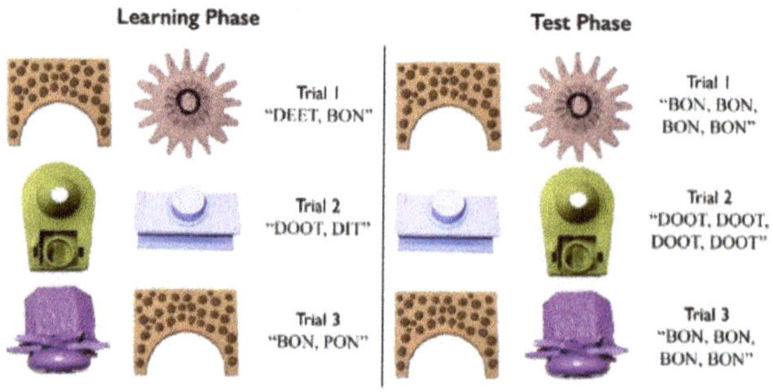

Figure 2. Example of a learning (**left**) and test (**right**) trial (figure from [27]).

2.4. Statistical Analysis

The data were analyzed using the statistical program R [48] with the brms package (using Stan; [48–50]). We used a multilevel Bayesian regression model to analyze participants' accuracy. We chose to use a Bayesian approach for the statistical analysis due to the advantage and flexibility of using probabilistic statistics with small sample sizes [51], as in the case of our L1 Mandarin L2 English group based in Sydney (N = 11). We have successfully used Bayesian modelling in previous studies where we provide further information and details of two other cases where probabilistic statistics are particularly useful [38,52–54].

For the multilevel Bayesian regression model reported below, we used dummy coding (the default in brms) for the factors of Language group and Pair type, with AusE and nonMP as reference levels. Approximate leave-one-out (LOO) cross-validation was used to find the best fitting models, which resulted in a model fitted with Language group and Pair type as fixed effects and Pair type and Trial number as random effects within-participants. We used weakly informative priors [55,56] with a Student-t's distribution with 3 degrees of freedom, a mean of 0 and a standard deviation of 2.5. We used a Bernoulli distribution to model accuracy responses (which consist of either 0′s and 1′s).

After fitting the model, we proceeded with hypothesis testing. Based on the predictions from the L2LP model, we hypothesized that the L1 Mandarin L2 English groups would perform less accurately than the AusE group for vMPs and cMPs but not for nonMPs. If experience with living in a country were English is spoken and in particular the English variety of the stimuli influences performance, we also expected the L2 group based in Sydney to perform better than the group based in Shanghai. We quantified the evidence for the tested hypotheses by using evidence ratios (ER), which are used to assess the likelihood of the test hypothesis against its alternative. To test our hypotheses, we only consider ERs above 30 (or of 1/30 or beyond) which qualify as "very strong evidence" and ERs of 10–30 (or of 1/10–1/30) which qualify as "strong" evidence (see [57] as cited by [58]). For readers unfamiliar with Bayesian statistics, an ER of >19 is approximately equivalent to an alpha of 0.05 in frequentist null-hypothesis testing [59]. In addition to the ERs, we also report the hypotheses' posterior probabilities (PP).

3. Results

Accuracy

Figure 3 shows the mean accuracy responses per language group and pair type. We first analyzed participants' accuracy (correct and incorrect responses) and tested whether participants were able to learn the word-object pairings for each pair type. Bayesian linear models on the Intercept, which are equivalent to frequentist one sample *t*-tests, revealed very strong evidence of above chance performance for all pair types in the AusE and in the L2 group from Shanghai, as indicated by posterior probabilities (PP) of > 0.98 and ERs of >3999. However, for the L2 group from Sydney we found very strong evidence of above chance performance only for the nonMPs (PP = > 0.999; ER = 3999), while evidence to support above chance performance was weaker for cMPs and vMPs (PPs = 0.92, 0.86; ERs = 11.82, 6.09, respectively). This is likely due to the higher response variability in this group (see Figure 3), which might be related to its smaller sample size (N = 11).

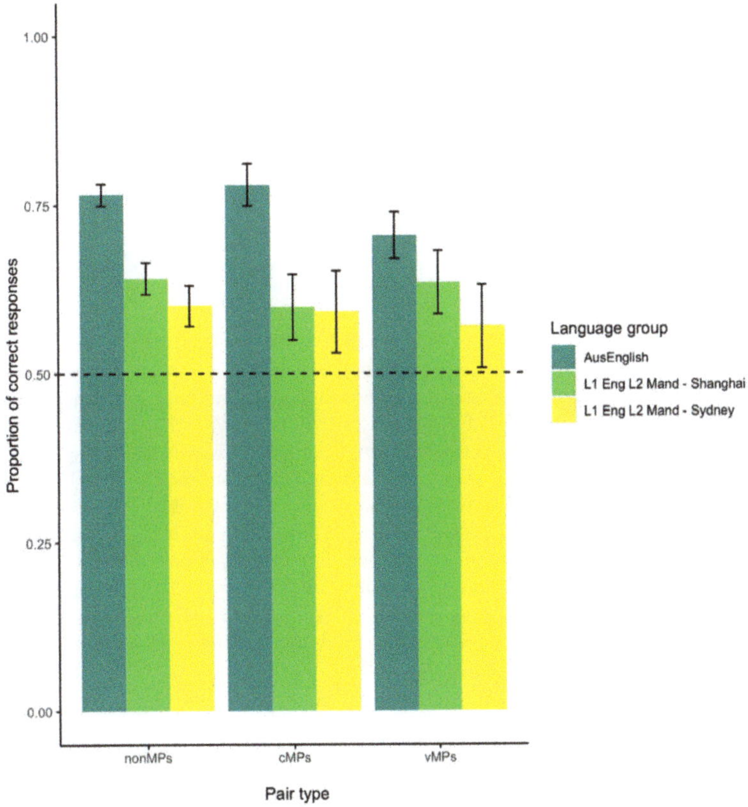

Figure 3. Mean accuracy in percentage per language group and pair type. Error bars indicate the standard error of the mean.

We then used a multilevel Bayesian regression model to estimate the interaction between Language group and Pair type on accuracy scores (see Table 1). We first tested whether there were differences in performance across the three pair types in each of the three language groups. For AusE, we found strong evidence that nonMPs and cMPs were more accurate than vMPs (PPs > 0.95; ERs of > 10), while no evidence for a difference between nonMPs and cMPs was found. These results replicate those reported in [26,27]) using frequentist statistics. Interestingly, no such difference between pair types was found for any of the L2 groups (PPs < 0.65 ERs < 10), suggesting their performance was similar across all three pair types.

Table 1. Hypothesis test results for the evidence ratios of the accuracy models' interaction between pair type and language group.

Hypothesis	Mean	90% CI	ER	PP
AusE				
nonMP > cMP	0.10	[−0.24, 0.45]	2.17	0.68
nonMP > vMP	0.35	[0.01, 0.68]	22.05	0.96
cMP > vMP	0.45	[0.01, 0.89]	21.06	0.95
L1 Mand L2 Eng–Shanghai				
nonMP > cMP	0.21	[−0.19, 0.60]	4.32	0.81
nonMP > vMP	0.06	[−0.34, 0.45]	1.48	0.60
cMP > vMP	−0.15	[−0.66, 0.36]	0.45	0.31
L1 Mand L2 Eng–Sydney				
nonMP > cMP	0.05	[−0.45, 0.54]	1.30	0.57
nonMP > vMP	0.18	[−0.32, 0.67]	2.65	0.73
cMP > vMP	0.13	[−0.50, 0.77]	1.72	0.63

Note: Mean = mean of the effect's posterior distribution. 90% CI = one-sided 90% credibility intervals. ER = evidence ratio = the odds that the effect is in the direction specified by the hypothesis. PP = the posterior probability of the tested hypothesis.

The results in Table 1 indicate no performance differences between the two L2 groups (residing in Sydney or Shanghai), which was confirmed by further hypothesis testing where we found no evidence of a between group difference for nonMPs (PP = 0.40; ER = 0.68), cMPs (PP = 0.40; ER = 1.23) or vMPs (PP = 0.32; ER = 0.47). We thus combined the data from the two L2 groups into one L1 Mandarin L2 English group to test our hypothesis that the L1 Mandarin L2 English learners should have lower performance on vMPs than on cMPs or nonMPs because the vowels in the vMPs are not contrastive in their L1 Mandarin.

A second multilevel Bayesian regression model was run to estimate whether monolingual AusE learners (N = 26) indeed performed better than L1 Mandarin L2 English learners (N = 29). As shown in Table 2, we found very strong evidence (for nonMPs and cMPs) and strong evidence (for vMPs) that the AusE group had higher accuracy than the L2 group for all three pair types, which runs contrary to the hypotheses that these L2 learners will have lower performance for vMPs than the other two pair types and that they would only differ from monolingual English speakers in the vMP trials.

Table 2. Hypothesis test results for the evidence ratios of the second accuracy models' interaction between pair type and language group.

Hypothesis AusEnglish > Mandarin	Mean	90% CI	ER	PP
nonMPs	0.78	[0.33, 1.22]	412.79	1.00
cMPs	1.03	[0.46, 1.60]	799.00	1.00
vMPs	0.51	[−0.03, 1.05]	15.20	0.94

Note: Mean = mean of the effect's posterior distribution. 90% CI = one-sided 90% credibility intervals. ER = evidence ratio = the odds that the effect is in the direction specified by the hypothesis. PP = the posterior probability of the tested hypothesis.

4. Discussion

This paper is the first to show that the mechanism of CSWL can be blocked by certain properties of the learner's L1, which go beyond segmental differences between L1 and L2 vowels and consonants. Overall, participants learned the word-object pairings for all pair types. In line with AusE monolinguals tested in [27], AusE monolinguals here were best at identifying words in a nonMP or cMP context compared to a vMP context. However, contradicting our prediction, this pattern was not found for either L1 Mandarin L2 English group, who were less accurate than the AusE group for all three pair types. This suggests that the phoneme inventory differences between Mandarin and English do not explain their L2 word learning performance. Experience with AusE did not impact performance either, as both L2 groups performed similarly.

In contrast with [26,27], where simultaneous English-Mandarin bilinguals outperformed AusE monolinguals and L2 learners with diverse L1 backgrounds performed similarly to AuE monolinguals, here we found that L1 Mandarin L2 English had lower accuracy than AusE monolinguals. Prior research suggests that bilinguals have an advantage in pseudo word learning due to enhanced phonological memory [60–63] and executive functioning (e.g., [64]). Conversely, L2 learners have been found to have low sensitivity to L2 phonological contrasts that are absent in their L1 [65]. This may be explained by the idea that L2 learners perceive the sounds of a new language through their native phonological categories (e.g., [9–11,33,66], which can lead to L2 word recognition problems and L2 representations that continue to be L1-like [13,20,67–71]. Instead, simultaneous bilinguals are able to fully inhibit each of their languages selectively, while L2 learners, despite their proficiency, have trouble doing so. This has been shown many times in previous studies where language dominance yields to interference, especially for the L2 in sequential bilinguals [43]. For example, the pseudo words making up the vMPs of the present study included vowels that are not present in Mandarin, namely /ɪ/ and /ʊ/, as mentioned in the Introduction. As predicted by the L2LP model [11,13,14], such vowel contrasts are likely to be perceived as the closest acoustically related native vowel, leading to problems with the recognition of words containing those L2 contrasts, as has been shown for similar L2 word recognition cases (e.g., [13,20–23,69]).

However, absent or L1-like L2 representations in the L1 Mandarin L2 English group cannot explain the current results because they found all pseudo pair types equally difficult to recognize. Rather, we propose that their general word learning difficulty may have resulted from the specific stimuli presented to them. As mentioned in the Methods, participants heard pseudo words produced in infant-directed speech (IDS), which is a speech style often used by mothers and caregivers when speaking to babies and infants and contains more variable pitch relative to adult-directed speech (ADS) [41]. Although many studies have shown that IDS can be beneficial for word learning in infants [42,44] and adults [41,45,46,72], IDS might negatively impact word learning for listeners who have heighted attention to pitch variation, such as tonal language speakers for whom pitch variations signify different lexical items [73].

We propose that heightened discrimination of pitch variation may have resulted in L2LP's Multiple Category Assimilation (MCA, L2LP; [74]), a scenario where an L2 category is acoustically similar to more than one L1 category, causing learners to perceive different tokens of a single L2 category as belonging to different categories in their L1 [17,22,66]. According to the L2LP proposal, this scenario results in listeners' perception of contrasts that do not exist in the L2 [12] and is referred to as a SUBSET problem [11,74]. When L2 sounds are a subset of what the learner can actually hear, there is no overt information from the target L2 that would allow the learner to stop hearing the extra category or stop activating irrelevant or spurious lexical items [11,22,46,74] resulting in higher lexical competition and overall less efficient L2 lexicalization and recognition. It is likely that MCA plays a role in the overall lower performance of Mandarin speakers in this study, specifically due to the use of IDS for the stimuli tokens. The IDS-induced pitch variations may have resulted in L1 Mandarin L2 English learners' perception of the two tokens of each word as two different words, challenging their ability to learn correct word-object pairs in the CSWL task. Importantly, the L2LP model is currently the only model of L2 perceptual and lexical developmental that can explain this type of L2 learning scenarios, starting from perceiving a single L2 category as more than one L1 category [75]. According to the L2LP model, this problem may not arise in simultaneous English-Mandarin bilinguals who may be able to de-activate their tonal language, succeeding at learning English words via CSWL and even surpassing monolingual English speakers, as reported in [26].

The L2LP explanation that the presence of additional pitch variation may be particularly problematic for speakers of a tonal language is further supported by findings that experience influences the perception of nonnative pitch variation. In many cases, tonal language experience is advantageous—for instance, Mandarin listeners outperform AusE listeners when learning a Thai tone distinction differing in pitch height contour [76,77]. In addition to tonal language learners, those who have experience with pitch via musical training have shown better tone discrimination [78], though not lexical tone learning [77]. However, in a recent study using the exact same CSWL paradigm and stimuli as in the present study [54] together with two standard music perception tests, we found that learners with high music perception abilities struggled most with IDS-produced words that had the highest pitch variability, i.e., vMPs. The tonal language speakers tested here struggled across all pair types, which may be due to them consistently using pitch information to discriminate between the exemplars of each word and across words for all pair types.

To confirm whether the additional pitch fluctuations induced by IDS indeed lead to a SUBSET problem and block CSWL in L1 Mandarin L2 English speakers, future research can use words produced in adult-directed-speech (ADS) with minimal pitch variation within and between exemplars of each word to examine whether word learning accuracy improves [54]. As vMPs naturally contain pitch variability, we expect the SUBSET problem to remain when tested with stimuli produced in ADS. If tonal language speakers indeed use suprasegmental information, such as pitch variations, when learning L2 words, their performance should thus improve more for nonMPs and cMPs than for vMPs when the variations in pitch are less prominent, as it is typical of ADS stimuli.

Lastly, we did not find a difference in experience with AusE, as both Mandarin groups performed similarly, providing no evidence that exposure to the L2 via an immersive environment mitigated L2 word learning difficulty. It could be that L2 exposure for the two L1 Mandarin groups is similar because while participants in Shanghai were not typically exposed to English in the community, both groups were students at English-speaking universities. Additionally, the two groups did not sufficiently differ in prior L2 exposure. However, a lack of group difference may also be due to a smaller sample size in the Sydney group, which resulted in higher variability in their results.

A limiting factor in this study is that by using self-reports as a measure of English proficiency, we may not be certain that participants have under- or overestimated their level of English. An English proficiency test administered alongside speech perception tasks may solve this issue. We also acknowledge that the missing details for each participants' English proficiency is a limitation. However, with the available demographic data, we replicated previous results showing that individual background does not play a role in performance. In a future study, more detailed information should be collected to confirm that this variable indeed does not affect CSWL. It is also important to note that individual differences in cognitive abilities may influence word learning in such word learning paradigms and in previous cross-situational word learning studies. Results from our lab show that cognitive skills, such as visuospatial memory, inhibition, or flexibility were not significant predictors of cross-situational and incidental word learning in four-year-old children [79,80], but this may be different for adults.

5. Conclusions

To conclude, although L1 Mandarin L2 English learners were able to learn the pseudo English words in an ambiguous word learning scenario, their performance was overall lower than that of monolingual English speakers. Given that their performance was equally low for all pair types regardless of L1-L2 phonological relationships, an explanation solely based on the absence of L2 representations in this L2 learners cannot adequately account for the results. The more nuanced L2LP model's explanation of a potential "subset problem," in which these L2 leaners may have perceived different tokens of the same word as separate words because of their L1 tonal language background, seems to a more adequate and accurate account. However, further research is needed to confirm this proposal.

Author Contributions: Conceptualization, P.E. and K.E.M.; methodology, P.E. and K.E.M.; formal analysis, E.A.S.; data curation, K.E.M. and E.A.S.; writing—original draft preparation, P.E., K.E.M. and E.A.S.; writing—review and editing, P.E., E.A.S. and K.E.M.; visualization, E.A.S.; supervision, P.E.; funding acquisition, P.E. All authors have read and agreed to the published version of the manuscript.

Funding: Data collection and K.M.'s work were funded by the Australian Research Centre of Excellence for the Dynamics of Language (CE140100041). P.E.'s and E.A.S.'s work were funded by an ARC Future Fellowship (FT160100514) awarded to P.E. Article publication fees were funded by Western Sydney University.

Institutional Review Board Statement: The study was conducted in accordance with the Declaration of Helsinki and was approved by the Western Sydney University Human Research Ethics committee (protocol code H11022 in 2017).

Informed Consent Statement: Informed consent was obtained from all subjects involved in the study.

Data Availability Statement: Anonymized data is available upon request to the first author.

Acknowledgments: We would like to thank Xiaoluan Liu and Nicole Traynor for their help with data collection in Shanghai and Sydney respectively and the participants for their time and participation.

Conflicts of Interest: The authors declare no conflict of interest.

References

1. DeKeyser, R. The robustness of critical period effects in second language acquisition. *Stud. Second Lang. Acquis.* **2000**, *22*, 499–533. [CrossRef]
2. Johnson, J.S.; Newport, E.I. Critical period effects in second language learning: The influence of maturational state on the acquisition of English as a second language. *Cogn. Pyschol.* **1989**, *21*, 60–99. [CrossRef] [PubMed]
3. Oyama, S. A sensitive period for the acquisition of a nonnative phonological system. *J. Psycholinguist. Res.* **1976**, *5*, 261–283. [CrossRef]
4. Piske, T.; MackKay, I.R.A.; Flege, J.E. Factors affecting degree of foreign accent in an L2: A review. *J. Phon.* **2001**, *29*, 191–215. [CrossRef]
5. Seliger, H.W.; Krashen, S.D.; Ladefoged, P. Maturational constraints in the acquisition of second language accent. *Lang. Sci.* **1975**, *36*, 20–22.
6. Tahta, S.; Wood, M.; Loewenthal, K. Foreign accents: Factors relating to transfer of accent from the first language to a second language. *Lang. Speech.* **1981**, *24*, 265–272. [CrossRef]
7. Jared, D.; Kroll, J.F. Do bilinguals activate phonological representations in one or both of their languages when naming words? *J. Mem. Lang.* **2001**, *44*, 2–31. [CrossRef]
8. Kroll, J.F.; Sunderman, G. Cognitive processes in second language learners and bilinguals: The development of lexical and conceptual representations. In *The Handbook of Second Language Acquisition*; Doughty, C.J., Long, M.H., Eds.; Blackwell Publishing: Oxford, UK, 2003; pp. 104–129.
9. Best, C.T.; Tyler, M.D. Nonnative and second-language speech perception: Commonalities and complementaries. In *Language Experience in Second Language Speech Learning: In Honor of James Emil Flege*; Bohn, O.-S., Munro, M.J., Eds.; John Benjamins: Amsterdam, The Netherlands, 2007; pp. 13–34.
10. Flege, J.E. Second language speech learning theory, findings and problems. In *Speech Perception and Linguistic Experience: Issues in Cross-Language Research*; Strange, W., Ed.; York Press: Timonium, MD, USA, 1995; pp. 229–273.
11. Escudero, P. *Linguistic Perception and Second Language Acquisition: Explaining the Attainment of Optimal Phonological Categorization*; Netherlands Graduate School of Linguistics: Amsterdam, The Netherlands, 2005.
12. Van Leussen, J.-W.; Escudero, P. Learning to perceive and recognize a second language: The L2LP model revised. *Front. Psychol.* **2015**, *6*, 1000. [CrossRef]
13. Escudero, P.; Benders, T.; Lipski, S.C. Native, non-native and L2 perceptual cue weighting for Dutch vowels: The case of Dutch, German and Spanish Listeners. *J. Phon.* **2009**, *37*, 452–465. [CrossRef]
14. Yazawa, K.; Whang, J.; Kondo, M.; Escudero, P. Language-dependent cue weighting: An investigation of perception modes in L2 learning. *Second Lang. Res.* **2020**, *36*, 557–581. [CrossRef]
15. Escudero, P.; Simon, E.; Mulak, K.E. Learning words in a new language: Orthography doesn't always help. *Billing. Lang. Cogn.* **2014**, *17*, 384–395. [CrossRef]
16. Escudero, P.; Vasiliev, P. Cross-language acoustic similarity predicts perceptual assimilation of Canadian English and Canadian French vowels. *J. Acoust. Soc. Am.* **2011**, *130*, EL277. [CrossRef]
17. Alispahic, S.; Mulak, K.E.; Escudero, P. Acoustic properties predict perception of unfamiliar Dutch vowels by adult Australian English and Peruvian Spanish listeners. *Front. Psychol.* **2017**, *8*, 52. [CrossRef]
18. Lengeris, A. Perceptual assimilation and L2 learning: Evidence from the perception of Southern British English vowels by native speakers of Greek and Japanese. *Phonetica* **2009**, *66*, 169–187. [CrossRef]

19. Boersma, P.; Escudero, P. Learning to perceive a smaller L2 vowel inventory. An optimality theory account. In *Contrast in Phonology: Theory, Perception, Acquisition*; Avery, P., Dresher, B.E., Rice, K., Eds.; De Gruyter Mouton: Berlin, Germany, 2008; pp. 271–301.
20. Escudero, P.; Broersma, M.; Simon, E. Learning words in a third language: Effects of vowel inventory and language proficiency. *Lang. Cogn.* **2013**, *28*, 746–761. [CrossRef]
21. Escudero, P. Orthography plays a limited role when learning the phonological forms of new words: The case of Spanish and English learners of novel Dutch vowels. *Appl. Psycholinguist.* **2015**, *36*, 7–22. [CrossRef]
22. Elvin, J.; Williams, D.; Escudero, P. Learning to perceive, produce and recognise words in a non-native language. In *Linguistic Approaches to Portuguese as an Additional Language*; Molsing, K.V., Perna, C.B.L., Ibaños, A.M.T., Eds.; John Benjamins: Amsterdam, The Netherlands, 2020; pp. 61–82.
23. Tuninetti, A.; Mulak, K.; Escudero, P. Cross-situational word learning in two foreign languages: Effects of native and perceptual difficulty. *Front. Commun.* **2020**, *5*, 602471. [CrossRef]
24. Yu, C.; Smith, L.B. Rapid word learning under uncertainty via cross-situational statistics. *Psychol. Sci.* **2007**, *18*, 414–420. [CrossRef]
25. Trueswell, J.C.; Medina, T.N.; Hafri, A.; Gleitman, L.R. Propose but verify: Fast mapping meets cross-situational word learning. *Cogn. Psychol.* **2013**, *66*, 126–156. [CrossRef]
26. Escudero, P.; Mulak, K.E.; Fu, C.S.; Singh, L. More limitations to monolingualism: Bilinguals outperform monolinguals in implicit word learning. *Front. Psychol.* **2016**, *7*, 1218. [CrossRef]
27. Escudero, P.; Mulak, K.E.; Vlach, H.A. Cross-situational word learning of minimal word pairs. *Cogn. Sci.* **2016**, *40*, 455–465. [CrossRef] [PubMed]
28. Chen, F.; Wong, M.L.Y.; Zhu, S.; Wong, L.L.N. Relative contributions of vowels and consonants in recognizing isolated Mandarin words. *J. Phon.* **2015**, *52*, 26–34. [CrossRef]
29. Jia, G.; Strange, W.; Wu, Y.; Collado, J. Perception and production of English vowels by Mandarin speakers: Age-related differences vary with amount of L2 exposure. *J. Acoust. Soc. Am.* **2006**, *119*, 1118–1130. [CrossRef] [PubMed]
30. Mi, L.; Tao, S.; Wang, W.; Dong, Q.; Jin, S.-H.; Liu, C. English vowel identification in long-term speech-shaped noise and multi-talker babble for English and Chinese listeners. *J. Acoust. Soc. Am.* **2013**, *113*, EL391. [CrossRef] [PubMed]
31. Tao, S.; Chen, Y.; Wang, W.; Dong, Q. English consonant identification in multi-talker babble: Effects of Chinese-native listeners' English experience. *Lang. Speech* **2019**, *62*, 531–545. [CrossRef]
32. Flege, J.E. Perception and production: The relevance of phonetic input to L2 phonological learning. In *Cross Currents in Second Language Acquisition and Linguistic Theory*; Huebner, T., Ferguson, A.C., Eds.; John Benjamins: Amsterdam, The Netherlands, 1991; pp. 249–290.
33. Flege, J.E.; Bohn, O.-S.; Jan, S. Effects of experience on non-native speakers' production and perception of English vowels. *J. Phon.* **1997**, *25*, 437–470. [CrossRef]
34. Cebrian, J. Input and experience in the perception of an L2 temporal and spectral contrast. In Proceedings of the 15th International Congress of the Ponetics Sciences, Barcelona, Spain, 3–9 August 2003; Recasens, D., Solé, M.J., Romero, J., Eds.; Universitat Autònoma de Barcelona/Causal Productions: Barcelona, Spain, 2003; pp. 2297–2300.
35. Flege, J.E.; Munro, M.J.; Fox, R.A. Auditory and categorical effects on cross-language vowel perception. *J. Acoust. Soc. Am.* **1994**, *95*, 3623–3641. [CrossRef]
36. Escudero, P.; Mulak, K.E.; Vlach, H.A. Infants encode phonetic detail during cross-situational word learning. *Front. Psychol.* **2016**, *7*, 1419. [CrossRef]
37. Curtin, S.A.; Fennell, C.; Escudero, P. Weighting of vowel cues explains patterns of word-object associative learning. *Dev. Sci.* **2009**, *12*, 725–731. [CrossRef]
38. Escudero, P.; Smit, E.A.; Angwin, A. Investigating orthographic versus auditory cross-situational word learning with online and lab-based research. *Lang. Learn.* 2022; *early view*.
39. Fikkert, P. Developing representations and the emergence of phonology: Evidence from perception and production. In *Laboratory Phonology 10: Variation, Phonetic Detail and Phonological Representation*; Fougeron, C., Kühnert, B., D'Imperio, M., Eds.; De Gruyter Mouton: Berlin, Germany, 2010; pp. 227–258.
40. Mulak, K.E.; Vlach, H.A.; Escudero, P. Cross-situational word learning of phonologically overlapping words across degrees of ambiguity. *Cogn. Sci.* **2019**, *42*, e12731. [CrossRef]
41. Kuhl, P.K.; Andruski, J.E.; Chistovich, I.A.; Chistovich, L.A.; Kozhevnikova, E.V.; Ryskina, V.L.; Stolyarova, E.I.; Sundberg, U.; Lacerda, F. Cross-language analysis of phonetic units in language addressed to infants. *Science* **1997**, *277*, 684–686. [CrossRef]
42. Graf Estes, K.; Hurley, K. Infant-directed prosody helps infants map sounds to meanings. *Infancy* **2013**, *18*, 797–824. [CrossRef]
43. Marian, V.; Spivey, M. Bilingual and monolingual processing of competing lexical items. *Appl. Psycholinguist.* **2003**, *24*, 173–193. [CrossRef]
44. Ma, W.; Golinkoff, R.M.; Houston, D.M.; Hirsh-Pasek, K. Word learning in infant- and adult-directed speech. *Lang. Learn. Dev.* **2011**, *7*, 185–201. [CrossRef]
45. Ellis, N.C. Salience, cognition, language complexity, and complex adaptive systems. *Stud. Second Lang. Acquis.* **2016**, *38*, 341–351. [CrossRef]
46. Golinkoff, R.M.; Alioto, A. Infant-directed speech facilitates lexical learning in adults hearing Chinese: Implications for language acquisition. *J. Child Lang.* **1995**, *22*, 703–726. [CrossRef]

47. Vlach, H.A.; Sandhofer, C.M. Retrieval dynamics and retention in cross-situational statistical word learning. *Cogn. Sci.* **2014**, *38*, 757–774. [CrossRef]
48. R Core Team. *R: A Language and Environment for Statistical Computing [Computer Software Manual]*; The R Project for Statistical Computing: Vienna, Austria, 2020.
49. Bürkner, P.-C. brms: An R package for Bayesian multilevel models using Stan. *J. Stat. Softw.* **2017**, *80*, 1–28. [CrossRef]
50. Bürkner, P.-C. brms: Advanced Bayesian multilevel modelling with the R package brms. *R J.* **2018**, *10*, 395–411. [CrossRef]
51. Van de Schoot, R.; Depaoli, S. Bayesian analyses: Where to start and what to report. *Eur. J. Health Psychol.* **2014**, *16*, 75–84.
52. Escudero, P.; Jones Diaz, C.; Hajek, J.; Wigglesworth, G.; Smit, E.A. Probability of heritage language use at a supportive early childhood setting in Australia. *Front. Educ.* **2020**, *5*, 93. [CrossRef]
53. Smit, E.A.; Milne, A.J.; Dean, R.T.; Weidemann, G. Perception of affect in unfamiliar musical chords. *PLoS ONE* **2019**, *14*, e0218570. [CrossRef] [PubMed]
54. Smit, E.A.; Milne, A.J.; Escudero, P. Music perception abilities and ambiguous word learning: Is there cross-domain transfer in nonmusicians? *Front. Psychol.* **2022**, *13*, 801263. [CrossRef] [PubMed]
55. Gelman, A.; Hwang, J.; Vehtari, A. Understanding predictive information criteria for Bayesian models. *Stat Comput.* **2014**, *24*, 997–1016. [CrossRef]
56. Gelman, A.; Lee, D.; Guo, J. Stan: A probabilistic programming language for Bayesian inference and optimization. *J. Educ. Behav. Stat.* **2015**, *40*, 530–543. [CrossRef]
57. Jeffreys, H. *The Theory of Probability*; OUP: Oxford, UK, 1998.
58. Kruschke, J.K. Rejecting or accepting parameter values in Bayesian estimation. *Adv. Methods Pract. Psychol. Sci.* **2018**, *1*, 270–280. [CrossRef]
59. Milne, A.J.; Herff, S.A. The perceptual relevance of balance, evenness, and entropy in musical rhythms. *Cognition* **2020**, *203*, 104233. [CrossRef]
60. Adesopa, O.O.; Lavin, T.; Thompson, T.; Ungerleider, C. A systematic review and meta-analysis of the cognitive correlates of bilingualism. *Rev. Educ. Res.* **2010**, *80*, 207–245. [CrossRef]
61. Kaushanskaya, M.; Rechtzigel, K. Concreteness effects in bilingual and monolingual word learning. *Psychon. Bull. Rev.* **2012**, *19*, 935–941. [CrossRef]
62. Majerus, S.; Poncelet, M.; Van der Linden, M.; Weeks, B.S. Lexical learning in bilingual adults: The relative importance of short-term memory for serial order and phonological knowledge. *Cognition* **2008**, *107*, 395–419. [CrossRef]
63. Service, E.; Simola, M.; Metsänheimo, O.; Maury, S. Maturational constraints in the acquisition of second language accent. *Eur. J. Cogn. Psychol.* **2002**, *14*, 383–403. [CrossRef]
64. Papagno, C.; Vallar, G. Phonological short-term memory and the learning of novel words: The effect of phonological similarity and item length. *Q. J. Exp.* **1992**, *44*, 47–67. [CrossRef]
65. Gor, K. Phonological priming and the role of phonology in nonnative word recognition. *Biling. Lang. Cogn.* **2018**, *21*, 437–442. [CrossRef]
66. Elvin, J.; Escudero, P. Perception of Brazilian Portuguese vowels by Australian English and Spanish listeners. In Proceedings of the International Symposium on the Acquisition of Second Language Speech (New Sounds 2013), Concordia Working Papers in Applied Linguistics. Montreal, Canada, 17–19 May 2013; pp. 15–156. Available online: http://doe.concordia.ca/copal/documents/12_Elvin_Escudero_Vol5.pdf (accessed on 1 May 2022).
67. Chrabaszcz, A.; Gor, K. Quantifying contextual effects in second language processing of phonolexically ambiguous and unambiguous words. *Appl. Psycholinguist.* **2017**, *38*, 909–942. [CrossRef]
68. Cutler, A.; Weber, A.; Otake, T. Asymmetric mapping from phonetic to lexical representations in second-language listening. *J. Phon.* **2006**, *34*, 269–284. [CrossRef]
69. Escudero, P.; Hayes-Harb, R.; Mitterer, H. Novel second-language words and asymmetric lexical access. *J. Phon.* **2008**, *36*, 345–360. [CrossRef]
70. Hayes-Harb, R.; Masuda, K. Development of the ability to lexically encode novel L2 phonemic contrasts. *Second Lang. Res.* **2008**, *24*, 5–33. [CrossRef]
71. Weber, A.; Cutler, A. Lexical competition in non-native spoken-word recognition. *J. Mem. Lang.* **2004**, *50*, 1–25. [CrossRef]
72. Houston-Price, C.; Law, B. How experiences with words supply all the tools in the toddler's word—Learning toolbox. In *Theoretical and Computational Models of Word Learning: Trends in Psychology and Artificial Intelligence*; Gogate, L., Hollich, G., Eds.; IGI Global: Hershey, PA, USA, 2013; pp. 81–108.
73. Han, M.; de Jong, N.H.; Kager, R. Lexical tones in Mandarin Chinese infant-directed speech: Age-related changes in the second year of life. *Front. Psychol.* **2018**, *9*, 434. [CrossRef]
74. Escudero, P.; Boersma, P. The subset problem in L2 perceptual development: Multiple-category assimilation by Dutch learners of Spanish. In Proceedings of the 26th Annual Boston University Conference on Language Development, Somerville, MA, USA, 2–4 November 2001; Skarabela, B., Fish, S., Do, A.H.-J., Eds.; Cascadilla Press: Somerville, MA, USA; pp. 208–219.
75. Escudero, P.; Hayes-Harb, R. The Ontogenesis Model may provide a useful guiding framework but lacks explanatory power for the nature and development of L2 lexical representation. *Biling. Lang. Cogn.* **2021**, *25*, 212–213. [CrossRef]
76. Ong, J.H.; Burnham, D.; Escudero, P. Distributional learning of lexical tones: A comparison of attended vs. unattended listening. *PLoS ONE* **2015**, *10*, e0133446. [CrossRef] [PubMed]

77. Ong, J.H.; Burnham, D.; Escudero, P.; Stevens, C.J. Effect of linguistic and musical experience on distributional learning of nonnative lexical tones. *J. Speech Lang. Hear. Res.* **2017**, *60*, 2769–2780. [CrossRef] [PubMed]
78. Ong, J.H.; Wong, P.C.M.; Liu, F. Musicians show enhanced perception, but not production of native lexical tones. *J. Acoust. Soc. Am.* **2020**, *148*, 3443–3454. [CrossRef]
79. Pino Escobar, G. Word Learning and Executive Functions in Preschool Children: Bridging the Gap between Vocabulary Acquisition and Domain-General Cognitive Processes. Ph.D. Thesis, Western Sydney University, Penrith, Australia, 2022.
80. Pino Escobar, G.; Tuninetti, A.; Antoniou, M.; Escudero, P. Understanding pre-schoolers' word learning success in different scenarios: Disambiguation meets statistical learning and eBook reading. *Dev. Psychol.* 2022; *manuscript submitted for publication*.

brain sciences

MDPI

Article

Native Listeners' Use of Information in Parsing Ambiguous Casual Speech

Natasha Warner [1,*], Dan Brenner [1], Benjamin V. Tucker [2] and Mirjam Ernestus [3]

1 Department of Linguistics, University of Arizona, Tucson, AZ 85721, USA; wobaidan@gmail.com
2 Department of Linguistics, University of Alberta, Edmonton, AB T6G 2E7, Canada; benjamin.tucker@ualberta.ca
3 Centre for Language Studies, Radboud University, 6500 HD Nijmegen, The Netherlands; mirjam.ernestus@ru.nl
* Correspondence: nwarner@arizona.edu

Abstract: In conversational speech, phones and entire syllables are often missing. This can make "he's" and "he was" homophonous, realized for example as [ɨz]. Similarly, "you're" and "you were" can both be realized as [jɚ], etc. We investigated what types of information native listeners use to perceive such verb tenses. Possible types included acoustic cues in the phrase (e.g., in "he was"), the rate of the surrounding speech, and syntactic and semantic information in the utterance, such as the presence of time adverbs such as "yesterday" or other tensed verbs. We extracted utterances such as "So they're gonna have like a random roommate" and "And he was like, 'What's wrong?!'" from recordings of spontaneous conversations. We presented parts of these utterances to listeners, in either a written or auditory modality, to determine which types of information facilitated listeners' comprehension. Listeners rely primarily on acoustic cues in or near the target words rather than meaning and syntactic information in the context. While that information also improves comprehension in some conditions, the acoustic cues in the target itself are strong enough to reverse the percept that listeners gain from all other information together. Acoustic cues override other information in comprehending reduced productions in conversational speech.

Keywords: reduced speech; conversation; comprehension; context; acoustic cues

Citation: Warner, N.; Brenner, D.; Tucker, B.V.; Ernestus, M. Native Listeners' Use of Information in Parsing Ambiguous Casual Speech. *Brain Sci.* **2022**, *12*, 930. https://doi.org/10.3390/brainsci12070930

Academic Editor: Yang Zhang

Received: 22 May 2022
Accepted: 12 July 2022
Published: 15 July 2022

1. Introduction

In normal daily-life conversations, humans convey information to each other efficiently, but do not produce all of the phones that they would in a careful speech version of the same sentences (although some of those phones may leave traces through coarticulation) [1–3]. This paper investigates what sources of information listeners use to understand reduced speech. For example, among normal, casual conversations we recorded, we have found tokens where speakers pronounced "gonna have to" as [gɔ̰ʔtə], or "a little" as a sonorant stretch of the waveform with low F2 and some change in formants and amplitude, but no distinguishable segments. (Audio examples are available at http://nwarner.faculty.arizona.edu/content/6 accessed on 13 July 2022) Such reduced speech clearly does not contain the same perceptual cues as a careful speech production does. Still, listeners usually perceive casual conversational speech with little difficulty, at least if they hear it in context and the speech is in their native language.

If listeners hear reduced speech out of the context it was produced in, they typically do not perceive the words of the reduced speech accurately. Koopmans-van Beinum [4] showed that listeners are very inaccurate at perceiving individual vowels that have been extracted from spontaneous speech. Ernestus et al. [5] showed that even very common words such as Dutch are recognized quite poorly (approximately 50% correct word identification) when highly reduced pronunciations such as [mok] are presented in isolation, extracted from the original context. Janse and Ernestus [6] further found that contextual information is more helpful to listeners if it is presented auditorily rather than orthographically. Arai [7], working on Japanese, similarly found that listeners badly misperceive strings of reduced speech out of context, for example hearing a recording of the five-mora string /kuiebaho/ as only the two morae /kebo/. These studies also confirmed that listeners perceive the same speech accurately when they hear it in context. Saerens et al. [8] investigated the perception of French voiceless stops with ambiguous voicing excised from conversational speech. They found that the lexical context of the rest of the words in the sentence helps listeners recover the intended voicing of the stops, but that secondary acoustic cues in or near the stop also play a role.

In some cases, reduction not only alters perceptual cues or segments, but it also creates ambiguity about what word was intended. In reduced speech, "he's" and "he was" can sound homophonous, with some of the reduced tokens sounding like [hɨz], and some so reduced that the only trace of the word(s) in the waveform is a [z] or [s]. (Audio examples are available on the website mentioned above.) Similarly, "we're" and "we were" can both sound like [wɚ] or possibly just [ɚ]. In both of these cases ("he's/he was" and "we're/we were"), reduction obscures the distinction between present and past tense, similar to how reduction and contextual speech rate can obscure the singular/plural distinction in sentences such as "The Petersons are looking to buy a brown hen/brown hens soon" [9]. Thus, reduction may not only inhibit how quickly and accurately listeners recognize content words in speech [10,11], it may also make function words with different meanings like "he's" and "he was" homophonous. Because function words are important for parsing the structure and meaning of the sentence, this makes reduction especially relevant for listeners' syntactic processing.

The speech signal includes several types of information that listeners might use to parse reduced speech and disambiguate reduced function words. One common assumption is that listeners perceive the other words of the utterance, and they retrieve the intended meaning through the syntactic or semantic context. For example, if one hears a highly reduced token of "We were supposed to see it yesterday" in which "we were" sounds like "we're," the word "yesterday" will allow the listener to realize the verb is past tense. This would likely be true even if the word "yesterday" is also reduced, since "yesterday" is more distinct from other lexical entries than "we're" and "we were" are from each other. When the authors play examples of reduced speech in classes or at conferences and ask the audience how they think listeners might be able to understand the speech, the first answer given is consistently "context," and when they are pressed to explain what they mean, they give some version of this explanation.

Several other types of information are also present in the signal. Most obviously, there is acoustic information in the reduced speech. It may not be sufficient for listeners to recognize the words, and it may even provide misleading perceptual cues. For example, our highly reduced recording of "gonna have to" mentioned above contains a creaky voice through a large part of the voicing, which may suggest a glottal stop. We find that listeners often misperceive this string as "got to," where a glottal stop is likely. If a token of "he was" out of context sounds to listeners like "he's," this means that the perceptual cues that are present are misleading. Any stretch of speech, no matter how reduced, contains acoustic information and hence perceptual cues, whether to the words the speaker intended or to something else.

Another type of information listeners use is speech rate. Listeners use speech rate within the same syllable to adjust the boundaries between aspirated and unaspirated stop categories [12,13]. Listeners use the speech rate of the surrounding utterance to help distinguish vowels such as /ɪ/ vs. /i/ (where /ɪ/ is intrinsically shorter), accepting a longer vowel as /ɪ/ if surrounding speech rate is slow [14]. At the word level, in phrases such as "leisure (or) time," altering the speech rate of either the surrounding context or the function word "or" can determine whether listeners perceive the function word at all [15]. If the function word is shorter than would be expected for the surrounding speech rate, either because the "or" was shortened or the surrounding speech was lengthened, listeners perceive "leisure time" instead of "leisure or time." Conversely, if the /r/ portion of the signal "leisure" is long relative to the surrounding speech rate, listeners may perceive an "or" that the speaker did not produce. Niebuhr and Kohler [16] showed a related result for German. This shows that listeners use the speech rate of the context to determine whether acoustic cues last long enough to constitute additional segments or words. Brown et al. [9] showed this effect through eye-tracking, establishing that it occurs in real time as part of how listeners develop expectations about upcoming words. Heffner et al. [17] investigated how listeners combine the information about context speech rate with acoustic cues within the word.

In the current work, if listeners hear [hiːz] with a relatively long duration, but the surrounding speech is very fast, they may hear "he was" instead of "he's." That is, the boundary between what counts as a good token of "he's" vs. a good token of "he was" may depend on duration, adjusted for the surrounding speech rate. One can imagine this as a subconscious process of "That was too long to be just 'he's' considering how fast this speech is going. Something must have been deleted. Maybe the speaker said 'he was.'"

In addition to syntactic and semantic information from other words in the utterance, listeners engaged in a conversation may also benefit from discourse information in the larger context, beyond the utterance. Knowing that one's interlocutor is discussing wedding plans, her part-time job, or a relative's health decisions may help the listener to adjust expectations for likeliness of words. However, this is probably less helpful for strings such as "he's" vs. "he was," which could both occur in most conversations. Another type of information available to listeners is information about a speaker's voice, both about properties such as vocal-tract conditioned vowel space (e.g., [18]) or the degree of habitual nasalization, and about idiosyncrasies or dialectal features [19]. Exposure to a longer sample of a speaker's voice allows listeners to adjust their expectations for the speaker's typical pronunciation. Furthermore, Brouwer et al. [20] found that if listeners have been hearing reduced spontaneous speech preceding a target, they penalize acoustic mismatches with lexical entries less strongly, so the speech style of the context also supplies information.

Van de Ven et al. [21] investigated how well semantically related words prime word recognition if the primes or targets are reduced vs. carefully pronounced. They found that the semantic information in reduced pronunciations of words does not help listeners to recognize subsequent words unless listeners are given more time than usual to fully process the reduced words before the related word is presented. This study used a priming methodology with words presented in isolation, so it did not test whether listeners use semantic information in the preceding parts of the sentence or discourse, but instead whether the activation of a related semantic concept outside of a discourse helps with isolated word recognition.

Using a different method, van de Ven et al. [22] showed that native speakers are able to use the syntactic and semantic information in a surrounding sentence to help them predict a missing adverb in the sentence at better than chance, but still low, rates. Native Dutch speakers in their experiment were able to predict the missing word "altijd" 'always' at better-than-chance rates in a sentence such as "Ik vertrouw altijd maar op mijn goede geluk" 'I always rely on my good luck.' The success rate at predicting such words was higher than would be expected based on n-gram probabilities, and was higher when listeners were able to hear the surrounding context auditorily rather than reading it. This indicates that there is some information in the phonetics, syntax, and semantics of the context, even for adverbs that are not predictable in the sentence. However, van de Ven et al. [22] also found that listeners obtain far more information about the words from hearing the word itself than from context. Van de Ven and Ernestus [23] presented various portions of the speech signal around and during reduced words in Dutch conversational speech, and found that listeners make less use of bigram probability based on the preceding or following word as they are given more acoustic cues from the target word to work with. Drijvers et al. [24] studied listeners' neuronal oscillations while hearing reduced vs. clearly pronounced word forms in various contexts, and found that reduced forms impose a higher cognitive load during recognition, which prevents lexical activation from spreading through the semantic network as quickly when words are reduced.

In order to determine which types of information (e.g., acoustic, speech rate, syntax/semantics) help listeners disambiguate forms such as "he's" vs. "he was" or "we're" vs. "we were" in casual, reduced speech, we conducted a series of experiments. We extracted utterances containing words/phrases such as "he's," "we were," "she was" etc., from recordings of spontaneous casual conversations that had been made for another experiment [25]. Some examples are "'Cuz he already told Steve he was in the wedding" or "When we were outside the bookstore..." (Underline indicates the target word/phrase.) A few items had a word other than a pronoun as the first word of the target, but these had the same potential tense ambiguity (e.g., "Katie was/Katie's"). We presented stimuli based on these utterances to participants with various types of context or information available, in order to determine how well listeners could disambiguate the reduced speech if given access to some types of information but not others. Experiments 1 and 2 provide baseline measures of how much information is available from the syntax and semantics of surrounding words, without the target words themselves. Experiment 3 turns to perception of the target word/phrase.

2. Experiment 1: Syntactic and Semantic Context without Acoustics (Orthographic Presentation)

Since the utterances we use as stimuli were taken from spontaneous, natural conversations, and were not constructed to be either semantically predictable or not, we need to establish a baseline of how much information about the target phrase/word one can gain just from the syntax and semantics of the rest of the utterance. In Experiment 1, we presented the utterances to participants written on a computer screen, with a blank for the target word/phrase, and asked participants to choose whether the present or past version of the target would be more likely to appear in the blank in the utterance. For example, participants would see "'Cuz he already told Steve ____ in the wedding" on the computer screen, with "he's/he is" and "he was" printed below. Participants pressed a button on a response box to indicate which alternative they thought was more likely to fill in the blank. Thus, in this experiment, participants had access to all the syntactic and semantic information in the surrounding context, but did not have access to any acoustic information, either about the target or about the context.

2.1. Methods

2.1.1. Materials

A total of 184 utterances containing words/phrases such as "he is," "he's," "she was," "she's," "we're," "we were," were chosen from recordings of 18 native speakers of American English who were originally recorded for a production study of spontaneous conversational speech (a superset of the speakers in [25]) (The study in [25] involved labor-intensive acoustic labeling that precluded measuring all of the participants who volunteered and were recorded at that time). Sample items appear in Appendix A (Table A1). Most of the speakers were completely monolingual in English until at least their teenage years, when they began taking language classes in school. Some had limited exposure to another language (e.g., Spanish, a Chinese language, Canadian French) in the home as children, but all were strongly English dominant and grew up in the U.S. The speakers were undergraduate students at the University of Arizona at the time the recordings were made (2005), and most were from the Southwestern U.S. or California. None spoke a dialect that was notably different from varieties typically heard in Arizona.

Speakers sat in a sound-protected booth and wore a high quality head-mounted microphone over the opposite ear from the one where they habitually held a telephone. Each speaker called a friend or family member and held a conversation of approximately 10 min on whatever topics they wished to discuss. Further details of the methods for obtaining this speech are available in [25]. Speech was recorded through the microphone, not the telephone.

The telephone was only used to allow a casual conversation with a well-known interlocutor, while still recording in a sound booth. This method succeeded in eliciting highly informal, casual speech, as shown by the range of topics discussed (including fraternities and drinking games as well as courses, part-time jobs, and family members).

During past work with these recordings, research assistants from a similar background to that of the speakers (undergraduate students at the same university a few years after the recordings were made) produced orthographic transcriptions of the recordings. We used these to locate sufficient numbers of tokens containing the strings "X is" (either contracted or not; both "he is" and "he's" were included), "X was," "X are" (including contracted forms, hence both "we're" and "we are"), and "X were." In each case, the longer past tense form had to be reducible to be homophonous with the shorter present tense form. For example, "Grammy's" could be used, because the words "Grammy was" could potentially be reduced to sound similar to "Grammy's," but "I was" could not be used, as there is no related shorter form "I'se" in English as spoken in Arizona with which it could become homophonous. All but 19 items had a personal pronoun (e.g., "he, she, it, we, they") in the X position. The remaining 19 included 7 items with "there," three with "who," two with "how," and one each of "parents, weekend, Katie, what, so, Grammy, everybody." The average number of words in the utterance before the target item was 3.08, and the average number after it was 4.98. The number of items drawn from each speaker's recording ranged from 2 to 42, and depended on how often the speaker used the target phrases. Using stimuli drawn from spontaneous conversation means that the stimuli are quite variable, for example in the sentence structure and focus in or near the target word/phrase. However, it has the advantage that the speech participants respond to is representative of what they hear in informal conversations in daily life, reducing the chance of task effects.

Each item was checked by the first author, as well as having been identified from the longer recording based on transcriptions made by the research assistants, whose age and dialect was a good match to the speakers'. Thus, each item was checked to determine whether the particular token was produced with "is" or "was" by at least two native speakers of American English who heard the entire discourse context of the longer recording and could listen to the utterance and any amount of context as many times as they wished. The first author agreed with the research assistants' perception of all items that were used. The orthographic transcriptions of these items form the materials for Experiment 1.

2.1.2. Participants

46 native speakers of English participated in Experiment 1. All were students in introductory Linguistics courses at the University of Arizona, and all had either been monolingual in American English until at least puberty, or had had some exposure to another language (e.g., German, Korean, Marathi, Gujerati, Spanish) in the home but were strongly English-dominant. All had grown up entirely in the U.S. Participants received extra credit in their Linguistics course as compensation.

2.1.3. Procedures

Participants sat in a sound-protected booth with a computer monitor outside the window of the booth. Participants saw each item in written form on the computer monitor, with a blank inserted for the target word/phrase, e.g., "'Cuz he already told Steve ____ in the wedding." Below the utterance the response options were printed, giving the present and past tense options, adjusted to use the correct word before the verb. That is, for this item the response options were "he's/he is" and "he was," while for the stimulus "And ____ huge houses too, it was weird, like" ("they're" deleted), the response options were "they're/they are" and "they were." The response options did not distinguish between contracted vs. full forms of present tense: participants were only asked to choose between present and past forms, not between "he is" vs. "he's," for example. Since for this experiment, participants did not hear the target or the context, they were instructed to choose which of the two response alternatives they thought would be more likely to occur in the blank. Participants were instructed that the sentences came from casual conversations.

The EPrime software (Psychology Software Tools [26]) was used to present stimuli and record responses. Participants pressed buttons on a response box to indicate whether they chose the left or right response on the monitor. The two response alternatives were randomly assigned to left and right position. Participants first responded to 6 practice items, followed by the full list of stimuli, in a different random order for each participant. After each stimulus appeared on the screen, the participant had up to 10 s to read it and respond, after which the program advanced to the next stimulus. On 1.2% of trials, participants failed to respond. The average reaction time was slightly over 4 s, reflecting time to read the stimulus. Six participants were mistakenly presented with a version of the experiment that omitted three items.

Approximately 90 additional items were included. Most were for an additional distinction ("X him" vs. "X them," as in "got 'em"), and some were additional items for the current conditions. These will not be discussed further, because the length of the subsequent experiments precluded the use of these additional items in the other experiments. The entire experiment, including the additional items, took approximately 35 min.

2.2. Results

Results for Experiment 1, as proportion correct, appear in Figure 1. (Using proportion correct as the dependent variable, instead of proportion present tense responses (or proportion past), focuses the analysis on investigating bias rather than whether listeners can distinguish the present from past. The d' analysis below focuses on the latter question.) During analysis, it became clear that participants' responses differed strongly depending on whether the target is followed by the quotative or discourse particle "like" or not (e.g., the stimuli "She's like, 'No! No more laptops!'", "And he was like, 'What's wrong?!'", and "Yeah he was like, ignoring me until he right, he, 'til right before he got on the bus."). This could be because speakers have the option of using the historical present to report a past conversation or situation, as in the first of these examples, making verb tense relatively uninformative before "like." Therefore, the presence of the word "like" after the target was included as a post hoc factor, as in Figure 1. Past research on "like" usage [27–30] suggests that these constructions may have properties that other usages of "he was, we were" etc. do not, confirming the need to include presence of following "like" as a factor. We did not attempt to distinguish among usages of "like," since it can be difficult to determine whether a given usage is quotative or not, and some stimuli ended with the "like" because the speaker made a long pause or stopped the utterance (e.g., "So like, she's like ... ").

"Like" occurs less often after plural subjects than singular subjects in our speakers' conversations (contrary to [27], perhaps suggesting a change in the intervening 15 years). The stimuli contained only two items with the target verb "are" followed by "like" and only seven with "were" followed by "like." This is too few items to provide reliable data, so we chose not to analyze the few items with "X are/were like" statistically. Since the singular conditions ("is, was") show a very strong difference in behavior depending on the presence of "like," we therefore analyzed the data in three subsets: "is" vs. "was" targets without a following "like," "is" vs. "was" targets with a following "like," and "are" vs. "were" targets without a following "like." Each analysis had the intended tense of the verb (present, past) as the fixed factor. (One could also analyze either both sets of "is" vs. "was" targets, or both sets of targets without "like," in a larger analysis with an additional factor. When tested, this type of higher-order analysis revealed an interaction that motivated testing each subset separately. (The higher-order LMEs generally showed significant interactions but also had failure to converge or singular fit warnings, and so we do not report numerical details of those models. However, to further motivate testing subsets of the data (simple effects tests) based on significant interactions, we performed by-subject ANOVAs (hence averaged over items). Because of the absence of plural "like" items, we performed an analysis on all of the singular data (with "like" and tense as factors). Both factors are within-subjects. A significant interaction motivated testing simple effects of tense of stimulus. Singular data: tense: $F(1,45) = 49.00$, like: $F(1,45) = 44.93$, interaction: $F(1,45) = 110.34$, all p's < 0.001. One could also analyze all of the data without "like" (with tense and number as factors), but the interaction in the singular data and absence of plural-like data already motivate testing simple effects of tense.).

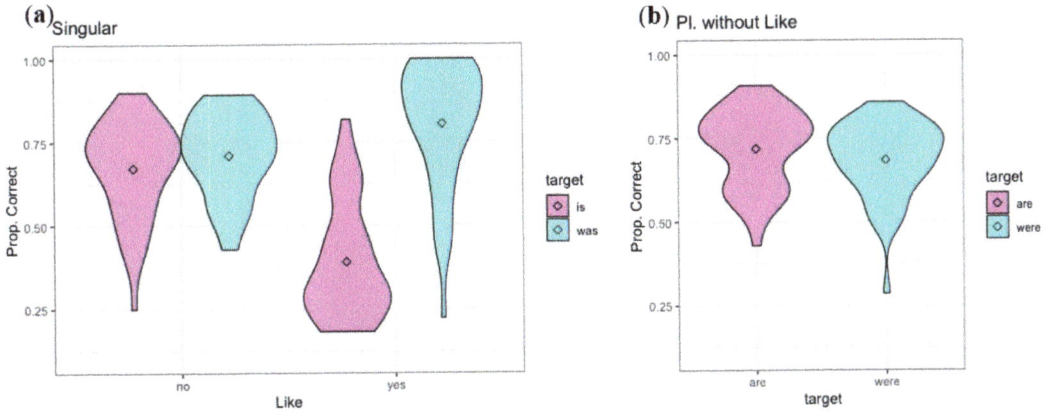

Figure 1. Results with orthographic presentation (distribution of listeners' averages over items). Dots indicate means for each condition. (**a**) "Is" and "was" targets. (**b**) "Are" and "were" targets.

We analyzed the data using generalized linear mixed effects models with a binomial link function as implemented in the lme4 package ([31] version 1.1–29) of R (using glmer), with the correctness of response as the dependent variable. The fixed factor tested for each present/past pair was the tense of the target word/phrase as produced by the speaker in each item (reference level: past). (For example, a participant sees "'Cuz he already told Steve ___ in the wedding," which was originally produced with "he was." The participant has a choice of "he's/he is" and "he was" as response options. If the participant selects "he's/he is" this is scored as incorrect, while selecting "he was" is scored as correct. The correctness of response is evaluated relative to what was originally produced in the target word/phrase, regardless of whether any other tensed verbs appear in the utterance. The independent variable of tense in the statistical analysis allows an analysis of how participants' accuracy on stimuli that originally contained present might differ from those that originally contained past.) (Models including larger subsets of the data at once, and more factors, showed significant interactions, so the tense factor had to be tested for each set separately.) Model selection was performed using an ANOVA comparison. Random intercepts for subject (participant) and item (sentence), as well as random slopes by subject for the tense factor, were included if the model converged and did not give singular fit warnings. Random intercepts for speaker (who produced the stimulus) were also tested and found not to improve the model's fit. Since participants did not hear the voices that produced the stimuli in this experiment, it is not surprising that the speaker random intercepts did not improve the models. (All three subsets (*is/was without like*, *is/was with like*, and *are/were without like*) used the model Correct ~ Tense + (1+Tense | Subject) + (1 | Item).)

The model for the "is/was" data not followed by "like" showed no significant effect of the tense of the stimulus ($\beta = -0.21$, $z = -0.78$, $p = 0.44$). The model for the "is/was like" data showed significantly more correct responses for "was" than "is" ($\beta = -2.63$, $z = -6.06$, $p < 0.001$). The model for the "are/were" data without "like" also showed no effect of tense ($\beta = 0.16$, $z = 0.62$, $p = 0.54$). The significance of the tense effect in the "is/was like" data does not indicate that "was like" is necessarily easier to perceive than "is like," but rather that participants are biased toward the "was" response. This could be because the quotative "like" is used to introduce reported speech, which must necessarily have been uttered in the past. Although one can use historical present to describe past speech with "he's like,...," participants seem to assume that speech uttered in the past will be reported in the past, and favor the "was" response. This leads to a high accuracy when the stimulus actually contains "was" and a low accuracy when it actually contains "is."

To further evaluate bias, we examined the average accuracy across past and present tense items, the detectability of the past/present distinction (d′), and bias (β) for each present/past verb pair (Table 1) (The corresponding results for later experiments of the paper are presented as well, and will be discussed below). The average accuracy, 58.3–70.2% for the various conditions, is substantially above chance. The d′ value for both pairs without following "like," at slightly more than one, indicates that listeners were able to extract some information about whether the verb was more likely to be present or past tense, but they were still far from being able to accurately recover tense. For the "is/was like" pair, the d′ is only 0.504, showing that this context offers only very weak information about which verb tense was intended. The value for bias confirms that participants were biased toward the past response for this pair.

Table 1. Signal detection measures d′ (detectability) and β (bias), and average proportion correct across the present and past verb of the pair, for the tense distinction for each pair of conditions. Positive β indicates bias toward the past response, negative toward the present response. All experiments are included here for ease of comparison. Number of items in each condition appears in Appendix A (Table A1).

Experiment/Condition	Context	d′	β	Avg. Prop. Correct
		Exper. 1 (Orthography)		
is/was, no "like"		1.005	0.054	0.692
is/was, with "like"		0.504	0.306	0.583
are/were, no "like"		1.062	−0.052	0.702
		Exper. 2 (Auditory, target replaced by beep)		
is/was, no "like"		0.889	−0.048	0.672
is/was, with "like"		0.414	0.078	0.581
are/were, no "like"		1.042	−0.338	0.690
		Exper. 3 (Target plus various contexts)		
is/was, no "like"	Isolation	1.927	−0.269	0.830
	Limited	2.162	−0.292	0.858
	Full	2.460	−0.820	0.878
is/was, with "like"	Isolation	0.910	−0.757	0.627
	Limited	0.852	−0.520	0.639
	Full	1.068	−0.653	0.672
are/were, no "like"	Isolation	1.443	−0.226	0.762
	Limited	1.943	−0.301	0.832
	Full	2.405	−0.868	0.871

2.3. Discussion

The results show that native speakers of English evaluate verb tense differently in phrases with a following "like" than in phrases without. In both cases, on average across all sentences used, they were able to extract at least some information about whether the verb is more likely to be present or past tense from the surrounding syntactic and semantic context. The sentential context in the sentences without "like" conveys more information about verb tense (69% correct) than in those with "like" (58% correct), as indicated by the higher d′. This is not surprising, since speakers have the option of using historical present to report past speech or events using the quotative "like." The bias toward "was" in this condition suggests that participants did not usually take the historical present option into account in reading the sentences. Instead, they seem to have assumed that a verb reporting past speech would be in the past tense, thus favoring the "was" response.

In the conditions without a following "like," participants had almost no bias, favoring neither present nor past verbs. Most of the items without "like" contain clearer syntactic or semantic information, as in "And so they were getting back to Desert [a school] right when Eric and I got there" or "'Cuz I can't hang out with anyone, 'cuz they're, everyone's gonna be studying." In both of these utterances, the opposite tense would be very unlikely. However, not all items without "like" contain tense information outside the target phrase itself, as in "But she's bored out of her mind," which could use either tense.

These results show how well native English speakers are able to recover the tense of the verb based solely on syntactic and semantic information. The average correct response rate of 69% across all stimuli without "like," and the d′ for these conditions of approximately one, show that participants are able to recover some information from the sentential context, but not enough to determine the verb tense with consistent accuracy. The two alternative forced choice task with an orthographic presentation gives a clear estimate of how much information is available in the syntax and semantics of the context of these particular utterances, without including any acoustic information, information from coarticulation with the target words, or any other auditory source. This provides a baseline for comparison with Experiments 2 and 3. It is possible that participants evaluate the verb tense differently than they would if they had heard the stimuli, even though they have been told that the material comes from conversations. Experiment 2 investigated how much information listeners can extract from syntactic, semantic, and prosodic context through the auditory modality, in order to provide an alternative baseline measure of how much information is available in the context.

3. Experiment 2: Syntactic and Semantic Context with Auditory Information

Experiment 2 replicated Experiment 1, but in the auditory modality. Instead of the target word/phrase being represented by a written blank, the portion of the speech signal corresponding to it was replaced with a beep sound similar to a square wave. The duration of the beep was standardized for all stimuli, so that duration of the target could not serve as a perceptual cue and was not confusing. Thus, listeners in this experiment had access to all of the information that participants in Experiment 1 did, plus the prosodic information in the rest of the utterance. Hence, they had access to all of the information except that of the target itself. Crucially, listeners in Experiment 2 still had no access to any acoustic cues during the target word/phrase itself. Using the auditory presentation modality may make the casual, conversational nature of the utterances more obvious to listeners. This task may also impose a higher processing load and may lead to more error and a less effective use of the information that is available, simply because the speech in most of the stimuli is fast and is presented just once, whereas participants in Experiment 1 could re-read the stimuli if they wished. Thus, we made no specific prediction about whether listeners could extract more information from the utterance context in Experiment 1 or Experiment 2, since Experiment 2 contains somewhat more information (prosody), but it is also a more difficult task.

3.1. Methods

3.1.1. Materials

The materials were made from the original conversational recordings of the 184 items of Experiment 1. Each item was extracted from the conversation from which it was recorded. The portion corresponding to the target word/phrase (e.g., "he's, she was, we were," etc.) was located and removed, and replaced by 262 ms of a beep sound. (This sound was a periodic wave with harmonics that are odd-numbered multiples of the fundamental frequency, approximating a square wave.) The 262 ms duration was the average duration of all the target portions. The boundaries for the portion to replace with a beep were adjusted to the nearest zero-crossing to avoid introducing artifacts.

Criteria for placing the boundaries at the edge of the items, and at the edge of the target word/phrase to be replaced by the beep, depended on the voicing and manner of the sounds at the boundaries and on how the sounds were realized phonetically in the particular token. All the boundaries were placed manually by inspecting the waveform and spectrogram. The complete item often began or ended at a pause, in which case the boundary between silence and voicing (for all voiced segments), frication noise, or bursts was identified as the edge of the item. When the item did not begin or end at a pause, the boundary criteria were the same as for the boundaries around the target word/phrase, described below.

For locating the boundaries of the target word/phrase (e.g., the outer edges of "he's," "you were"), the boundary between a vowel or sonorant and the preceding voiceless obstruent was placed at the onset of voicing. For example, in "I guess he was on the phone," (target underlined) there was strong frication noise for the /s/ of "guess" and no change in the frequency of the noise that would indicate an [h]. The "h" of "he was" was absent in this token. Therefore, the boundary between "guess" and the target "he was" was placed at onset of voicing. Because voicing frequently continues well into the closure of a post-vocalic voiceless stop, the boundary between a vowel or sonorant consonant and a following voiceless obstruent phoneme was placed at offset of F2, rather than the cessation of voicing (e.g., "you were telling me"). Boundaries between a vowel and a voiced obstruent were placed at the onset/offset of F2 (for example in "he was on the phone," with the /z/ fully voiced, the boundary between "he was" and "on" was placed at onset of F2). For boundaries between a vowel and a nasal, the boundary was placed at the sudden change in frequency distribution of energy visible in the spectrogram. For a voiceless stop burst with a following fricative (as in "like he's"), the boundary was placed at the change in quality of frication noise from broadband burst noise to frication noise.

In these spontaneous speech recordings, many sounds one would normally expect to find in the words were not present. For example, in one stimulus containing "and you're gonna," this string was realized as [ĩ:jɚɰɨ̃ nə̃], with the "and" reduced to a nasalized vowel assimilated to the following "j", and the following "g" was realized as a weak velar glide. Boundaries between a vowel and a glide, whether these were the expected segments or a result of reduction as in "you're gonna" here, were placed at the most sudden change in amplitude of formants for the glide, or if there was no change in formant amplitude, at the most sudden acoustic change of any sort visible in the spectrogram. If no acoustic change was present at all (as in the boundary between "and you're" [ĩ:jɚ] in this case), then the boundary was placed in the middle of that sound. This was also the case if the first/last segment of the target was adjacent to another instance of the same phoneme, as in "they're recording." Boundary placement was based on the phonetic realization of the particular production, not on what segments would be expected. For example, in "guess you're gonna" realized as [gɨsɨginʌ] (Figure 2), with a central vowel as the only realization of "you're", the boundaries for "you're" were placed at onset of voicing for the "s"-vowel boundary and the offset of F2 for the vowel-"g" boundary. Because the placement of such boundaries can be difficult in spontaneous speech, all boundaries were also verified auditorily to make sure that segments of adjacent words were not included within the target portion, so that the target portion itself would not contain excessive cues to its neighboring words. The placement of boundaries was conducted by hand labeling, using Praat [32].

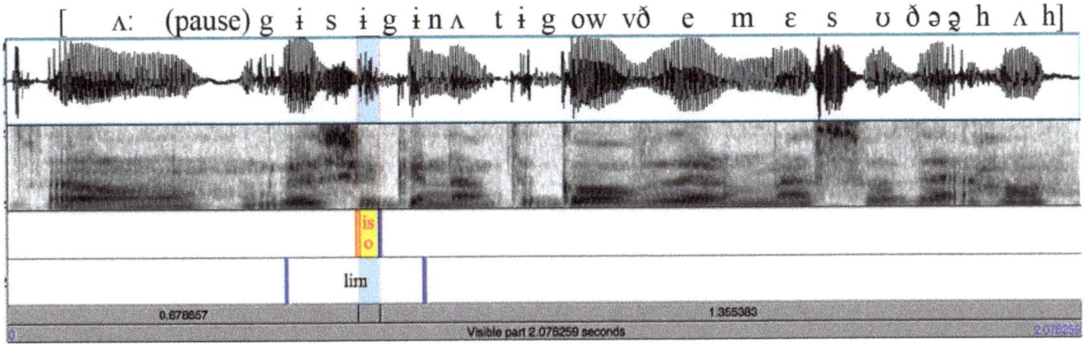

Figure 2. Waveform and spectrogram of a stimulus "Oh, guess you're gonna hafta go over there and mess with it, huh?" (referring to repairing a computer), containing highly reduced speech, with the target "you're" realized as a single central vowel. The portion marked "iso" is the portion corresponding to the target "you're," and was replaced by a beep in Experiment 2. The portions "iso" and "lim" are explained for Experiment 3 below.

3.1.2. Participants

For this experiment, 111 native speakers of American English participated. They were drawn from the same population as the participants in Experiment 1 in a different semester and did not participate in Experiment 1. (The higher number of participants reflects solely the larger number of students wishing to participate at that time.) As in Experiment 1, all participants were either monolingual in English until at least puberty, or had received some exposure to another language in the home but were English-dominant and had grown up in the U.S. No listeners reported any speech or hearing problems.

3.1.3. Procedures

Listeners were seated in a sound-protected booth and heard the stimuli over headphones. The E-Prime software was used to present stimuli and record responses. Listeners first heard 5 practice items similar to the test items, and then heard the 184 test items (and an additional 3 items that were later eliminated for comparability across experiments). The stimuli were blocked by speaker, so that listeners would be able to adjust to the phonetic features of a given speaker, as happens in the perception of conversation in daily life. At the beginning of each speaker block, a filler item by the same speaker was inserted to give listeners a chance to adjust to the new voice before data were collected. These acclimation items were not indicated to the listeners in any way, and listeners responded to them just as for test items. Each listener received the speaker blocks and the items within each speaker block in a different random order. Each stimulus was presented only once.

For each item, the listener heard the entire utterance with the target word/phrase replaced by a beep, as described above (e.g., "'Cuz he already told Steve [beep] in the wedding"). The response options were the same as for Experiment 1, but the utterances themselves were not orthographically displayed on the monitor, only the response options were (e.g., "he's/he is" and "he was" or "they're/they are" and "they were" and so on as appropriate to the item were displayed on the screen). Items were randomly assigned to have the correct response appear on the left vs. the right side of the screen. Listeners responded by means of the E-Prime response box, as in Experiment 1. If the listener did not respond, the program advanced to the next stimulus 9 s after onset of the stimulus; this occurred for 352 trials (1.7% of trials were excluded from the data below). The median length of time listeners took to respond on all other trials was approximately 3 s from the onset of the stimulus. The experiment took approximately 25 min. This task was difficult. Anecdotally, when hearing these fast and casual stimuli, it was often difficult even to be sure where in the sentence the rather short beep occurred.

3.2. Results

The results for Experiment 2 (auditory, with beep replacing target) appear in Figure 3. The data were analyzed using the same designs as for Experiment 1, again with proportion correct as the dependent variable. As in Experiment 1, an interaction of the tense and "like" factors motivated examining the fixed factor of tense for each present–past pair separately ("is/was like," "is/was" without "like," and "are/were" without "like"). (By-subject ANOVAs for details of significant interactions, as in Experiment 1 above: Singular data: tense: $F(1,110) = 5.05$, $p < 0.03$, like: $F(1,110) = 105.56$, $p < 0.001$, interaction: $F(1,110) = 53.17$, $p < 0.001$.). The same methods for model selection and choice of random effects structure were used.(For *is/was without like* and *are/were without like*, the model was Correct ~ Tense + (1+Tense | Subject) + (1 | Item); *is/was with like* used Correct ~ Tense + (1 | Subject) + (1 | Item). Models including speaker random intercepts gave either singular fit or failure to converge warnings.). For the "is/was like" items, the effect of tense was significant, with more accurate responses for "was" than "is" items ($\beta = -0.63$, $z = -3.23$, $p < 0.005$), while tense had no significant effect for "is/was" without "like" ($\beta = 0.20$, $z = 0.92$, $p = 0.36$). Detectability and bias measures appear in Table 1 above. For the "is/was like" items, listeners were somewhat biased toward the past tense response, but less so than in the orthographic task of Experiment 1. They showed greater detectability for the tense distinction if the following word was not "like.".

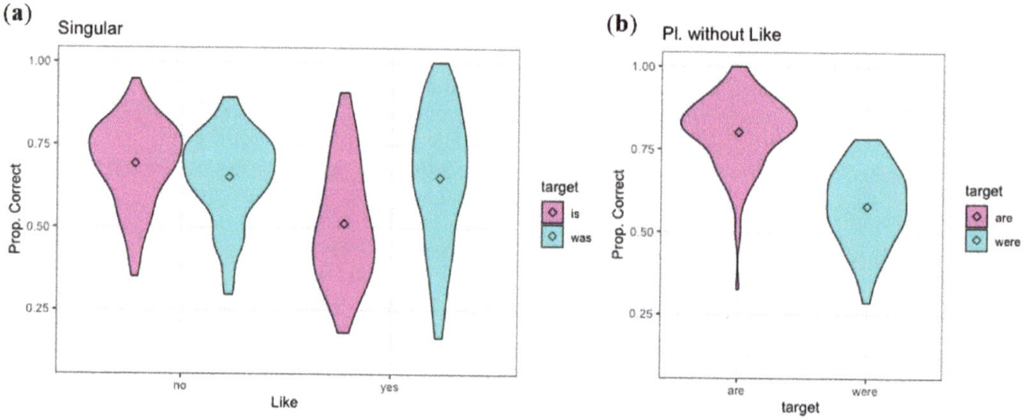

Figure 3. Results for auditory presentation of context with the target replaced by a beep sound (distribution of listeners' averages over items). Dots indicate means for each condition. (**a**) "Is" and "was" targets. (**b**) "Are" and "were" targets.

For the "are/were" pair (without "like") listeners showed significantly more accurate perception of "are" (present) than "were" ($\beta = 1.24$, $z = 5.44$, $p < 0.001$). This reflects a bias toward the present tense response, as well as some ability to hear the distinction (Table 1).

3.3. Discussion

These results show that listeners are able to extract some information about whether the verb is present or past tense from the surrounding syntactic and semantic information in the auditory modality, as well as the visual modality (Experiment 1 above). As with the visual modality, they find more information about verb tense in the sentential context if the following word is not "like." Before "like," listeners are biased toward the past tense option, although not as strongly when the information is presented auditorily as when it is presented orthographically. This suggests that listeners take the possibility of the historical present into account more when they can hear that the utterance comes from a casual conversation. That is, the tendency to assume that the verb must be in the past tense because it is reporting a past conversation is weaker if they hear the speech than if they see the written sentence.

Overall, the results of Experiment 2 confirm those of Experiment 1, but show a weaker bias toward past tense before "like" (more chance of taking historical present into account) in the auditory modality. Experiment 2 confirms that listeners are able to gather some information from the surrounding syntactic and semantic context about the verb tense when hearing the speech: across the four conditions without following "like," listeners averaged 68.2% correct answers. Clearly, the surrounding syntactic and semantic context provides some information even when it is presented auditorily, a single time, at the fast rate of spontaneous speech, but it does not provide enough to fully disambiguate the verb tense, as 68.2% correct is far from 100%.

Experiments 1 and 2 provide two types of baseline that reveal how much information about the verb tense native speakers can obtain from the syntactic, semantic, and prosodic context around that verb. When we give conference talks or class presentations on reduced spontaneous speech, we find that a common informal assumption about how listeners succeed in understanding reduced conversational speech is that they do it by understanding the surrounding syntactic and semantic information, which may include some more clearly pronounced words, and making inferences based on it. Shockey [33], section 4.2.2 suggests that hearing several words of an utterance's context after a reduction may sometimes allow listeners to suddenly recognize the reduced word, which could not be recognized until then. A large amount of past literature generally references context outside of the word as being helpful to the perception of reduced words and sounds, although not necessarily syntactic and semantic context, e.g., [4–9,15,17,22]. Experiments 1 and 2 show that there is useful information present in the rest of the sentence, but not enough for listeners to identify the verb tense very accurately. One might expect that native listeners would be more than 68–69% accurate in perceiving verb tense. Therefore, the acoustic signal within the target word/phrase itself may be providing considerable information, even when the speech is reduced. Experiment 3 addresses this issue.

4. Experiment 3: Auditory Targets with and without Context

In Experiment 3, we investigated how much information about the reduced speech of the verb native listeners can obtain from the acoustics of the target itself, with or without context. In some stimulus tokens, the portion of the signal corresponding to "you're" consists only of a single central vowel (e.g., Figure 2), or the portion corresponding to "you were" when heard in isolation sounds like an excellent example of "you're." When hearing reduced tokens in isolation, it can be tempting to infer that listeners cannot possibly be using the small amount of acoustic information in the target word/phrase itself to perceive the content. However, listeners may be relying on the acoustic information even when it is very reduced, perhaps even if this acoustic information leads them to misperceive the word, for example, if a reduced token of "you were" is misperceived as "you're" because it contains only one vowel. Thus, one question is how much use listeners make of the acoustic information within the target itself, even if it may lead to the wrong answer. We can answer this by presenting listeners with just the target word/phrase, in isolation.

Hearing the target word/phrase in isolation and hearing the rest of the sentence without the target (Experiment 2) are not simply two separable parts of perceiving the whole utterance. Coarticulation between the target word/phrase and the sounds just outside of it may be helpful to listeners. Furthermore, context provides information about the speech rate and speech style of the utterance, and listeners may use this to normalize their expectations about the duration of words, as shown in [12,15] and related work. If the surrounding speech is very fast, the boundary between what counts as "he's" vs. "he was" may fall at shorter durations than if the surrounding speech is slow, because the listener expects the speech in the target word/phrase to be fast as well. If the information the context provides is about speech style rather than just rate, knowing that the surrounding speech is spontaneous and reduced could have the same effect of causing the listener to expect shorter, more reduced pronunciations of the longer possible parse "he was."

In this experiment, we presented listeners with the target word/phrase and three levels of context (the same levels as in [5]), blocked by amount of context. This methodology is similar to that in [23]. In one condition, listeners hear only the target (e.g., "we're," "he was"), with no additional context (isolation condition). This condition thus provides listeners with whatever acoustic information occurs within the target word/phrase itself, but nothing more. In the limited context condition, listeners hear from the onset of the vowel preceding the target through the offset of the vowel following it (the stretch including the target and out to the edges of its surrounding vowels). For example, for the utterance "'Cuz he already told Steve he was in the wedding," the listener hears whatever portion of the signal corresponds to the phoneme string /iv hi wʌz ɪ/ ("-eve he was i-"). Any consonants intervening between the nearest vowel and the target are also included in this condition (e.g., /v/ in "Steve"). The speech out to the edge of the surrounding vowels should be enough to give listeners some information about speech rate independently of the target itself, but not enough to allow them to recognize the neighboring words with certainty in most cases. When the following word is "like," it is usually recognizable, since the vowel includes coarticulation with the following "k" and "like" is a very probable word after many of the targets. However, most other surrounding words cannot be recognized with certainty based on the limited context. For example, in the limited stimuli, the syllable after the target in both "I thought you were asking me" and "and we were outside the bookstore" sounds like "at" rather than "asking" or "outside." We believe the limited level of context does provide some information about speech rate, because [34,35] show that listeners do not assume that they might be hearing only part of a segment when a segment is cut off; instead they parse whatever acoustic cues they have heard as a segment, so in this case listeners were unlikely to assume that the neighboring vowels could be longer than what they heard. Furthermore, the intervening consonants such as /v/ in "Steve" in this case also provide some speech rate information. This amount of context could be confusing to listeners, since it includes incomplete words, but it is important to test whether context is useful independent of lexical information.

The third level of context allows the listener to hear the entire utterance, including the target (full context condition). This is the same acoustic signal as in Experiment 2, but with the target presented as well, not obscured in any way. Thus, this condition provides all possible types of information that occur within the utterance: the acoustics of the target word/phrase itself, speech rate, the syntactic and semantic context, and cues to the speech style of the utterance. The only additional source of information this condition does not provide is the long-term discourse context: information about what the speaker has been discussing up to this point in the conversation, or long-term acoustic cues such as speech rate beyond the single utterance (which shows an effect as speech rate across the experiment in [36]). Thus, in this experiment, listeners can use the acoustic cues in the target word/phrase itself and can also use the information present in various amounts of surrounding context. This differs from Experiments 1 and 2, where only the context information was presented, without the target itself.

4.1. Methods

4.1.1. Materials

The materials were the same recordings used in Experiment 2 except for the portion of the signal presented. For the full context condition, the stimuli were identical to those of Experiment 2 except that the target word/phrase was not replaced by a beep. These materials were simply extracted from the surrounding speech stream using the boundary criteria described for Experiment 2 and were not further manipulated. For the isolation condition, the same portion that was removed for Experiment 2 (the target word/phrase) formed the stimuli for the isolation condition of Experiment 3.

For the limited context condition, the portion from onset of the preceding vowel through the target word/phrase and up through the end of the following vowel was presented, e.g., for the utterance "you know, you were telling me about his roommate," the stimulus was the portion of the signal corresponding to /oʷ ju wɚ tɛ/. If a pause occurred in between the target and its nearest vowel, that was also included, as in this token, which had a short pause between "you know" and "you were." In some stimuli, the target word/phrase was at the beginning or end of the full context utterance, as in "He was totally making fun of me today." In such cases, the limited context stimulus included the neighboring vowel on the side that had one, e.g., /hi wʌz toʷ/.

The criteria for placing the boundary at the outer edge of the neighboring vowel were the same as described in Experiment 2 above, relying on onset/offset of F2, onset of voicing in the case of a voiceless obstruent followed by a vowel, the most sudden change in amplitude of formants for glide-vowel or vowel-glide boundaries, sudden change in the distribution of energy for nasal–vowel or vowel–nasal boundaries, etc. If the neighboring vowel was one of a string of vowels, as in "we were doing it" with the /uɪŋɪ/ portion of "doing it" realized only as a string of nasalized vowels, then the boundary was placed halfway through the F2 transition from the vowel adjacent to the target to the next vowel if there was an F2 transition (/uɪ/ in this case), and at the end of the vocalic stretch if the vowels were merged into a single vowel. If the consonant after the neighboring vowel was realized entirely as a creaky voice in place of a glottal stop (e.g., "they're not recording" with the /t/ of "not" as creaky voice), the onset of the creaky voice was considered to be the boundary between the vowel and consonant. If a target's neighboring vowel was absent, leaving a syllabic sonorant (e.g., neighboring "and" realized as [n̩]), then the end of the sonorant consonant was used as the end of the neighboring "vowel." However, if a target's neighboring vowel was absent and there was no sonorant consonant present, as in deletion of the vowel of "the," then the limited context extended to the outer edge of the next vowel that was phonetically present. Since the word "the" sometimes has very little acoustic content at all, this is necessary to have a neighboring vowel present. All boundary points were adjusted to the nearest zero-crossing to avoid adding click artifacts.

4.1.2. Participants

74 native speakers of American English who had not participated in the previous experiments participated. They were drawn from the same population as the participants for Experiments 1 and 2 and had similar language backgrounds to those participants. These participants also received extra credit in their Linguistics course for participation.

4.1.3. Procedures

The procedures were the same as for Experiment 2 above, except that listeners received the three conditions (full context, limited context, isolation) in blocks, with a break between each condition. The conditions were presented in that order (from most to least information) for all listeners, because the full context condition is most similar to the material presented in Experiments 1 and 2. Thus, having listeners respond to the full context block first, so that they cannot be influenced by having heard the other context conditions, makes the full context results directly comparable with the results of Experiments 1 and 2. Furthermore, among the 184 items, most with targets such as "he's, we're, we were" etc., it is unlikely that listeners would be able to remember specific items and apply knowledge from having heard the full context condition when hearing the corresponding item in other conditions, especially since the other conditions did not present enough context for surrounding words to generally be recognizable.

Within each condition (full, limited, and isolation), listeners first heard five practice items made from the same tokens as were used for practice items in Experiment 2. Experimental items were blocked by speaker (within each condition block), as for Experiment 2, and as in that experiment one acclimation item by the same speaker was presented before the experimental items, but was not indicated to participants as being different from the experimental items. Data from acclimation items were not analyzed. In total, within each context block, listeners heard five practice items, 18 acclimation items, 184 test items, and 4 additional test items that were later excluded for comparability across experiments. While all listeners received the context blocks (full, limited, isolation) in the same order, within each context block, the order of the speaker's voices and the order of items within each voice (after the acclimation item) was a different randomization for each listener.

The display on the computer monitor and the response alternatives for this experiment were identical to those in Experiment 2. Listeners were instructed to press the correct button to show whether the word/phrase within the sentence was, for example, "he's/he is" or "he was." None of the utterances contained the same target word/phrase twice, so there was no ambiguity as to which target was intended. The entire experiment took approximately 50 min. All other aspects of procedures were the same as Experiment 2. The time-out, after which the computer advanced to the next item if the listener failed to respond, was 9 s from onset of the stimulus. This occurred only on 320 trials, 0.8% of the data. These trials were excluded from further analysis. Median reaction time across all blocks was approximately 1200 ms from stimulus onset.

4.2. Results

The results for Experiment 3 appear in Figure 4. This experiment had context (isolation, limited, full) as an additional factor beyond those used in the experiments above, and context was the factor of primary interest, to answer which types of information the listener uses in perceiving potentially homophonous reduced speech forms. Limited context was used as the reference level for all analyses in order to reveal whether limited context allows listeners to perceive the target more accurately than the absence of context does, and whether they are able to extract more information from the full context than the limited. Initial models showed significant interactions between context and the other factors (tense and presence/absence of "like"), which motivated testing just the context factor for six subsets of data ("is like", "was like", "is" without "like", "was" without "like", "are" without "like", "were" without "like"). (By-subject ANOVAs for details of significant interactions, as in Experiment 1 above: Singular data: tense: $F(1,73) = 163.81$, like: $F(1,73) = 2141.84$, context: $F(2,146) = 18.57$, tense x like: $F(1,73) = 257.14$, tense x context: $F(2,146) = 5.37$, $p < 0.01$, like x context: $F(2,146) = 1.07$, $p > 0.10$, tense x like x context: $F(2,146) = 15.35$, all p's < 0.001 unless otherwise specified.). Model selection and choice of random effects structure was done in the same way as for Experiments 1 and 2. (The model for *is without like*: Correct ~ Context + (1 | Subject) + (1+Context | Item); *was and are without like*: Correct ~ Context + (1+Context | Subject) + (1 | Item) + (1 | Speaker); *is with like*: Correct ~ Context + (1 | Subject) + (1 | Item); was with like: Correct ~ Context + (1+Context | Subject) + (1+Context | Item); *were without like*: Correct ~ Context + (1+Context | Subject) + (1 | Item).). We predicted before beginning the analysis that the perception of "are" vs. "were" would be very different from the perception of "is" vs. "was", because of the segmental content of the words, and hence the acoustic cues, are so different. Therefore, the data cannot be analyzed when pooled over the singular and plural verbs, as was confirmed by the significant interactions with context. The "like" factor was added post hoc, as discussed for Experiment 1 above, because it had a large interaction with other factors, making it impossible to conduct a meaningful single analysis over items with and without "like."

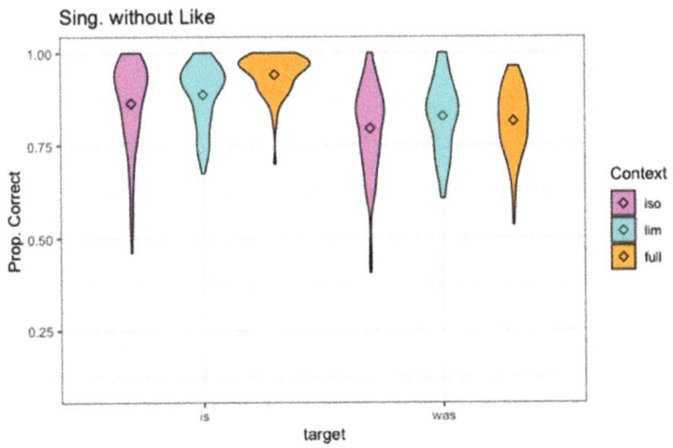

(**a**)

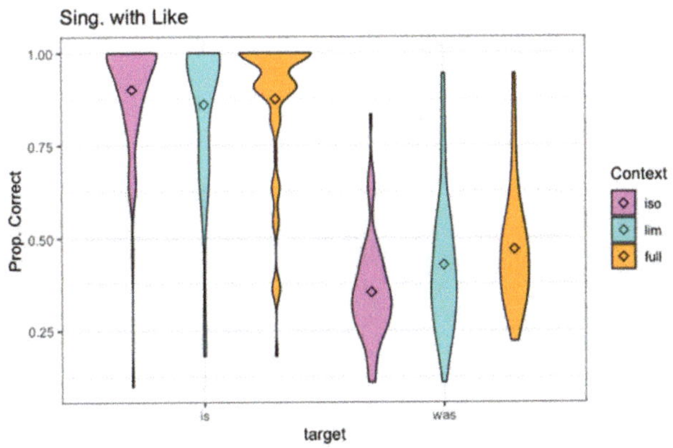

(**b**)

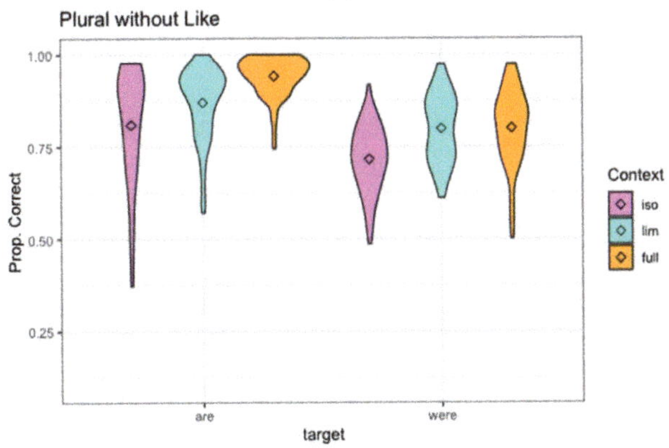

(**c**)

Figure 4. Results for listeners hearing targets with various amounts of context (iso = Isolation, lim = Limited, full = Full utterance context, distribution of listeners' averages over items). Dots indicate means for each condition. (**a**) "Is" and "was" targets not followed by "like." (**b**) "Is" and "was" targets followed by "like." (**c**) "Are" and "were" targets not followed by "like.".

For the singular "is" targets not followed by "like," the limited context was perceived significantly better than the tokens in isolation ($\beta = -0.55$, $z = -3.52$, $p < 0.001$), and full context was perceived better than the limited context ($\beta = 0.74$, $z = 4.33$, $p < 0.001$). For a singular "was" not followed by "like," isolation was perceived less accurately than limited context ($\beta = -0.29$, $z = -2.42$, $p < 0.02$), but full context provided no significant additional benefit ($\beta =-0.17$, $z = -1.54$, $p = 0.124$). Thus, for "is/was" targets not followed by "like," both types of context facilitated the perception of "is" but only limited context helps listeners recognize "was." Hearing the content of the rest of the utterance (full context) did not provide any additional benefit when listeners are hearing "was." The implications of this for listeners' use of various types of information will be discussed in Sections 4.3 and 5 below.

For the "is" targets followed by "like," listeners performed significantly better in isolation than with limited context ($\beta = 0.59$, $z = 3.56$, $p < 0.001$). This apparent negative effect of limited context on perception will be discussed below. Full context led to no difference relative to limited context ($\beta = 0.26$, $z = 1.67$, $p = 0.094$). For "was" targets followed by "like," just as for "was" without "like" above, limited context was perceived significantly better than isolation ($\beta = -0.57$, $z = -3.16$, $p < 0.005$), but full context provided no additional benefit ($\beta =0.29$, $z = 1.41$, $p = 0.159$).

Turning to the plural "are/were" pair, of which only those without "like" are analyzed as noted above, the "are" targets showed significant improvement in perception with each additional level of context (limited vs. isolation: $\beta =-0.39$, $z = -4.00$, $p < 0.001$; full vs. limited: $\beta =0.94$, $z = 7.59$, $p < 0.001$). For the "were" targets, limited context led to improved perception relative to isolation ($\beta =-0.62$, $z = -7.34$, $p < 0.001$), but Full context provided no additional benefit ($\beta =0.05$, $z = 0.54$, $p = 0.588$). This is the same pattern as for the "is/was" pair without "like": both types of context improved the perception of the shorter present tense form, while only limited context improved the perception of the longer past form. The longer past tense forms "was" and "were" (including "was" both with and without "like") showed no additional benefit when listeners hear the entire surrounding utterance in full context.

Detectability and bias measures appear in Table 1 above. The detectability results show that listeners are able to distinguish present and past tense verbs much better if the stimulus sentence was not produced with a following "like" (whether that following word is presented or not). The bias results in Table 1 for this experiment indicate that the listeners are biased toward the shorter present tense response in all conditions, unlike in Experiments 1 and 2.

4.3. Discussion

The results of Experiment 3 show that when potentially reduced function words are not followed by "like," listeners are able to extract considerable information about the intended function word from the acoustics of the target word/phrase itself. The relatively high accuracy for "is/was" and "are/were" without following "like" in the isolation condition (average of 83 and 76% correct, respectively) provide evidence of this. The availability of context, whether the speech rate and coarticulation context afforded by the limited condition or the syntactic, semantic, and prosodic context contained in the full condition, does lead to better perception. However, this improvement is somewhat modest, with accuracy improving only to 88% ("is/was" without "like") and 87% ("are/were" without "like") in full context. Information in the surrounding utterance is not enough to fully disambiguate the target words, even with the combination of bottom-up and top-down processing that contextual information allows. The acoustic information in the target word/phrase itself seems to contribute more than either type of context.

With the data in this experiment, we could examine closely which types of context contribute to listeners' perception. For all of the past tense target conditions ("was" with "like," "was" without "like," "were" without "like"), limited context improves accuracy relative to isolation, but full context leads to no significant further improvement. The past tense targets are always the longer linguistic form in the number of phonemes relative to their present tense counterpart (e.g., "was" vs. "is/'s"), and in careful speech, the past forms must constitute a syllable, whereas the present forms can be contracted to a single consonant. The results indicate that when listeners hear a reduced production of a longer past form that sounds ambiguous or sounds like the corresponding shorter present form, the information in the limited context helps them to reconstruct the longer form from the reduced acoustics, but the semantic and syntactic information of the rest of the sentence does not help them with this aspect of processing. The limited context condition, extending only to the outer edges of the nearest vowels to the target, provides listeners with information about the speech rate of the utterance and coarticulation with neighboring sounds, but is not enough to provide consistent or accurate syntactic or semantic information. When the following word is "like," it is recognizable in limited context. Other surrounding words are not consistently recognizable from the limited context stimuli, and may be misperceived as other words or not recognized as words, although some of the neighboring words may be correctly perceived as well. It seems that when listeners hear a reduced form in spontaneous speech, part of the process of perceiving it is evaluating it relative to the speech rate of the surrounding speech, as also shown in [5,15]. If the surrounding speech is fast, then a given duration might be too long to be a good candidate for "we're," but would be a better candidate for "we were." If the listener does not have access to information about the surrounding speech rate, as in the isolation condition, they might assume the same production was "we're" instead. This is similar to Miller & Volaitis' [12] finding that listeners adjust their category boundaries for aspirated vs. unaspirated stops depending on the surrounding speech rate. Here, listeners are adjusting their category boundary for "we're" vs. "we were" or "he's" vs. "he was" when they have information about the surrounding speech rate.

It is possible that the limited context supplies perceptual information other than speech rate, for example simply because including the target's neighboring segments provide information through coarticulation. That is, there could be additional perceptual cues to the segments of the target word/phrase in the adjacent segments. For example, in /iv hi wʌz ɪ/ extracted from "told Steve he was in the wedding," it is possible that the final /ɪ/ could contain perceptual cues to the preceding word "was." However, it is unlikely that coarticulation rather than speech rate is the primary source of perceptual improvement in the limited condition. For all target words/phrases, the past and present target forms begin and end with the same segments (e.g., "he's" and "he was" both begin with /hi/ and end with /z/). Coarticulation between the final /z/ of the target and its following vowel may make the /z/ more perceptible, but this would not help listeners distinguish "he's" from "he was." It is more likely that the crucial information in the limited condition is speech rate. When listeners hear the fast surrounding speech, they realize that the duration of the "was/were" targets is too long to be the shorter present tense form at that speech rate. This allows them to hypothesize that segments have been deleted and reconstruct the longer past tense form.

For the shorter present tense targets, the presence/absence of a following "like" affects which type of context improves perception. For both "is" and "are" without "like," each type of context leads to significant improvement. Listeners are already biased toward the shorter present tense responses even in isolation in these conditions. Having either speech rate or syntactic and semantic information available further strengthens their judgement that these forms are the present tense option. For "is" followed by "like," the only effect of context is to reduce the accuracy of perception rather than improve it, only with the addition of limited context. This unexpected negative effect may reflect context, leading listeners to rely less on bias (which is strongly toward "is" in this condition). That is, proportion correct dropped not because the listeners became worse at realizing they have heard "is," but rather because limited context gives them enough information to rely less on bias toward "is" and more on the ambiguous cues they hear. This direction of bias toward the "is" response before "like" is notably different from the bias in the "is/was like" conditions of Experiments 1 and 2, as will be discussed below.

This bias toward "is" in utterances with "like" was present even in the isolation condition, where listeners cannot hear the "like." (We verified by listening that coarticulation during the verb is not sufficient to perceive that "like" follows.) To verify statistically (beyond the bias value toward "is" in Table 1) that responses are quite different if the stimulus was extracted from before "like" than if it was not, we performed a post hoc comparison of "was" in isolation in the "like" vs. non-"like" conditions. The proportion correct is significantly lower in stimuli where a "like" (unheard by the listener) had originally followed the "was" ($\beta = -3.18$, $z = -5.20$, $p < 0.001$). The fact that this bias toward "is" occurs even when the listeners do not hear the "like" (isolation) suggests that the collocation with "like" causes a difference in the acoustics of the preceding target word itself, which is what influences listeners' behavior. For speakers who use quotative or discourse adverb "like," phrases such as "he was like," "he's like," "I was like" are extremely common [28,30], and thus especially subject to reduction [37–40]. Since reduction makes forms shorter and makes the /w/ of "was" or "were" less distinct [4,25,41], more reduced productions of "he was" will sound more like "he's." Thus, speakers reduce the entire phrase pronoun-is/was-like because of its high frequency, and this gives the "is/was" before "like" perceptual cues expected for "is" rather than "was." This leads to listeners' strong preference for the present tense "is" response for stimuli before "like" in all context conditions of Experiment 3. The bias becomes somewhat weaker in limited and full context conditions as listeners gain more ability to detect the difference between "is like" and "was like."

5. General Discussion

These three experiments test what types of information listeners use when perceiving reduced, potentially homophonous function words such as "he's" vs. "he was" in spontaneous, conversational speech. The types of information available include the acoustic information present in the target words themselves, the rate of surrounding speech, coarticulation with nearby sounds, cues that inform the listener that the speech style is conversational and reduced, and syntactic and semantic cues in the rest of the utterance, such as tense of other verbs or presence of a time adverb like "yesterday." Across four comparisons enumerated below, the current results suggest that listeners make more use of acoustic cues than of anything else, while using both bottom-up and top-down processing to reach a percept. This finding of dominance of acoustic cues over meaning is consistent with findings of [6,22].

First, comparison of Experiments 1 and 2 to the isolation condition of Experiment 3 shows that listeners are able to extract more information about "is" or "was" and "are" or "were" from the brief, reduced acoustic cues in the target word/phrase itself than from the entire surrounding context, no matter whether it is presented auditorily or in writing. (Experiment 2 provided listeners with acoustic cues to the context, but not to the target itself.) For this, we examined the conditions without "like," to avoid the other factors influencing those utterances. The average proportion correct and the d′ measure of detectability in Table 1 were considerably higher for the isolation condition of Experiment 3 than for either Experiments 1 or 2, both for "is/was" and "are/were" without "like." To confirm this statistically beyond the d′ values, we calculated each listener's average proportion correct for each word pair (is/was vs. are/were without "like," averaged over items first since number of present and past items is not equal), and used a linear mixed effects analysis to confirm that proportion correct was significantly lower in each of Experiments 1 and 2 than in Experiment 3′s isolation condition. (CorrectNum ~ Exper * Wordpair + (1 | Subject); the interaction of Wordpair by Experiment was significant, so the Wordpair was releveled to test with both *is/was* and *are/were* as the reference level. With *is/was* as reference level, Exper. 1 shows significantly lower proportion correct than Exper. 3 Isolation (set to reference level): $\beta = -0.15$, $t = -9.25$, $p < 0.001$; Exper. 2 also does: $\beta = -0.16$, $t = -12.56$, $p < 0.001$. The same is true with *are/were* as reference level: Exper. 1: $\beta = -0.06$, $t = -3.91$, $p < 0.001$, Exper. 2: $\beta = -0.07$, $t = -5.10$, $p < 0.001$.). Examining the average proportion correct for a present/past pair avoids the issue of bias in either direction in order to focus on how well listeners can hear the difference. As discussed in the Methods section of Experiment 2, the average duration of the isolation stimuli was only 262 ms, and many were severely reduced, as in Figure 2. Out of context, in isolation, these words/phrases can be very difficult to perceive. However, listeners perceived the targets more accurately from just the acoustic information in the target itself than they did when they were presented with all the context information in the utterance, including semantic and syntactic cues such as time adverbs, even if the context is presented in writing with ample time to read it several times. These results suggest that listeners can gain more information about reduced function words in conversation from the acoustics of just the word itself than from the entire total of all information in the context.

Second, in all three past tense conditions of Experiment 3 ("was" without "like," "was like," and "were" without "like"), the only type of context that improved listeners' accuracy significantly was the limited context, as compared to the absence of context (isolation condition). In all three cases, the full context condition led to no significant additional improvement. The limited context is not enough to allow the accurate perception of most words adjacent to the target except for the word "like." The limited context extends only as far as the outer edge of the neighboring vowels, as in /iv hi wʌz ɪ/ ("-eve he was i-") from "Cuz he already told Steve he was in the wedding." Because listeners often could not recognize the context in the limited condition as words, the addition of limited context might make the stimuli somewhat confusing relative to the isolation condition. Still, listeners were able to use the limited context to help them partially recover from the reduction of the was/were forms and to correctly parse them as the longer past tense (was/were) forms. Hearing the entire utterance did not provide significant additional benefit.

As discussed above in Experiment 3, the most likely explanation for what information listeners use from the limited context is speech rate information. Hearing that the surrounding speech is fast could shift the listeners' boundary between "he's" vs. "he was" to a shorter duration, because in fast speech, the longer past form "he was" is expected to take less time than in slower speech. Thus, listeners can use the speech rate information (or perhaps information about speech style and degree of reduction) in the limited context to help them recover the longer form from its reduced, shorter pronunciation. This is similar to how a listener adjusts the range of expected values for VOT depending on surrounding speech rate [13,17]. It is also similar to the finding [15] that the surrounding speech rate influences listeners' perception of whether function words such as "or" are present at all ("leisure or time"/"leisure time") and to the finding [36] that listeners also use long-term speech rate over the context of the experiment for this type of normalization. Notably, the addition of syntactic and semantic information from the full context did not help them significantly more beyond the limited context in the perception of the past tense conditions. This also suggests that acoustic cues (in this case speech rate and/or style) are more important than the larger context of meaning. Still, it is likely that listeners are combining bottom-up and top-down processing to perceive speech, even when the acoustic cues dominate the percept.

Third, the direction of bias in the "like" conditions provides another argument that acoustic cues outweigh other cues. In Experiment 1, where no acoustic cues were available and participants received only syntactic and semantic information, the "is/was like" conditions showed a strong bias toward "was." This is evident from the positive β value in Table 1 and the high proportion correct for "was like" stimuli and significantly lower proportion correct for "is like" stimuli. (If the participants decided entirely based on bias toward "was," with no detectability, we would see 100% correct for "was" and 0% correct for "is.") In Experiment 2, where the same context information was presented auditorily, the bias is in the same direction, but is not as strong. Experiments 1 and 2 suggest that when participants have only the syntactic and semantic information available, they assume that these sentences containing "like" are likely to be in the past tense because they are often reporting speech that happened in the past or describing a past situation. Given only the syntactic and semantic context, listeners do not take the possibility of historical present usage into account well, and instead assume past events have past tense verbs. The reason that the bias toward "was" before "like" is stronger in Experiment 1 than 2 may be that seeing the written form leads participants toward a more prescriptive judgment of which verb tense would be "correct" in a sentence.

However, in Experiment 3, where listeners can hear the target word/phrase itself, the direction of bias in the "like" conditions reverses, to a rather strong bias toward "is." This is true even in the isolation condition, where the listeners could not hear the "like." It is also true in the full context condition, which differs from Experiment 2 only in that the target word/phrase itself was also played. Thus, the acoustic information in that target word/phrase is sufficient to override all of the syntactic and semantic information that is available in the utterance context (the same information available in Experiment 2) and reverse the direction of bias. Both the target word and the utterance context cause a bias, but in opposite directions. If both are available (full condition of Experiment 3), the acoustic cues of the target word override the syntactic and semantic information of the context.

Fourth, the direction of bias in all conditions of Experiment 3 (with and without "like") is toward the present response ("is" or "are", reflected in negative β values for all conditions). Acoustic information in the stimulus target words themselves can be a source of bias. Because the stimuli are from spontaneous, casual conversations, many of the productions of target words are rather reduced, and reduction makes the past tense forms "was, were" sound more like "is, are," by making them shorter with less distinct segments. Thus, if listeners rely strongly on the acoustic cues in the targets themselves, and if many tokens contain reduction, we would expect to see bias toward the present tense responses. The acoustic cues listeners rely on may mislead them into choosing the present tense response more often than it was actually produced. While in much research bias is something undesirable to be removed in the analysis, in this case, it provides evidence for listeners' use of the acoustic cues in the target words/phrases, since reduction shifts these acoustic cues toward the present tense end of the distinction. Listeners favor the acoustic information in the targets over other information even when it misleads them.

Our findings that acoustic cues outweigh syntactic and semantic cues in the utterance context relate to the findings [21,42] that listeners make less use of semantic information in reduced pronunciations of words when recognizing subsequent words, and that they need more time to process reduced speech before they can use the semantic information in a reduced word. Drijvers et al. [24] found that reduction makes it more difficult for listeners to activate semantic information. Acoustics also outweigh word bigram probabilities in context as more acoustic information becomes available in [23], but in that case, words did not have homophones. In the current potentially homophonous short phrases such as "he's/he was" and "we're/we were," we found that the acoustics outweigh the meaning of the utterance context.

6. Conclusions

Overall, these results show that native listeners of English integrate several types of information during the process of perceiving function words in reduced spontaneous speech. They use the acoustic information within a word itself, the surrounding speech rate, and syntactic and semantic information from the rest of the utterance, as well as potentially other types of information not tested here. For example, we sometimes found that it is easier to understand a highly reduced utterance if one has heard the preceding conversation and knows what topics the speakers are discussing, but we were unable to test this here.

The listeners showed a stronger reliance on acoustic cues than on any other type of information. This does not mean that listeners fail entirely to use syntactic and semantic information in the utterance context: the significant improvement from limited to full context for "is" and "are" without "like" in Experiment 3 show use of that information. However, four comparisons across various parts of the three experiments all lead to the conclusion that acoustic cues dominate: (1) participants perceived the targets "is," "are," "was," "were" more accurately based on just the very short recording of the target phrase in isolation than they did based on the entire utterance context; (2) acoustic information in the immediately surrounding few sounds (limited context) helps listeners to recover from reduction to recognize the longer past tense forms was/were, while the addition of the entire meaning of the rest of the utterance does not provide any further benefit; (3) for "is like/was like," the utterance context biases participants toward the "was" response, but just the addition of the acoustic cues in the target is sufficient to reverse that to a bias toward "is," even when the utterance context is also heard; (4) whenever listeners hear the acoustics of the target, they show bias toward the shorter "is" or "are" response, consistent with following the acoustic cues of this reduced speech. This dominance of acoustic cues is especially interesting because it contradicts one potential explanation for how listeners understand reduced speech: that other words in the utterance are pronounced more clearly, and listeners use those instead to avoid having to parse the reductions. Instead, what we find is that listeners favor whatever acoustic information is available.

Author Contributions: Conceptualization, M.E., B.V.T. and N.W.; methodology, all authors; software, D.B.; data collection, D.B.; data curation, D.B. and N.W.; data analysis: all authors; writing—original draft preparation, N.W.; writing—review and editing, all authors; visualization, N.W. and B.V.T. All authors have read and agreed to the published version of the manuscript.

Funding: This research received no external funding.

Institutional Review Board Statement: The study was conducted in accordance with the Declaration of Helsinki, and approved by the Institutional Review Board of the University of Arizona (projects B03.195 (approved 10/20/03) and 03-0704-00 (approved 2/10/09)).

Informed Consent Statement: Informed consent was obtained from all subjects involved in the study.

Data Availability Statement: The human subjects permission and approved consent form included a statement that data would only be available to the researchers and their assistants or others collaborating with them. If you wish to collaborate on a future project involving the data, please contact the corresponding author.

Acknowledgments: We wish to acknowledge the help of Anna Woods, Carolien Schieke, Jessica Robins, Robert Henshaw, Jaycie Martin, and Amelia Zurn, who ran participants in the experiment and maintained data files. We also wish to thank all of the participants in the experiments, as well as the speakers who produced the stimulus recordings. We thank the two anonymous reviewers for their helpful comments.

Conflicts of Interest: The authors declare no conflict of interest.

Appendix A

Table A1. Examples of the 184 target items used in all of the experiments, by condition. The target word is underlined, and the phrase or sentence given here is what was used for the "full" context.

Present Tense target verbs	Past Tense target verbs
"Is" without "like" (n = 40)	"Was" without "like" (n = 30)
He's pretty closed off to everybody. But he's still s- you know. It's not like she's gonna be there all day, you know? But, either way there's I mean, I can't . . . Grammy's a grown, a grown woman.	Did you think he was ugly? She was really hyper earlier. No, it was probably last Tuesday. The other night he was at, um, like, his fraternity house. He was totally making fun of me today.
"Is" with "like" (n = 11)	"Was" with "like" (n = 16)
And she's like, "Yay! I'm so excited for you!" Yes, he's like superhyper. She's like, "No! No more laptops!" So, I don't, she's like, she has an older computer at home. Maybe my Dad could help you, he's like . . .	He was like . . . And he was like, "What's wrong?!" I called Dad and asked him about the internet, and like, he was like . . . 'Cuz she was like, "I'm gonna, you know, be screwed if I don't." It was like, the words of a giant.
"Are" without "like" (n = 43)	"Were" without "like" (n = 35)
Oh, you're going to, uh, Gymboree with Sam? And then they, like, read it to see how good of a writer you are too, so . . . I don't even know what we're gonna do. So they're gonna have like a random roommate. He's like, "You are very lucky, 'cuz that's not how it works."	You know, you were telling me about his roommate. We were gonna go out to dinner so he could see her, I dunno. He said they were dating. That was, my parents were so happy when I didn't get a bid to a sorority. Plans got changed, 'cuz like we were supposed to leave Thursday.
"Are" with "like" (excluded, n = 2)	"Were" with "like" (excluded, n = 7)
I was like, "What are you guys doing," and they're like . . . But, they're like, "It's cheaper that way!"	They were like, "Oh my God," that's, it's like almost . . . Um, Kaibab, but they were like talking to a girl and there were ambulances there and stuff. So he was so funny, you were like . . . Yeah, when I talked to you, you sounded like you were like dying. But now it's so far past Easter that we were like . . .

References

1. Greenberg, S. Speaking in shorthand—A syllable-centric perspective for understanding pronunciation variation. *Speech Commun.* **1999**, *29*, 159–176. [CrossRef]
2. Johnson, K. Massive reduction in conversational American English. In *Spontaneous Speech: Data and Analysis, Proceedings of the 1st Session of the 10th International Symposium*; Yoneyama, K., Maekawa, K., Eds.; The National International Institute for Japanese Language: Tokyo, Japan, 2004; pp. 29–54.
3. Ernestus, M.; Warner, N. An introduction to reduced pronunciation variants. *J. Phon.* **2011**, *39*, 253–260. [CrossRef]
4. Koopmans-Van Beinum, F.J. Vowel Contrast Reduction: An Acoustic and Perceptual Study of Dutch Vowels in Various Speech Conditions. Ph.D. Thesis, Universiteit van Amsterdam, Amsterdam, The Netherlands, 1980.
5. Ernestus, M.; Baayen, H.; Schreuder, R. The recognition of reduced word forms. *Brain Lang.* **2002**, *81*, 162–173. [CrossRef] [PubMed]
6. Janse, E.; Ernestus, M. The roles of bottom-up and top-down information in the recognition of reduced speech: Evidence from listeners with normal and impaired hearing. *J. Phon.* **2011**, *39*, 330–343. [CrossRef]
7. Arai, T. A case study of spontaneous speech in Japanese. In Proceedings of the International Congress of Phonetic Sciences (ICPhS), San Francisco, CA, USA, 1–7 August 1999; Department of Linguistics, University of California: Berkeley, CA, USA, 1999; pp. 615–618.
8. Saerens, M.; Serniclaes, W.; Beeckmans, R. Acoustic versus contextual factors in stop voicing perception in spontaneous French. *Lang. Speech* **1989**, *32*, 291–314. [CrossRef]

9. Brown, M.; Dilley, L.C.; Tanenhaus, M.K. Real-time expectations based on context speech rate can cause words to appear or disappear. In Proceedings of the 34th annual conference of the Cognitive Science Society, Sapporo, Japan, 1–4 August 2012; pp. 1374–1379.
10. Tucker, B.V. The effect of reduction on the processing of flaps and /g/ in isolated words. *J. Phon.* **2011**, *39*, 312–318. [CrossRef]
11. Ranbom, L.J.; Connine, C.M. Lexical representation of phonological variation in spoken word recognition. *J. Mem. Lang.* **2007**, *57*, 273–298. [CrossRef]
12. Miller, J.L.; Volaitis, L.E. Effect of speaking rate on the perceptual structure of a phonetic category. *Percept. Psychophys.* **1989**, *46*, 505–512. [CrossRef]
13. Volaitis, L.E.; Miller, J.L. Phonetic prototypes: Influence of place of articulation and speaking rate on the internal structure of voicing categories. *J. Acoust. Soc. Am.* **1992**, *92*, 723–735. [CrossRef]
14. Gottfried, T.L.; Miller, J.L.; Payton, P.E. Effect of speaking rate on the perception of vowels. *Phonetica* **1990**, *47*, 155–172. [CrossRef]
15. Dilley, L.C.; Pitt, M.A. Altering context speech rate can cause words to appear or disappear. *Psychol. Sci.* **2010**, *21*, 1664–1670. [CrossRef] [PubMed]
16. Niebuhr, O.; Kohler, K.J. Perception of phonetic detail in the identification of highly reduced words. *J. Phon.* **2011**, *39*, 319–329. [CrossRef]
17. Heffner, C.C.; Dilley, L.C.; McAuley, J.D.; Pitt, M.A. When cues combine: How distal and proximal acoustic cues are integrated in word segmentation. *Lang. Cogn. Process.* **2013**, *28*, 1275–1302. [CrossRef]
18. Ladefoged, P.; Broadbent, D.E. Information conveyed by vowels. *J. Acoust. Soc. Am.* **1957**, *29*, 98–104. [CrossRef]
19. Labov, W.; Ash, S. Understanding Birmingham. In *Language Variety in the South Revisited*; Bernstein, C., Nunnally, T., Sabino, R., Eds.; University of Alabama Press: Tuscaloosa, AL, USA, 1997; pp. 508–573.
20. Brouwer, S.; Mitterer, H.; Huettig, F. Speech reductions change the dynamics of competition during spoken word recognition. *Lang. Cogn. Process.* **2012**, *27*, 539–571. [CrossRef]
21. Van de Ven, M.; Tucker, B.V.; Ernestus, M. Semantic context effects in the comprehension of reduced pronunciation variants. *Mem. Cogn.* **2011**, *39*, 1301–1316. [CrossRef]
22. Van de Ven, M.; Ernestus, M.; Schreuder, R. Predicting acoustically reduced words in spontaneous speech: The role of semantic/syntactic and acoustic cues in context. *Lab. Phonol.* **2012**, *3*, 455–481. [CrossRef]
23. Van de Ven, M.; Ernestus, M. Segmental/durational cues in the processing of reduced words. *Lang. Speech* **2018**, *61*, 358–383. [CrossRef]
24. Drijvers, L.; Mulder, K.; Ernestus, M. Alpha and gamma band oscillations index differential processing of acoustically reduced and full forms. *Brain Lang.* **2016**, *153*, 27–37. [CrossRef]
25. Warner, N.; Tucker, B.V. Phonetic variability of stops and flaps in spontaneous and careful speech. *J. Acoust. Soc. Am.* **2011**, *130*, 1606–1617. [CrossRef]
26. Schneider, W.; Eschman, A.; Zuccolotto, A. *E-Prime (Version 2.0). [Computer Software and Manual]*; Psychology Software Tools Inc.: Sharpsburg, PA, USA, 2002.
27. Blyth, C.; Recktenwald, S.; Wang, J. I'm like, "say what?!": A new quotative in American oral narrative. *Am. Speech* **1990**, *65*, 215–227. [CrossRef]
28. Dailey-O'Cain, J. The sociolinguistic distribution of and attitudes toward focuser *like* and quotative like. *J. Socioling.* **2000**, *4*, 60–80. [CrossRef]
29. Drager, K.K. Sociophonetic variation and the lemma. *J. Phon.* **2011**, *39*, 694–707. [CrossRef]
30. Podlubny, R.G.; Geeraert, K.; Tucker, B.V. It's All about, Like, Acoustics. In Proceedings of the 18th International Congress of Phonetic Sciences, Glasgow, UK, 10–14 August 2015; pp. 1–4. Available online: https://www.internationalphoneticassociation.org/icphs-proceedings/ICPhS2015/Papers/ICPHS0477.pdf (accessed on 13 July 2022).
31. Bates, D.; Mächler, M.; Bolker, B.; Walker, S. Fitting Linear Mixed-Effects Models Using lme4. *J. Stat. Softw.* **2015**, *67*, 1–48. [CrossRef]
32. Boersma, P. Praat, a system for doing phonetics by computer. *Glot Int.* **2001**, *5*, 341–345.
33. Shockey, L. *Sound Patterns of Spoken English*; John Wiley and Sons: Hoboken, NJ, USA, 2008.
34. Smits, R.; Warner, N.; McQueen, J.M.; Cutler, A. Unfolding of phonetic information over time: A database of Dutch diphone perception. *J. Acoust. Soc. Am.* **2003**, *113*, 563–574. [CrossRef]
35. Warner, N.; McQueen, J.M.; Cutler, A. Tracking perception of the sounds of English. *J. Acoust. Soc. Am.* **2014**, *135*, 2995–3006. [CrossRef] [PubMed]
36. Baese-Berk, M.M.; Heffner, C.C.; Dilley, L.C.; Pitt, M.A.; Morrill, T.H.; McAuley, J.D. Long-term temporal tracking of speech rate affects spoken-word recognition. *Psychol. Sci.* **2014**, *25*, 1546–1553. [CrossRef]
37. Pluymaekers, M.; Ernestus, M.; Baayen, R.H. Lexical frequency and acoustic reduction in spoken Dutch. *J. Acoust. Soc. Am.* **2005**, *118*, 2561–2569. [CrossRef]
38. Pluymaekers, M.; Ernestus, M.; Baayen, R. Articulatory planning is continuous and sensitive to informational redundancy. *Phonetica* **2006**, *62*, 146–159. [CrossRef]
39. Bell, A.; Brenier, J.; Gregory, M.; Girand, C.; Jurafsky, D. Predictability effects on durations of content and function words in conversational English. *J. Mem. Lang.* **2009**, *60*, 92–111. [CrossRef]

40. Bybee, J.; Scheibman, J. The effect of usage on degrees of constituency: The reduction of don't in English. *Linguist. Interdiscip. J. Lang. Sci.* **1999**, *37*, 575–596. [CrossRef]
41. Warner, N.; Fountain, A.; Tucker, B.V. Cues to perception of reduced flaps. *J. Acoust. Soc. Am.* **2009**, *125*, 3317–3327. [CrossRef] [PubMed]
42. Van de Ven, M.; Tucker, B.V.; Ernestus, M. Semantic facilitation in bilingual everyday speech comprehension. In Proceedings of the Interspeech, Makuhari, Chiba, Japan, 26–30 September 2010; pp. 1245–1248.

brain sciences

MDPI

Article

Adaptation to Social-Linguistic Associations in Audio-Visual Speech

Molly Babel

Department of Linguistics, University of British Columbia, Vancouver, BC V6T 1Z4, Canada; molly.babel@ubc.ca

Abstract: Listeners entertain hypotheses about how social characteristics affect a speaker's pronunciation. While some of these hypotheses may be representative of a demographic, thus facilitating spoken language processing, others may be erroneous stereotypes that impede comprehension. As a case in point, listeners' stereotypes of language and ethnicity pairings in varieties of North American English can improve intelligibility and comprehension, or hinder these processes. Using audio-visual speech this study examines how listeners adapt to speech in noise from four speakers who are representative of selected accent-ethnicity associations in the local speech community: an Asian English-L1 speaker, a white English-L1 speaker, an Asian English-L2 speaker, and a white English-L2 speaker. The results suggest congruent accent-ethnicity associations facilitate adaptation, and that the mainstream local accent is associated with a more diverse speech community.

Keywords: perceptual adaptation; linguistic expectations; social stereotypes; speech in noise; intelligibility

1. Introduction

Listeners' experiences in the linguistic world contribute to the formation and reinforcement of associations between language, people, and social structures. Listeners learn that, for example, females being, on average, smaller in stature, have smaller vocal tracts than men, and thus generally have higher frequency boundaries between, for example, vowels [1] and sibilant fricatives [2]. Such expectations about the relationship between talker size and phonetic realizations arguably assist in processing spoken language more efficiently and adeptly. While phonetic associations related to gender or sex are at least partially rooted in physiological differences [3]—as opposed to being wholly culturally-specific learned patterns, see Johnson [4]—between women and men, listeners also connect pronunciation patterns with completely arbitrary social groups. Drawing upon learned associations, listeners can categorize a speaker on a number of different social identities based on speech samples (e.g., ethnicity [5]) and use inferred social characteristics to guide the categorization of spoken language (e.g., [6]). For overviews of the evidence in support of listeners' vast sociophonetic knowledge space, see Drager [7] and Hay and Drager [8]. Supported by decades of empirical evidence that listeners jointly track social and linguistic information (see overviews in [9,10]), Kleinschmidt et al. [11] build upon their Bayesian ideal adaptor framework for phoneme identification [12] and present a computational model of how listeners can leverage probabilistic co-patterning of socio-indexical characteristics and linguistic features. This join tracking models how listeners are able to infer both linguistic judgments (e.g., given what is known about this talker socially, was that a /p/ or a /b/?) and social-indexical judgments (e.g., given what is known about the identified linguistic category, what dialect region is that talker from?).

The associations and joint probabilities listeners possess may (typically) originate in veridical experiences, but these expectations and the sociolinguistic knowledge listeners carry can warp their perception of the speech stream. Listeners' ultimate percepts or decisions about what they heard of a given utterance are influenced by what they *expect* a talker

Citation: Babel, M. Adaptation to Social-Linguistic Associations in Audio-Visual Speech. *Brain Sci.* **2022**, *12*, 845. https://doi.org/10.3390/brainsci12070845

Academic Editors: Richard Wright and Benjamin V. Tucker

Received: 25 May 2022
Accepted: 25 June 2022
Published: 28 June 2022

from a particular social category to produce. For example, given acoustically identical perceptual stimuli, New Zealand listeners perceive speakers who seem younger as having a more complete NEAR/SQUARE merger, consistent with younger speakers being probabilistically more likely to have merged the sounds Hay et al. [6]. Niedzielski [13] found that listeners from Michigan, USA assumed that an apparent speaker from Ontario, Canada had a different accent from their own (despite this lack of difference) and categorized vowels accordingly. Niedzielski also showed that these Michigan listeners perceived their own accent as patterning more with a mainstream American one, indicating a disconnect between actual and perceived pronunciation in their speech community. This indicates that listener expectations about accents and speech patterns, including their own, affect their perceptual or recognition space.

The current research is focused on listener associations between accent and ethnicity. Associations between accent and ethnicity in English-speaking countries with histories as colonizers present a particular challenge, as the associations are frequently shown to be fallible. The fallibility of accent and ethnicity associations is, of course, not a problem that is unique to English-speaking countries. For example, despite multicultural and diverse non-white demographics in the United States, to be considered maximally "American", one must be white [14]. This association is implicated in speech studies that tap expectations or stereotypes about who is expected to speak "unaccented" English. For example, an influential set of studies paired photos of a white face and a East Asian face with voices representing native and non-native accents [15,16]. Recent scholarship makes a convincing case for the abandonment of the vague label *native speaker* [17]. The term is used in this manuscript as a short-hand for a perceptibly mainstream accent for a local speech community. This phrasing of "perceptibly mainstream" is intended to signal that what is crucial within the current work is that a talker's accent is *perceived* as being a member of a particular category or speech community, in spite of an individual having, for example, multiple native languages, as is typical in the local speech community. Mainstream language use, however, often masks this multilingual upbringing. In Kang and Rubin's work, when the voices were paired with the East Asian face, they were perceived as more accented and were associated with lower accuracy on a cloze task. Kang and Rubin call this outcome reverse linguistic stereotyping, which results in evaluations of low social status that negatively affect speech comprehension. Kang and Rubin's theory hinges on the listener-valued social prestige, riding on some aspect of volition. However, experience and stereotypes may affect speech processing instead of or in addition to a listener's inclinations. McGowan [18] pursued an exemplar-theoretic explanation of accent and ethnicity associations that is based on experience and not a listener's willingness to comprehend. In support of his approach, McGowan found that Mandarin-accented English was more intelligible when paired with an East Asian face than with a white face [18]. This facilitating effect in comprehending L2-accented speech is consistent with expectations about the phonetic patterns associated with a given social group. Using speech from a larger set speakers of Canadian English, Babel and Russell [19] demonstrated a similar effect in a speech in noise task that compared audio-only trials with ones pairing audio with white Canadian or Chinese Canadian faces. They found lower accuracy in the transcription of Chinese-Canadian speech only in combination with Chinese-Canadian faces. This effect was greater for listeners who reported spending more time with Chinese Canadians, suggesting that the findings may not be about negative social associations, but instead involve erroneous ethnicity/accent expectations. Similarly, in another English-speaking context, Gnevsheva [20] found that New Zealand English listeners rated a white German-L1/English-L2 speaker as least-accented when presented with a video, middlingly-accented in an audio-only condition, and most-accented in an audio-visual condition. This contrasts with an ethnically Korean Korean-L1/English-L2 speaker being rated as consistently highly accented in all three conditions. Gnevsheva reasons that while listeners expect an ethnically Korean individual to speak English with a non-native accent, their expectations of a white talker are that they will exhibit a native English accent. The mismatch between this expected native accent and reality in an audio-video condition prompts an increase in perceived accentedness.

Gnevsheva [20] examined perceived accentendness and not intelligibility. This distinction is both theoretically and empirically important. Through a series of experiments, Zheng and Samuel [21] demonstrate the changes in perceived accentedness are better characterized as a change in "interpretation" and not "perception" of the speech. These changes in interpretation may be more malleable than changes in perception, as they may tap into stereotypes more directly. Indeed, presenting listeners with native Dutch passages, Hanulíková [22] found that co-presenting the native speech with a photo of an ethnically Moroccan face did not change intelligibility and only increased ratings of accentedness in adverse listening conditions.

Not all accents are equivalent in their ability to elicit ethnicity and accent associations in speech processing. This may be because particular accents are simply more intelligible (e.g., the signal quality of a mainstream accent may be more robust than an accent with which one has less experience) or because particular accents are socially associated with a more diverse group of talkers (e.g., mainstream accents versus rural regional accents). Evidence for the latter interpretation comes from infants. Infants raised in highly multilingual communities develop ethnicity and language associations from an early age. May et al. [23] demonstrated that 11 month old English-acquiring infants in Vancouver, British Columbia associate Cantonese more strongly with Asian faces than white faces. Eleven month old English-acquiring white infants' looking times at Asian versus white (static) faces are equivalent when infants are presented with English, demonstrating that infants consider both of the faces equally likely to produce (natively-accented) English. On Cantonese language trials, however, infants looked longer at the Asian faces, suggesting an expectation that the Asian face would be more likely to speak Cantonese. Through a series of experiments, May and colleagues suggest that their results are not simply due to an association of an unfamiliar language with a less familiar face.

Using three groups of listeners—teens, young adults, and elderly adults—Hanulíková [24] assessed the effect of face primes on the intelligibility and perceived accentedness of Standard German, Korean-accented German, and Palatinate German. Perceived accentedness was significantly higher for all speech samples for the elderly adult group when accompanied by an Asian face. Only the standard accent was perceived as more accented in the presence of an Asian face for the teen and younger adult age groups. There was some evidence that listeners found the Korean-accented German more intelligible when co-presented with the Asian face and that the Palatinate German accent was more intelligible when accompanied by a white face. Hanulíková [24] found no effect of intelligibility on the standard German accents. These results also align with there being more diverse associations with mainstream accents compared to regional or non-native accents. Such results reiterate that listeners' experiences shape and guide their social expectations (see also [19]). In Montréal, Québec, Canada, where bi/multi-lingualism is an embedded aspect of the local culture, listeners appear to be more impervious to the effects of white and South Asian face primes paired with American English, British English, and Indian English voices compared to listeners from Gainesville, Florida, USA, where bi/multi-lingualism is a less valued asset [25].

Whether listeners specifically tailor their expectations to a particular accent—that is, adjusting the anticipated phonetic distributions to align with the pronunciation patterns of particular non-native accent (e.g., Mandarin-accented English)—or engage a more global relaxation mechanism is a matter of debate. Like the targeted adaptation to Mandarin-accented English when presented with an image of an Asian talker found in McGowan [18], Vaughn [26] found that giving listeners information about the identity of an upcoming L1-Spanish/L2-English talker improved transcription accuracy. These results suggest a targeted adaptation mechanism that improved American English listeners' ability to parse Mandarin- and Spanish-accented English in those respective studies. Melguy and Johnson [27] find evidence that supports a more global adaptation mechanism, showing that listeners who believe the talker exhibits any non-native accent show higher transcription accuracy.

The majority of the literature exploring ethnicity and accent or language associations relies on static photos. The reason for this is likely to allow for more convincing applications of matched-guise techniques in the experimental design. The use of static photos, however, removes a layer of ecological validity, as voices are most often accompanied by moving faces, not static ones. Those moving faces are an important source of phonetic information. Generally, audio-visual speech receives a boost in performance compared to audio-only speech (e.g., [28]). This audio-visual benefit, however, has been shown to be larger for natively accented talkers [29]. Yi and colleagues tested listeners using native and Korean-accented English in audio-only and audio-visual conditions. A greater audio-visual boost was found for the native English speakers. For the Korean-accented speakers, listeners' performance was predicted by the strength of an association between the categories "Asian" and "foreign". They conclude less experience with Korean faces inhibits listener ability to exploit the facial movements that are known to aid alignment and boost intelligibility.

The summarized literature suggests that listeners use experiences and stereotypes to buffer expectations that help and hinder the processing of accents. The quality of the evidence is mixed, however, with some finding support for expectations exerting influence in both intelligibility and accentedness (e.g., [19]), only accentedness (e.g., [22]), or a mixed bag (e.g., [24]) when intelligibility and accentedness are investigated in tandem. Whether adaptation to accent and ethnicity associations is targeted or global is also mixed [26,27]; recent evidence in support of both targeted and global adaptation mechanisms at work in lexically-guided perceptual adaptation is presented in Babel et al. [30]. The conflicting results within this body of literature may be expected due to the uniqueness of the subject population—rarely are the social and linguistic experiences of the listener population described at length—and the specific social associations and demographic facts of a speech community. As a case in point, Babel and Russell [19] recruited talkers from a particular suburb with a historic and well-established Cantonese-speaking population, and also informed listeners that the talkers were from this particular suburb. We were leveraging locally-held social associations. Regardless, the conflicting results may also be due to spurious findings in either direction—that is, either in support of the role of social expectations in speech perception or against such a mechanism. These considerations warrant an analysis strategy that offers nuance to interpretation, a focus on effect size, and a side-lining of null-hypothesis significance testing. Bayesian data analysis satisfies these desiderata and is deployed for the current set of research questions.

Those research questions are focused on how listeners adapt to accent and ethnicity associations in naturally-produced audio-visual speech. While varied, the literature generally suggests that listeners should be better at adapting to accent and ethnicity associations that match local stereotypes. We test this with a speech in noise sentence transcription task using naturally produced audio-visual stimuli with speech embedded in −5 dB signal-to-noise ratio (SNR) pink noise. The speech samples come from four talkers who vary in terms of self-identified ethnicity—white and Asian—and whether they speak English as a first or second language. Comparing high predictability training sentences and low predictability test sentences, we expect listeners to adapt more easily to the talkers who match accent and ethnicity stereotypes. Local for the current study is the same urban area as Babel and Russell [19] and May et al. [23], which means that we expect to see a reduction in transcription accuracy between training and test trials for the Asian English-L1 and the white English-L2 speakers due to assumptions that ethnically Asian individuals should be non-native English speakers and ethnically white individuals should be native speakers of English. Being trained on an accent and ethnicity pairing counter to local stereotypes is predicted to make adaptation more difficult due to a mismatch between predicted and perceived signals. The white English-L1 and the Asian English-L2 talkers conform to local stereotypes, and we predict that listeners will adapt more to these talkers, showing generalization from the high predictability training sentences to the low predictability test set.

2. Methodology

2.1. Materials: Audio-Visual Stimuli

Four female talkers in their twenties were recorded reading high and low predictability sentences from Bradlow and Alexander [31]. Example sentences are provide in Table 1. The full sentence list is available at an OSF repository (accessed on 24 May 2022). The talkers included two first language speakers of Canadian English and two second language English speakers. For both the L1 and L2 pairs, one talker was Asian and the other was white. The Asian English-L2 talker was a native speaker of Mandarin and the white one was a native speaker of Spanish; these speakers were chosen out of convenience. The social demographic labels applied to these speakers—female and either Asian or white—are self-identified categories for each talker.

Table 1. Example high and low predictability sentence stimuli.

Sentence	Predictability
The opposite of hot is cold.	High
For your birthday, I baked a cake.	High
In the spring plants are full of green leaves.	High
He pointed at his hair.	Low
Mom thinks that is yellow.	Low
She talked about the leaves.	Low

Audio recordings were digitized at 44.1 kHz using a Sennheiser MKH-416 shotgun microphone connected to a USB Pre-2 amplifier and a PC. Video recordings were made using Panasonic HC-V700M high definition video camera, which also recorded audio. The video recordings included the talkers from the neck up against a white background. The high quality audio recordings were RMS-amplitude normalized and embedded in pink noise at a −5 dB SNR. The video and high-quality audio streams were synced using Adobe Premier Pro using the lower quality audio recorded from the video recorder to guide the audio alignment. Sentences with speech errors were eliminated, leaving 120 unique sentences. Participants were always presented with simultaneous audio-video stimuli.

2.2. Participants

A total of 83 listeners were recruited from undergraduate linguistics courses and received partial course credit in exchange for their participation. There were 66 female and 17 male participants between 18 and 26 years of age (Mean = 20). Listeners were either first language or early learners of English, which we operationalize as before the age of 5. Listeners self-reported their ethnicities (33 = Asian, 23 = White, 9 = South Asian, 3 = Indian, 2 = Asian and White, 1 = Asian Pacific Islander, 1 = Filipino, 1 = Japanese Canadian, 1 = Middle Eastern, 1 = First Nations, 1 = South East Asian and White; ethnicity information was missing for 7 participants).

2.3. Procedure

Participants were seated in front of a computer in sound-attenuated cubicles for the duration of the experiment. Listeners heard each sentence over headphones at approximately 65 dB while watching accompanying video of the talker on the screen. They were asked to type sentences on a keyboard and told to focus on being as accurate as possible while not worrying about minor spelling errors.

To facilitate adaptation to each individual talker, the task was blocked by talker. In each block, listeners heard 30 sentences from each talker. In order to control for talker order, there were 24 different permutations of the experiment. These orders were implemented cyclically, such that one participant would have order A and the next order B, resulting in approximately three to four participants for each. The 30 sentences were separated into 15 high predictability and 15 low predictability blocks, randomly selected for each listener. The high and low predictability blocks are thus designed and analyzed as training and test blocks, respectively. There were breaks between talkers, but not between sentence types within a talker.

This within-subject design for all talkers allows us to ignore talker-specific differences in intelligibility and focus on change—improvement or decline in performance—between high and low predictability blocks for each of the four talkers.

3. Results

3.1. *Analyses*

The measure of interest in this study is the change in listeners' accuracy in transcribing a talker's speech in noise between the set of high predictability sentences and the set of low predictability sentences. Transcription accuracy was automatically scored using the Token Sort Ratio, which is a fuzzy logic matching metric Bosker [32].

Data were analyzed with a Bayesian multilevel regression model using *brms* [33] in R using *cmdstanr* on the back end [34]. The model syntax was: TSR Sentence Predictability * Talker + (Sentence Predictability * Talker | Subject) + (1 | Sentence), phi Sentence Predictability * Talker), family = Beta(). The TSR score is similar to proportion correct, and is bounded between 0 and 1, so a beta regression was used [35]. The actual 0 and 1 values were "squeezed" following the formula provided by Smithson and Verkuilen [35]. These squeezed TSR scores were the dependent measure. For the mean parameter, sentence predictability and Talker were the population-level (the "fixed effects" in frequentist mixed effects modeling jargon) effects, the interaction of which was also included in the model. Both sentence predictability and talker were dummy coded with high predictability sentences and the white English-L1 speaker as the reference levels. Listener and sentence were the group-level effects (i.e., the "random effects" in a frequentist model). There was a random intercept for sentence, while the listener-level effects included a random intercept and random slopes for sentence predictability, talker, and their interaction. Beta regression models include a phi parameter, which models the variance of the TSR scores. Sentence predictability and talker (without their interaction) were included as the population-effects for the phi parameter. Priors for all population-level effects were weakly informative priors of normal distributions with a mean of 0 and standard deviations of 2 and 1 for the intercept and population-level parameters, respectively, for both the mean and the phi components of the analysis. The standard deviations for the group-level effects had an exponential distribution of rate 1 as priors, and correlations used an LKJ prior of concentration 1. The model was fit using 4 Markov chains and 4000 samples each with 1000 warm-up samples per chain.

There were no divergent transitions and the $\hat{R}$ values were all <1.01, suggesting well-mixed chains. Inspection of the graphical posterior predictive check indicated that the model fit the data well.

Bayesian analysis allows for more nuance in evaluation evidence. When the 95% Credible Interval (CrI) for a given parameter excludes 0, this is considered strong evidence for an effect. The evidence for an effect is described as weak if the CrI includes 0, but the probability of direction is more than 95%. These choices follow Nicenboim and Vasishth [36].

3.2. *Empirical Observations*

The empirical data are presented as a box-and-whisker plot in Figure 1, with intelligibility—the Token Sort Ratio (TSR) score on the y-axis and the four talkers along the x-axis. Separate boxes visualize the distribution of responses for the high predictability training sentences (dark purple boxes) and the low predictability test sentences (yellow boxes). The empirical data suggest a generalization from the high to low predictability sentences for the Asian English-L2 and white English-L1 talkers, the talkers with stereotypically congruent race/accent associations. The median intelligibility score for these talkers is maintained across the two sentence types or, in the case of the Asian English-L2 increases across the testing and training, suggesting robust adaptation and generalization. The empirical data for the stereotypically incongruent Asian English-L1 and white English-L2 talkers shows a loss of intelligibility across the high predictablity training sentences and the low predictability test sentences, suggesting that listeners are less able to adapt to the noise for the talkers with stereotypically incongruent race/accent pairings.

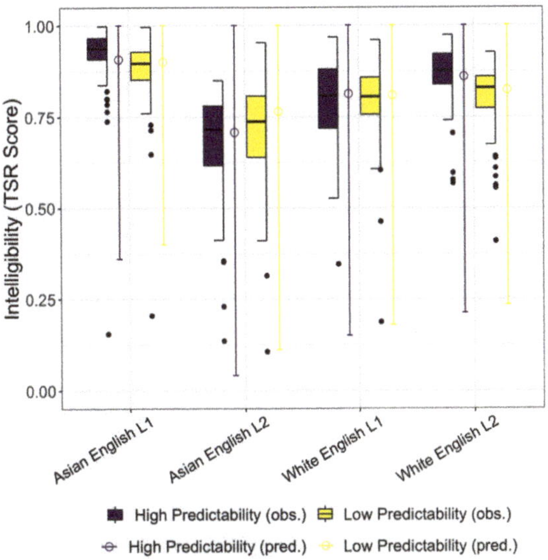

Figure 1. A box-and-whisker plot of the empirical results and the range of the posterior predictive distributions for intelligibility, plotted as TSR scores, for the four talkers separated by high and low predictability sentences.

3.3. Bayesian Regression

The empirical observations are quantitatively assessed through a Bayesian regression model. The $\hat{\beta}$ Estimate, standard error, 95% CrI, and the Probability of Direction of the fixed effects for this model are summarized in Table 2. The range of the posterior predictive distributions are plotted alongside the empirical data in Figure 1.

The high predictability utterances provided strong training for the low predictability utterances for the white English-L1 talker, who is the reference level in the model. There is no evidence for a loss of intelligibility across the training and test sentences [$\hat{\beta} = 0, CrI = [-0.2, 0.2], Pr(\hat{\beta} > 0 = 0.52$]. The white English-L1 talker's high predictability sentences were lower intelligibility than those of the Asian English-L1 talker [$\hat{\beta} = 0.82, CrI = [0.71, 0.94], Pr(\hat{\beta} > 0 = 1$] and the white English-L2 talker [$\hat{\beta} = 0.37, CrI = [0.26, 0.48], Pr(\hat{\beta} > 0 = 1$]. The white English-L1 talker's high predictability sentences had higher intelligibility than the Asian English-L2 talker's high predictability sentences [$\hat{\beta} = -0.57, CrI = [-0.69, -0.45], Pr(\hat{\beta} < 0 = 1$]. Of primary interest, however, are the interactions between predictability and the talkers. While the empirical data and the point estimate suggest a loss of intelligibility for the Asian English-L1 talker across the test and training sentences compared to the white English-L1 talker, the CrI contains 0 and the probability of direction is relatively low [$\hat{\beta} = -0.07, CrI = [-0.2, 0.05], Pr(\hat{\beta} < 0 = 0.88$]. The results for the L2 speakers of English are in line with predictions, however. There is strong evidence that the high predictability sentences provided robust training for the stereotypically congruent Asian English-L2 talker [$\hat{\beta} = 0.27, CrI = [0.14, 0.41], Pr(\hat{\beta} > 0 = 0.99$]. Likewise, the evidence is strong the the high predictability training sentences do not generalize to test performance with the low predictability sentences for the stereotypically incongruent white English-L2 talker [$\hat{\beta} = -0.29, CrI = [-0.42, -0.16], Pr(\hat{\beta} < 0 = 1$].

Table 2. Population-level or fixed-effect predictors for the beta regression model. The $\hat{\beta}$ estimate, standard error, and 95% Credible Interval (CrI) for TSR are reported for the means and phi parameters.

Mean

	$\hat{\beta}$ Estimate	Standard Error	95% CrI	Probability of Direction
Intercept	1.4553	0.0857	[1.29, 1.62]	1
Low Predictability	−0.0030	0.1013	[−0.2, 0.2]	0.52
Asian English-L1	0.8227	0.0619	[0.71, 0.94]	1
Asian English-L2	−0.5695	0.0612	[−0.69, −0.45]	1
white English-L2	0.3731	0.0594	[0.26, 0.48]	1
Low:Asian English-L1	−0.0734	0.0633	[−0.2, 0.05]	0.88
Low:Asian English-L2	0.2717	0.0690	[0.14, 0.41]	0.99
Low:white English-L2	−0.2901	0.0594	[−0.42, −0.16]	1

phi

	$\hat{\beta}$ Estimate	Standard Error	95% CrI	Probability of Direction
Intercept	0.4101	0.0340	[0.34, 0.48]	1
Low Predictability	0.2043	0.0318	[0.14, 0.27]	1
Asian English-L1	0.2289	0.0490	[0.13, 0.32]	1
Asian English-L2	−0.1530	0.0410	[−0.23, −0.07]	0.99
white English-L2	0.0768	0.0444	[−0.01, 0.16]	0.96

Precision is inversely related to variance. Positive distributions for the phi parameter indicates an increase in precision, meaning that listeners were more consistent in the accuracy of their responses, while negative distributions indicate less precision and more variance. The positive distribution for the Low Predictability sentences with a CrI that does not contain 0 provides strong evidence that listeners were more consistent in the accuracy of their responses for the low predictability sentences compared to the high predictability sentences for the white English-L1 talker, who was the reference level [$\hat{\beta} = 0.2, CrI = [0.14, 0.27], Pr(\hat{\beta} > 0 = 1]$. Recall the analysis of the mean indicated that the Asian English-L1 and the white English-L2 talkers' high predictability sentences were overall more intelligible than the white English-L1 talker's. For the Asian English-L1 talker, this is accompanied by strong evidence for high precision in her high predictability sentences [$\hat{\beta} = 0.23, CrI = [0.13, 0.32], Pr(\hat{\beta} > 0 = 1]$. Listeners were more consistent in their accuracy to her high predictability utterances. There is weak evidence that listeners were more precise in transcribing the high intelligibility sentences for the white English-L2 talker compared to the reference level [$\hat{\beta} = 0.08, CrI = [-0.01, 0.16], Pr(\hat{\beta} > 0 = 0.96]$. However, there was strong evidence that the Asian English-L2 talker, who had the least intelligible voice, elicited lower precision and, thus, more variance in response accuracy from listeners compared to the white English-L1 talker [$\hat{\beta} = -0.15, CrI = [-0.23, -0.07], Pr(\hat{\beta} < 0 = 0.99]$.

4. Discussion

In parts of North America, stereotypes about ethnicity and accent associations present a socio-linguistic landscape where white individuals are licensed to be native speakers of English, whereas non-white individuals are presumed to be second language speakers of English. In the local context, individuals of Asian descent may be stereotypically associated with non-English language, an expectation that is developed in infancy [23]. Any individual in the local context—infant or otherwise—has the opportunity for rich and diverse input. Nearly 50% of individuals in the Greater Vancouver Area identify as a visible minority [37], and nearly 45% of individuals report an "immigrant" language as their mother tongue, which includes all languages other than Aboriginal languages (First Nations languages, Inuktitut, and Métis), English, and French [38]. Certainly, it is not the case that all individuals who speak a Canadian-census-labelled "immigrant" language identify as visible minorities. However, the sheer amount of diversity presents listeners with a wide range of ethnicity and accent associations. Some local suburbs have historically coupled ethnicity and language associations that are particularly strong. And, indeed, previous work has shown that when local adults are cued to consider local stereotypes about particular suburbs that are strongly associated with Chinese Canadian culture, they find white Canadian individuals' speech more intelligible and less accented than the speech from Chinese Canadians *only* when they are aware they are listening to individuals who self-identify as Chinese Canadian [19]. In this previous work, individuals with stronger Asian Canadian social networks showed these effects more strongly, suggesting that negative bias towards Asian Canadians is unlikely to underlie the drop of intelligibility and increase in perceived accentedness. These results, along with others' findings in the literature on ethnicity and accent associations (e.g., [18,24,29]), suggest that experiences may undergird ethnicity and accent associations more than negative social attitudes [16]. This scholarship led to the hypothesis that listeners should more readily adapt to degraded audio-visual speech when the talker's ethnicity and accent align with local stereotypes. Specifically, this hypothesis offers the prediction that listeners will more readily adapt to a white L1-English speaker and an Asian L2-English (Mandarin-accented, in this case) speaker. The white L2-English speaker contradicts local expectations about white speakers, and listeners were predicted to struggle in their adaptation to her speech. While a diverse population is expected to speak the local mainstream variety of English [23], listeners may also implicitly carry the expectation that Asian individuals speech English with a non-native accent, which would entail listeners not adapting to the speech of the Asian L1-English speaker (e.g., [15,16,18,19]).

The current experiment tested this hypothesis space in an experiment that compared the change in intelligibility—measured by listeners' accuracy in transcribing audio-visual speech in noise—between high predictability training sentences and low predictability test sentences. The data were analyzed using a Bayesian mixed effects beta regression model, which allows nuance in interpretation and a joint consideration of estimates for means and variance (the phi parameter). The four talkers varied in their respective baseline levels of intelligibility in the high predictability test sentences. The white English-L1 speaker had relatively low baseline intelligibility, and the Asian English-L1 speaker and the white English-L2 speaker were both more intelligible than the white English-L1 speaker in the high predictability test sentences. Anecdotally, the white English-L1 speaker's speech style exhibited a relatively small amount of head and jaw motion, the movement of which is known to be beneficial for intelligibility [39]. The, perhaps, surprisingly low intelligibility of the white English-L1 speaker underscores the importance of using a paradigm that allows each individual and their own speech idiosyncrasies as their own control. The Asian English-L2 speaker was less intelligible than the white English-L1 speaker for the high predictability sentences.

It is the interactions between the sentence type and talker, however, that provide insight to the research question: is adaptation affected by accent and ethnicity associations? Starting with the clear results, listeners showed the predicted adaptation to the Asian English-L2 speaker, generalizing their experiences in the high predictability training sentences to the more challenging low predictability test sentences. This is in line with the stereotyped expectation that Asian individuals will have a non-native accent; with this expectation, listeners were able to leverage the expected non-native accent to more robustly learn and generalize from the high predictability sentences. Also in accordance with the prediction that listeners anticipate a white individual to speak English as a first language, listeners' exhibited a loss of intelligibility across training and test sentences for the white English-L2 speaker. This indicates that despite the white English-L2 speaker being a clear talker, as evidenced by the high intelligibility in the high predictability test sentences, listeners were not able to adapt to her speech patterns. There is little-to-no evidence that listeners are unable to adapt to the Asian English-L1 speaker. While the mean point estimate is negative, suggesting a tendency towards a challenge to adapt, which would suggest an expectation of an Asian individual being a non-native English speaker, the 95% credible interval crosses 0 and the probability of direction is 88%. The effect size would be small and the spread of the 95% credible interval is wide, though not as wide as it is for the interaction between sentence type and the white English-L1 talker and the Asian English-L2 talker. This lack of an effect at the population-level suggests that listeners are prepared for an Asian individual to speak the local accent, an expectation that is well-suited to the multi-ethnic landscape of their speech community. These results are in line with infants' developing expectations about the local native accent within this particular speech community [23], in addition to listeners' flexibility with "standard" German accents in the German-speaking context [24].

As noted, the spread of the credible interval for the English-L2 talkers is wide, indicating a large amount of variation in response to these non-native accents. Indeed, as the listeners were sampled from the same speech community as the speakers, all parties are exposed to a wide variety of first and second (and beyond) language accents in the local university setting and metropolitan area. Individuals who have more direct or regular experience with Mandarin-accented and Spanish-accented Englishes, the varieties spoken by the Asian and white talkers in this study, would have representations that are better equipped to parse these talkers. Note, however, that despite experience, the expectations that faces and voices presented in a university laboratory would conform to stereotypes about ethnicity and accents were generally maintained.

A beta regression model requires the modeling of precision, which is inversely related to variance. This phi parameter suggested that listeners were more consistent, showing less variance, in their transcriptions of the Asian English-L1 talker, who also had higher baseline intelligibility than the white English-L1 talker. These results simply suggest that an English-L1 speaker with clear speech patterns is more uniformly intelligible to listeners. Speaking to the wide credible interval in the estimate for the means, the variance associated with transcription accuracy for the Asian English-L2 speaker was higher compared to the white English-L1 talker (the reference level for the statistical model). Listeners likely have more varied levels of experience with Mandarin-accented English, and this is evidenced in the range of credible intervals and the lower precision values. The evidence for less variance for the white English-L2 talker compared to the white English-L1 reference level was weak, which may be due to her higher baseline intelligibility coupled with her non-native accent, to which listeners will have varied experience.

While the age distribution of the listeners in the current study is quite constrained (18–26 years of age), the self-identified ethnicity of the participants is varied. This was not the result of targeted recruitment, but a representation of the local university setting and speech community in which this study was conducted. This participant diversity aligns with other studies on the role of stereotypes on speech intelligibility at the same university [19]. Note that other scholars who report listener self-reported ethnicity generally have a much less diverse sample in terms of ethnicity (e.g., [24,25,27]). With respect to age, however, Hanulíková et al. [40] used aged-diverse listeners (teens, young adults, and older adults) and found that accent-ethnicity stereotypes were more pronounced in older listeners. Given this, it may be that older listeners in the local speech community would exhibit stronger effects than the young adult population used in the current study. Such a prediction, however, would depend on older listeners' having the requisite experiences to generate such predictions about accent and ethnicity associations.

The current results provide empirical behavioural support for a model of speech processing where listeners buffer their linguistic expectations about the incoming speech stream based on socio-cultural information about the speaker. Support for the notion that listeners have expectations about speakers also exists, however, in neurolinguistic measures. Listeners make socio-demographic assessments about an individual based on their voice (e.g., deducing from the speech signal that a talker is an adult or child) and parse the pragmatic appropriateness of their utterances (e.g., a child's or adult's voice saying "Every evening I drink some wine before I go to sleep."). Van Berkum et al. [41] found an N400 effect when the speaker and their linguistic message were pragmatically inconsistent.

Listeners also have knowledge and expectations about linguistic forms as they relate to an individual's accent. Hanulíková et al. [40], for example, found that L1 Dutch listeners exhibited a P600 effect in response to a grammatical error when the voice producing the error was a Dutch-L1 accent, but not when the voice had a Turkish-accented Dutch accent. They reason that since the grammatical error is typical for Turkish-L1 Dutch-L2 speakers, listeners anticipated the grammatical error, hence the lack of a P600. Recently, Zhou et al. [42] demonstrated that the expectations about the grammatical patterns of natively accented and non-natively accented from imagined speech — i.e., auditory perceptual simulation. (Zhou et al. [42] presented participants with a photo of a white female and an Asian female, and they report presenting these photos with recordings from "native" and "non-native" female speakers. The voices and photos were further accompanied by an English name and a Chinese name, so presumably the "non-native" accent was Mandarin-accented English.) Listeners exhibited a P600 effect when imagining the L1 English voice producing a sentence with a grammatical error, but showed no P600 effect when the same utterance was imagined in a Mandarin-accented English voice.

Given the local linguistic diversity, it is assumed that for the current population the stereotypes about accent and ethnicity are at least partially formed by experience and not media-mediated linguistic stereotypes, though such input may indeed play a critical role in seeding associations. The observed ethnicity-accent associations are certainly overgeneralizations in the sense that they are often erroneous. Melguy and Johnson [27] discuss, for example, how a generic association of white talkers with native English accents is incompatible in English-speaking communities with a large number of white immigrants from non-English-speaking countries (e.g., their example is immigration in the dissolution of the Soviet Union). We can add to this, of course, that in many countries, particularly those with long-standing colonial histories or those with large amounts of immigration, non-white individuals are native speakers of a majority language like English. While expectations may be part and parcel of the organization of the linguistic processing system, we need awareness of these affects to mitigate their social costs. It is simply unfair for linguistic stereotyping to impose an intelligibility cost on particular ethnicity-accent associations.

Lastly, while audio-visual methods are far from novel, their application in social linguistic inquiries is, arguably, under-utilized. The use of audio-visual speech to probe the role of previous experience and the use of phonetic and phonological expectations in the processing of speech offers the distinct advantage of being more ecologically valid. Behaviours gathered in the context of audio-visual speech, as opposed to static photos, may also be less susceptible to strategies and task effects [21].

5. Conclusions

In a multicultural and multilingual speech community, listeners exhibit accent and ethnicity associations. The local variety of English is spoken by a diverse demographic, and listeners are flexible with English-L1 accents, whether spoken by white or Asian English-L1 speakers. For non-native accents, listeners adapted to the accent and ethnicity association that conforms to local stereotypes (an Asian English-L2), but not to an incongruent association (a white English-L2). These results provide support for accounts where intelligibility is supported by the alignment of phonetic and phonological expectations with the apprehended phonetic signal.

Funding: This research was funded by a SSHRC grant awarded to the author.

Institutional Review Board Statement: The study was conducted in accordance with the Declaration of Helsinki, and approved by the Behavioural Ethics review Board at the University of British Columbia (protocol code H12-02739 and 5 October 2012).

Informed Consent Statement: Informed consent was obtained from all subjects involved in the study.

Data Availability Statement: Data and code are available at https://osf.io/pwj6b/?view_only=290b33be56f648eaa26f51fea27e5ee2 (accessed on 24 May 2022).

Acknowledgments: Thank you to Gloria Mellesmoen and Sophie Bishop for assistance in data collection and stimuli preparation. Thank you to Roger Lo for guidance on the statistical approach. Earlier versions of this work were presented at the International Congress of the Phonetic Sciences 2019.

Conflicts of Interest: The authors declare no conflict of interest

Abbreviations

The following abbreviations are used in this manuscript:

SNR	signal-to-noise ratio
TSR	token sort ratio
CrI	credible interval

References

1. Johnson, K.; Strand, E.A.; D'Imperio, M. Auditory–visual integration of talker gender in vowel perception. *J. Phon.* **1999**, *27*, 359–384. [CrossRef]
2. Strand, E.A.; Johnson, K. Gradient and visual speaker normalization in the perception of fricatives. In Proceedings of the KONVENS, Bielefeld, Germany, 1 October 1996; pp. 14–26.
3. Munson, B.; Babel, M. The Phonetics of Sex and Gender. In *Routledge Handbook of Phonetics*; Katz, W., Assmann, P., Eds.; Routledge: London, UK, 2019; pp. 499–525.
4. Johnson, K. Resonance in an exemplar-based lexicon: The emergence of social identity and phonology. *J. Phon.* **2006**, *34*, 485–499. [CrossRef]
5. Thomas, E.R.; Reaser, J. Delimiting perceptual cues used for the ethnic labeling of African American and European American voices. *J. Socioling.* **2004**, *8*, 54–87. [CrossRef]
6. Hay, J.; Warren, P.; Drager, K. Factors influencing speech perception in the context of a merger-in-progress. *J. Phon.* **2006**, *34*, 458–484. [CrossRef]
7. Drager, K. Sociophonetic variation in speech perception. *Lang. Linguist. Compass* **2010**, *4*, 473–480. [CrossRef]
8. Hay, J.; Drager, K. Sociophonetics. *Annu. Rev. Anthropol.* **2007**, *36*, 89–103. [CrossRef]
9. Sumner, M.; Kataoka, R. Effects of phonetically-cued talker variation on semantic encoding. *J. Acoust. Soc. Am.* **2013**, *134*, EL485–EL491. [CrossRef]
10. Pierrehumbert, J.B. Phonological representation: Beyond abstract versus episodic. *Annu. Rev. Linguist.* **2016**, *2*, 33–52. [CrossRef]
11. Kleinschmidt, D.F.; Weatherholtz, K.; Florian Jaeger, T. Sociolinguistic perception as inference under uncertainty. *Top. Cogn. Sci.* **2018**, *10*, 818–834. [CrossRef]
12. Kleinschmidt, D.F.; Jaeger, T.F. Robust speech perception: Recognize the familiar, generalize to the similar, and adapt to the novel. *Psychol. Rev.* **2015**, *122*, 148. [CrossRef]
13. Niedzielski, N. The effect of social information on the perception of sociolinguistic variables. *J. Lang. Soc. Psychol.* **1999**, *18*, 62–85. [CrossRef]
14. Devos, T.; Banaji, M.R. American = white? *J. Personal. Soc. Psychol.* **2005**, *88*, 447. [CrossRef] [PubMed]
15. Rubin, D.L. Nonlanguage factors affecting undergraduates' judgments of nonnative English-speaking teaching assistants. *Res. High. Educ.* **1992**, *33*, 511–531. [CrossRef]
16. Kang, O.; Rubin, D.L. Reverse linguistic stereotyping: Measuring the effect of listener expectations on speech evaluation. *J. Lang. Soc. Psychol.* **2009**, *28*, 441–456. [CrossRef]
17. Cheng, L.S.; Burgess, D.; Vernooij, N.; Solís-Barroso, C.; McDermott, A.; Namboodiripad, S. The problematic concept of native speaker in psycholinguistics: Replacing vague and harmful terminology with inclusive and accurate measures. *Front. Psychol.* **2021**, *12*, 715843. [CrossRef]
18. McGowan, K.B. Social expectation improves speech perception in noise. *Lang. Speech* **2015**, *58*, 502–521. [CrossRef]
19. Babel, M.; Russell, J. Expectations and speech intelligibility. *J. Acoust. Soc. Am.* **2015**, *137*, 2823–2833. [CrossRef]
20. Gnevsheva, K. The expectation mismatch effect in accentedness perception of Asian and Caucasian non-native speakers of English. *Linguistics* **2018**, *56*, 581–598. [CrossRef]
21. Zheng, Y.; Samuel, A.G. Does seeing an Asian face make speech sound more accented? *Atten. Percept. Psychophys.* **2017**, *79*, 1841–1859. [CrossRef]
22. Hanulíková, A. The effect of perceived ethnicity on spoken text comprehension under clear and adverse listening conditions. *Linguist. Vanguard* **2018**, *4*. [CrossRef]
23. May, L.; Baron, A.S.; Werker, J.F. Who can speak that language? Eleven-month-old infants have language-dependent expectations regarding speaker ethnicity. *Dev. Psychobiol.* **2019**, *61*, 859–873. [CrossRef] [PubMed]
24. Hanulíková, A. Do faces speak volumes? Social expectations in speech comprehension and evaluation across three age groups. *PLoS ONE* **2021**, *16*, e0259230. [CrossRef] [PubMed]
25. Kutlu, E.; Tiv, M.; Wulff, S.; Titone, D. Does race impact speech perception? An account of accented speech in two different multilingual locales. *Cogn. Res. Princ. Implic.* **2022**, *7*, 7. [CrossRef] [PubMed]

26. Vaughn, C.R. Expectations about the source of a speaker's accent affect accent adaptation. *J. Acoust. Soc. Am.* **2019**, *145*, 3218–3232. [CrossRef] [PubMed]
27. Melguy, Y.V.; Johnson, K. General adaptation to accented English: Speech intelligibility unaffected by perceived source of non-native accent. *J. Acoust. Soc. Am.* **2021**, *149*, 2602–2614. [CrossRef] [PubMed]
28. Sumby, W.H.; Pollack, I. Visual contribution to speech intelligibility in noise. *J. Acoust. Soc. Am.* **1954**, *26*, 212–215. [CrossRef]
29. Yi, H.G.; Phelps, J.E.; Smiljanic, R.; Chandrasekaran, B. Reduced efficiency of audiovisual integration for nonnative speech. *J. Acoust. Soc. Am.* **2013**, *134*, EL387–EL393. [CrossRef]
30. Babel, M.; Johnson, K.A.; Sen, C. Asymmetries in perceptual adjustments to non-canonical pronunciations. *Lab. Phonol.* **2021**, *12*. [CrossRef]
31. Bradlow, A.R.; Alexander, J.A. Semantic and phonetic enhancements for speech-in-noise recognition by native and non-native listeners. *J. Acoust. Soc. Am.* **2007**, *121*, 2339–2349. [CrossRef]
32. Bosker, H.R. Using fuzzy string matching for automated assessment of listener transcripts in speech intelligibility studies. *Behav. Res. Methods* **2021**, *53*, 1945–1953. [CrossRef]
33. Bürkner, P.C. brms: An R package for Bayesian multilevel models using Stan. *J. Stat. Softw.* **2017**, *80*, 1–28. [CrossRef]
34. Gabry, J.; Češnovar, R. cmdstanr: R Interface to 'CmdStan'. 2021. Available online: https://mc-stan.org/cmdstanr, https://discourse.mc-stan.org (accessed on 1 March 2022).
35. Smithson, M.; Verkuilen, J. A better lemon squeezer? Maximum-likelihood regression with beta-distributed dependent variables. *Psychol. Methods* **2006**, *11*, 54–71. doi: 10.1037/1082-989X.11.1.54. [CrossRef] [PubMed]
36. Nicenboim, B.; Vasishth, S. Statistical methods for linguistic research: Foundational Ideas—Part II. *Lang. Linguist. Compass* **2016**, *10*, 591–613. [CrossRef]
37. Statistics Canada. Statistics Canada 2016 Immigration and Ethnocultural Diversity Highlight Tables. 2016. Available online: https://www12.statcan.gc.ca/census-recensement/2016/dp-pd/hlt-fst/imm/Table.cfm?Lang=E&T=43&geo=59&vismin=2&age=1&sex=1&SP=1&SO=13D (accessed on 16 February 2020).
38. Statistics Canada. Focus on Geography Series, 2016 Census. Statistics Canada Catalogue no. 98-404-X2016001. 2016. Available online: https://www12.statcan.gc.ca/census-recensement/2016/as-sa/fogs-spg/Facts-PR-Eng.cfm?TOPIC=5&LANG=Eng&GK=PR&GC=59 (accessed on 2 March 2022).
39. Munhall, K.G.; Jones, J.A.; Callan, D.E.; Kuratate, T.; Vatikiotis-Bateson, E. Visual prosody and speech intelligibility: Head movement improves auditory speech perception. *Psychol. Sci.* **2004**, *15*, 133–137. [CrossRef] [PubMed]
40. Hanulíková, A.; Van Alphen, P.M.; Van Goch, M.M.; Weber, A. When one person's mistake is another's standard usage: The effect of foreign accent on syntactic processing. *J. Cogn. Neurosci.* **2012**, *24*, 878–887. [CrossRef]
41. Van Berkum, J.J.; Van den Brink, D.; Tesink, C.M.; Kos, M.; Hagoort, P. The neural integration of speaker and message. *J. Cogn. Neurosci.* **2008**, *20*, 580–591. [CrossRef] [PubMed]
42. Zhou, P.; Garnsey, S.; Christianson, K. Is imagining a voice like listening to it? Evidence from ERPs. *Cognition* **2019**, *182*, 227–241. [CrossRef] [PubMed]

brain sciences

Article

The Role of the Root in Spoken Word Recognition in Hebrew: An Auditory Gating Paradigm

Marina Oganyan and Richard A. Wright *

Department of Linguistics, University of Washington, Seattle, WA 98195, USA; marina0@uw.edu
* Correspondence: rawright@uw.edu; Tel.: +1-206-616-2426

Abstract: Very few studies have investigated online spoken word recognition in templatic languages. In this study, we investigated both lexical (neighborhood density and frequency) and morphological (role of root morpheme) aspects of spoken word recognition of Hebrew, a templatic language, using the traditional gating paradigm. Additionally, we compared the traditional gating paradigm with a novel, phoneme-based gating paradigm. The phoneme-based approach allows for better control of information available at each gate. We found lexical effects with high-frequency words and low neighborhood density words being recognized at earlier gates. We also found that earlier access to root-morpheme information enabled word recognition at earlier gates. Finally, we showed that both the traditional gating paradigm and gating by phoneme paradigm yielded equivalent results.

Keywords: spoken word recognition; morphology; Hebrew

Citation: Oganyan, M.; Wright, R.A. The Role of the Root in Spoken Word Recognition in Hebrew: An Auditory Gating Paradigm . *Brain Sci.* **2022**, *12*, 750. https://doi.org/10.3390/brainsci12060750

Academic Editors: Yang Zhang and Heather Bortfeld

Received: 5 April 2022
Accepted: 1 June 2022
Published: 7 June 2022

1. Introduction

In this study, we investigate online morphological processing of spoken Hebrew, a templatic language, using the traditional gating paradigm and a novel phoneme-based adaptation of the paradigm. In particular, we investigate online lexical processing of templatic words and online morphological processing and the role of the root in spoken word recognition.

When investigating the role of morphological complexity in spoken word recognition, it is important to take into account the relative structure of the language's morphology. The structure of concatenative languages, such English, Japanese, or Swahili (and many other languages), differs from that of templatic languages, such as Hebrew and Arabic (and a handful of other languages), in the distribution of lexical and derivational morphemes. In concatenative languages, complexity is largely expressed through affixation, with morphemes occurring sequentially, while in templatic languages, lexical and derivational morphemes are interleaved (the differences are illustrated in Table 1). In Semitic languages, the combination of the *root*, the semantic/lexical part, and *template*, the derivational part, form the stem for all Semitic words (excluding non-Semitic loanwords). This makes all Semitic words inherently morphologically complex. While affixation is used in Semitic languages, it is restricted to inflectional morphology.

Morphological processing of affixed words in concatenative languages has been widely studied in visual word recognition, resulting in strong evidence for (at least some) decomposition of words into composite morphemes during word recognition, e.g., [1] (English), [2] (French), [3] (Finnish), [4] (Japanese). As with concatenative words, there is evidence that templatic words are parsed into their morphological units (namely the root and template) during visual word recognition [5–10]. A smaller but also substantial body of research exists for word auditory word recognition in concatenative languages with similar findings (e.g., [11,12]).

Table 1. Examples of Concatenative vs. Templatic derivational morphology.

	Concatenative (Swahili)				
	Prefix	**Root**	**Suffix**	**Word**	**Meaning**
verb	*ku- (inf.)*	-pend-	*-a (verb)*	kupenda	to love
noun	*u- (nom.)*	-pend-	-o *(nom)*	upendo	love (n.)
	Templatic (Hebrew)				
		Root	**Template**	**Word**	**Meaning**
verb		/x/-/k/-/ʁ/	_a_a_ (verbal)	/xakaʁ/	investigated (v. m. past)
noun		/x/-/k/-/ʁ/	mi_ _a_ (nominal)	/mixkaʁ/	research (n.)

In Hebrew and Arabic, the root morpheme has been shown to be key to word recognition in a way that is different from concatenative languages. For example, in concatenative words, letter transpositions do not inhibit priming, while in templatic words, transposition of root letters inhibits priming ([7] (Hebrew), [8] (Arabic)). Additionally, while in concatenative languages, priming occurs at the stem or affix level, templatic words can be primed by words with shared roots regardless of semantic relationship (e.g., [13]).

While a small amount of research has been conducted on auditory word recognition in Semitic languages, it has primarily used offline paradigms such as priming and auditory masking. For example, Geary and Ussishkin [10] found that root priming in templatic words extends to the auditory domain. Additionally, Oganyan, Wright, and Herschensohn [14] found that noise-masking root morpheme sounds in auditory stimuli makes a word more difficult to recover than noise-masking template sounds. Extending word recognition research to the auditory domain is important because it strengthens our understanding of the role of morphology by reducing the potential for orthographic interference. In this study, we use gating paradigms to investigate real-time auditory processing of words where full root information (the *root completion point* RCP) is presented either earlier or later in the signal to test the relative importance of the root and template in the timing of word recognition.

An important aspect of auditory perception is the linearity of the signal and the ability to amend perception of the word as more of the signal becomes available. For example, when a Hebrew listener hears the onset of a word beginning with a/k/, as in the Hebrew word (/katav/כתב), there are a large number of /k/-initial lexical competitors; however, as the auditory word progresses, the number of competitors narrows. This aspect of auditory perception has given rise to models of spoken word recognition such as Trace [15] and Cohort [16,17] and, more recently, cognitive network approaches (e.g., [18]). To explore this aspect of spoken language, we employ an auditory gating paradigm [19].

The *gating paradigm*, originally developed by Grosjean in 1980 [19], exposes increasing information from the speech signal. This paradigm mirrors the temporal unfolding of speech information in the auditory perception process while permitting the experimenter to probe the time course of auditory perception and word recognition at different time points. One finding in Grosjean's [19] study that is relevant to the current investigation is that word duration, measured in number of syllables, affects word recognition time, where the greater the syllable count, the later word recognition takes place. Additionally relevant to this investigation is his finding that words with high usage frequency are recognized earlier than their less-frequent counterparts. A later study by Metsala in 1997 [20] used the same gating paradigm to extend the study of lexical effects to include phonological-neighborhood density, the number of phonological competitors, and its interaction with usage frequency. In his study, he used a two-by-two design crossing *frequency* (low and high) with *density* (low and high). He found that for high-frequency sets, low-density words were recognized earlier than high-density words, but for low-frequency sets, high-density words were recognized sooner than low-density words.

The first set of goals of this paper is to replicate and extend findings of lexical effects for usage frequency and phonological-neighborhood density, observed using the gating paradigm in concatenative languages [19,20], to Hebrew and to test the effect of templatic morphology on the process. In particular, we evaluate whether earlier access to complete root information (RCP) relative to the uniqueness point (UP), the point in the word where there are no possible auditory competitors, leads to earlier word identification. We hypothesize that lexical effects will extend to spoken Hebrew in a way that is analogous to studies of English, with high-frequency words being recognized sooner than their low-frequency counterparts and with interactions between neighborhood density and word frequency.

A second goal of this paper is a methodological one: to test a novel alternative version of the gating paradigm with gates set by perceptual phoneme boundaries rather than traditional fixed 20–60 ms windows. One drawback of traditional gating methods with fixed-window durations is that stimuli have to be very carefully matched to avoid consonant-manner effects interfering with observations. The reason for this is that different consonant manners have different acoustic time courses, and therefore, a fixed-window duration will reveal very different amounts of lexical information if the stimuli are not matched. This severely limits the number and variety of stimulus words since they have to have the same manner sequences within comparison groups. On the other hand, a phoneme-gating paradigm allows for the use of words that are not matched in this manner. A second drawback to the traditional fixed-window paradigm is testing time; with short, fixed windows, a word is broken into a large number of gates, and many of the gates are redundant with previous ones in terms of phonemes. If our novel phoneme-gating paradigm is equivalent to the traditional gating paradigm, as we hypothesize, it will greatly increase the number of possible stimuli, reduce testing time, and increase the kinds of research questions which can be addressed. One important limitation to the novel approach is that the phonemic gates have to be very carefully applied by thoroughly trained acoustic phoneticians to avoid revealing information about preceding or following phonemes. The necessary expertise will limit who can conduct research with this method.

2. Materials and Methods

2.1. Stimuli

All stimuli were Hebrew words read from randomized wordlists by a male native Hebrew speaker. The recordings were made using a Zoom H4n professional recorder with an AKG C520 head-mounted condenser microphone in a sound-treated recording booth at the Phonetics Laboratory on the University of Washington campus.

For each word, we calculated the uniqueness point (UP) and the root completion point (RCP). The UP refers to the point in the acoustic signal where a word has no lexical competitors. The RCP refers to the point in the acoustic signal at which all Semitic root information has been completed.

The wordlists included two sets of spoken-word recognition stimuli: (1) *lexical*, which were used to test the effects of usage-frequency and phonological-neighborhood density, and (2) *morphological*, which were used to test the effects of Hebrew morphology. Acoustic stimulus duration ranged from 547 ms to 999 ms with an average of 736 ms. Within stimuli, phone duration ranged from 17 ms to 394 ms with an average of 127 ms. A full list of stimuli used can be found in the Appendix A. Results data are available upon request from the authors.

2.1.1. Lexical Stimuli

Forty nouns were selected in a 2 × 2 stimulus matrix design for neighborhood density and usage frequency. Usage-frequency was taken from the database by Frost and Plaut [21], which is a database of written word-usage frequency based on newspapers. Neighborhood density (ND) is defined using the method established by Charles-Luce and

Luce [22] as the edit distance of one phoneme (addition, subtraction, deletion, or substitution). Neighborhood density was calculated using a modified version of the MILA corpus lexicon [23] with phonological transcriptions. The MILA ("word" in Hebrew) corpus lexicon of Hebrew words contains more than 25,000 lexicon items. Half of the words were high-frequency (>16 per million), and half were low-frequency (<4 per million). Half of the words were high-density (>12 neighbors), and half were low-density (<3 neighbors). This resulted in a total of 10 words for each combination: high-frequency, high neighborhood density (HF-HND); high-frequency, low neighborhood density (HF-LND); low-frequency, low neighborhood density (LF-LND); and low-frequency, high neighborhood density (LF-LND) (see Table 2). Words were all equal in length (5 phones), beginning with a root sound and for each word the UP and RCP coincided.

Table 2. Lexical Stimuli Properties.

	Phones	Freq	ND	Initial Sound Manner
HF-HND	5	34 (16–64)	14.7 (12–16)	Fricative (6), Stop (4)
HF-LND	5	30.2 (16–52)	1.8 (0–3)	Fricative (3), Nasal (3), Stop (4)
LF-LND	5	1.9 (1–4)	15.1 (12–23)	Fricative (6), Nasal (1), Stop (3)
LF-HND	5	1.7 (1–3)	1.8 (1–2)	Fricative (3), Nasal (4), Stop (3)

2.1.2. Morphological Stimuli

Thirty nouns were selected for their relative position of RCP and UP to form three conditions split across the stimuli with ten words in each. Words either had root completion precede uniqueness point (RCP < UP), uniqueness point precede root completion (UP < RCP), or the two occurring at the same point (RCP = UP). Words were balanced initially for manner of initial phoneme, frequency, and density (see Table 3). All words began with a root sound.

Table 3. Morphological Stimuli Properties.

	Phones	Freq	ND	Initial Sound Manner
RCP < UP	5.9 (5–8)	2.6 (1–7)	3.3 (0–8)	Fricative (4), Stop (4), Liquid (2)
RCP = UP	5.7 (5–7)	2.8 (1–7)	3.3 (0–10)	Fricative (4), Stop (4), Liquid (2)
UP < RCP	5.7 (5–7)	2.9 (1–7)	3.3 (0–7)	Fricative (4), Stop (4), Liquid (2)

2.2. *Gating Paradigms*

In the first gating paradigm, which we refer to as the *traditional paradigm*, words were cut into 50 ms segments increasing in length with each gate and with the final segment being the full length of the word. This is the traditional gating paradigm first developed by Grosjean [19].

In the second paradigm, which we refer to as the *phoneme paradigm*, words were cut at perceptual phoneme boundaries with the first segment containing the first phoneme, the second the first two phonemes, and continuing until the last contained the full word. Gates were assigned by two trained acoustic phoneticians and were tested on a native Hebrew speaker to ensure that there was insufficient coarticulatory information for the following speech sound to be recovered.

There are several reasons for exploring a phoneme-based alternative to the traditional gating paradigm. The first is that phonemes vary in their intrinsic duration, so a fixed-gate duration exposes different amounts of acoustic information when words differ in their segmental makeup. Thus, an initial gate of 50 ms for a word starting with a stop followed by a vowel, for example, קיטור(kituʁ), will expose significantly more information than for one with a fricative followed by a vowel, for example שחק(ʃaχak). This difference is illustrated in Figure 1, which shows a pair of spectrograms with the two gating paradigms marked out (50 ms gates and phoneme gates).

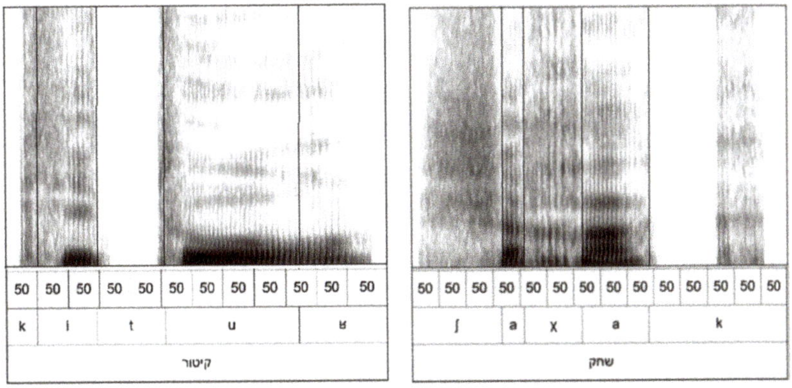

Figure 1. Spectrograms of stimuli קיטור(kituʁ) (**left**) and שחק(ʃaχak) (**right**) illustrating the traditional 50 ms gates and the phoneme gates.

As can be seen in comparing the two spectrograms, the initial 50 ms gate in the stimuli reveal very different amounts of information about the word. For קיטור(kituʁ) (left), the initial 50 ms gate reveals the consonant's release burst and a portion of the following vowel, whereas for שחק(ʃaχak) (right), the onset of the vowel is not revealed until the fourth 50 ms gate. To avoid this problem, traditional gating paradigms must restrict the manner of articulation of consonants to have comparable information flow across time. Therefore, gating by phonemes greatly increases the number of possible stimuli available to the researcher. Moreover, the number of gates needed per word greatly decreases because the duration of most consonants and vowels is longer than 50 ms. This can be seen in שחק(ʃaχak) (right), where the traditional paradigm with 50 ms gates requires 15 gates, whereas when gating by phoneme, only 5 are needed. This results in a threefold reduction in testing time. Finally, the traditional gating paradigm is difficult to apply for a research question where control of access to phonemic information is important, such as in the morphological research in this study. This is because the arbitrary nature of the gates means that the information at each gate is difficult to control. A particular gate may contain only partial information about a speech sound, while a different gate may contain information about more than one speech sound if it straddles a boundary. Having more control over the type of information presented at each gate, as the phonemic paradigm does, allows for these types of questions to be addressed.

2.3. Participants and Procedure

The experiment was run using an online version of Psychopy (version 2020.1.3) [24] running on the Pavlovia platform (https://pavlovia.org/ accessed from 1 June 2020 to 7 Febuary 2022). All participants were recruited using the Prolific platform (https://www.prolific.co/ accessed from 1 June 2020 to 7 Febuary 2022). Using Prolific's screening, participants were screened for being native speakers of Hebrew who had grown up in a monolingual household. They were also screened for reporting having normal hearing. Participants who reported living outside of Israel for more than two years were excluded from the study. All instructions were in Hebrew.

Participants wore headphones of their choosing, reporting the brand and model as part of an initial survey to ensure headphone use. Headphone use was set as a technical requirement on the Prolific platform for participation in the study. Participants chose a comfortable listening level on their own devices.

The two gating paradigm experiments were run as independent experiments. Within each gating paradigm experiment, all stimuli were divided into five lists with a balanced sampling from each of the different stimulus types (the four conditions for lexical and three for morphological). Each list was posted on Prolific as a separate task within its relevant experiment. Participants were restricted to participation in only one of the two experiments and were able to complete 1 to 5 of the tasks in their chosen experiment.

For the lexical (frequency by density) experiments, a total of 130 participants took part. For the traditional gating paradigm, 57 participants took part (35 male, 22 female). Participant ages ranged from 18 to 42 years with an average age of 26.5 years (5.7 standard deviation). Each word was responded to between 26 and 35 times (avg. 31). In the gating by phoneme paradigm, there were 73 participants (34 male, 39 female). Participant ages ranged from 18 to 59 years with an average age of 29.5 years (7.5 standard deviation). Each word was responded to between 32 and 40 (avg. 36) times.

In the morphological experiments, a total of 128 participants took part. For the traditional gating paradigm, 56 participants took part (34 male, 22 female). Participant ages ranged from 18 to 42 years with an average age of 26.4 years (5.9 standard deviation). Each word was responded to between 25 and 36 (avg. 31) times. For the phoneme gating paradigm, 72 participants took part (34 male, 38 female). Participant ages ranged from 18 to 59 years with an average age of 29 years (7.7 standard deviation). Each word was responded to between 31 and 39 (avg. 36) times.

In both paradigms, words were presented incrementally increasing in duration with each gate. At each gate, participants were asked to guess the identity of the word and give a confidence value for their guess.

2.4. *Analysis*

Each stimulus was analyzed using both the *recognition point* (RP) and the *isolation point* (IP). The term recognition point here is the point when the word was first guessed correctly, while the isolation point is the point at which the participant guessed the word without changing the guess at subsequent gates. Both RP and IP were used to have a more thorough comparison between gating paradigms since different researchers used one or the other of these with the traditional gating paradigm (e.g., [19,25] IP, [20] RP). Because the IP and RP are largely equivalent, the IP results are reported in the body of the text (RP results are reported in separate tables). Any differences are discussed in the results and discussion sections.

2.4.1. Preprocessing

In preprocessing the results, we established inclusion criteria for responses. Participants were excluded if they stated that Hebrew was not their first language or if they had more than one first language. Participants were also removed if a valid participant ID was missing, indicating an improper submission; all such entries contained no valid guesses. Valid guesses were those with at least one Hebrew letter in the guess. If a participant had no valid guesses for any of the gates of a particular word, responses to that word were omitted for that participant. See Table 4 for a summary of all omitted data as raw counts. Four stimulus words were excluded from the lexical experiment: two for not meeting criteria of having identical gating and uniqueness points and two for having close homonyms. Results were processed with a script, which removed (1) all entries by participants not meeting inclusion criteria, (2) erroneous stimuli, and (3) a participant's responses to any word with no valid guesses. In addition, the script marked as correct all non-ambiguous typos or misspellings. Due to the vowelless nature of the Hebrew spelling system, incorrectly typed words could only be allowed if there was a clear typo (e.g., inclusion of a non-letter key such as a number or shift key) or confusion between two letters with the same sound, which did not form a different word.

Table 4. Omitted Data, reported as raw counts.

	By Time		By Phoneme	
	Lex.	**Morph.**	**Lex.**	**Morph.**
Participants —Invalid ID	8	8	7	7
Participants —Language	10	10	10	10
Entries —No Valid Guesses	35	33	29	30
Stimuli	LF HND-רתך,שכירה LF LND-נחיל HF LND-בדיחה	n/a	LF HND-רתך,שכירה LF LND-נחיל HF LND-בדיחה	

2.4.2. Statistical Analysis

Responses to lexical stimuli were analyzed both in terms of absolute IP, as has been traditionally done in the gating paradigm (e.g., [19,26]), and in terms of difference between IP and UP (IP–UP). This difference measure is useful for controlling for variations in acoustic-word duration. That is, while there may be identical numbers of letters in a written word and therefore no durational difference, different speech sounds may exhibit small differences in duration, introducing noise into the estimation of word recognition point. For the lexical stimuli, no statistically reliable difference was expected between the results for the two measures (IP, IP–UP) because stimulus word length was relatively easy to control for. The research question for the lexical stimuli also lends itself to both IP and IP–UP measure analyses. However, for the morphological stimuli, the stimulus design resulted in word length differences (see Table 3). Furthermore, the research question about the relative ordering of the UP and the RCP did not lend itself to analysis in terms of absolute IP. Therefore, for morphological responses, data were analyzed using only the IP–UP difference measure. For the traditional paradigm, the IP–UP difference was measured in ms, while in the gating by phoneme paradigm, it was measured by gates.

The results of the lexical experiments, IP and IP–UP, were submitted to 2 × 2 linear mixed effects (LMER) models with *density* and *frequency* as fixed effects and *participant* as a random intercept (R formula = IP or IP-UP~Freq * Density + (1|Participant). Two additional comparisons were made, one for frequency (high vs. low) and the other for neighborhood density (high vs. low), using linear mixed-effects regression (LMER) models with *type* as the independent variable, IP or IP–UQ as the dependent variable, and participant as a random intercept (R formula = IP or IP-UP~Freq or Density + (1|Participant). To compensate for potential interactions between neighborhood density and frequency, where a frequency effect can mask neighborhood density effects (e.g., [20]), comparisons for neighborhood density were also made within low-frequency and high-frequency sets using an analogous LMER Model.

In the morphological experiments, the IP–UP differences were submitted to an ANOVA with three condition types: RCP < UP, RCP = UP, and UP < IP. An LMER model was additionally used to compare the three conditions (RCP < UP, RCP = UP, and UP < IP). A second LMER model was used to compare only two condition types: RCP < UP and UP < RCP. In both LMER models, the dependent variable was the IP–UP difference, the independent variable was condition type (RC < UP, RC = UP, UP < IP, or RC < UP, UP < IP), and participant was a random intercept.

3. Results

3.1. Gating by Time: Lexical

3.1.1. Isolation Points

The results of the lexical experiments are summarized in Table 5. Average isolation points by type were HF-HND 413 ms, HF-LND 385 ms, LF-HND 470 ms, and LF-LND 456 ms. The 2 × 2 LMER revealed an effect for *frequency* (t = 5.781, $p < 0.001$) and for *density* (t = −3.164, $p < 0.01$) but not the *frequency* by *density* (t = 1.127, $p < 0.26$) interaction. Additional LMER models were run separately for *frequency* (H vs. L) and for *density* (H vs. L), with participant as a random intercept. An additional LMER model was run for *density* (H vs. L) within high and low frequencies. LF words were identified on average 63 ms slower than HF words (t = 9.253, $p < 0.001$). Overall, LND words were identified on average 17 ms sooner than HND ones (t = 2.38, $p < 0.05$). The effect was carried by differences for low-frequency words, with no significant effect for high-frequency words (t = −1.42, $p < 0.156$). Low-frequency words with LND words were recognized 28 ms sooner relative to UP than HND on average (t = −2.948, $p < 0.01$).

Table 5. Lexical Results Summary Table (* indicates significance at $p < 0.05$).

Gating by Time						
	2 × 2 Freq	2 × 2 ND	2 × 2 Freq:ND	Freq	ND in HFreq	ND in LFreq
IP	*	*	NS	H < L *	NS	L < H *
IP-UP	*	NS	NS	H < L *	NS	L < H *
Gating by Phoneme						
	2 × 2 Freq	2 × 2 ND	2 × 2 Freq:ND	Freq	ND in HFreq	ND in LFreq
IP	*	*	NS	H < L *	NS	L < H *
IP-UP	*	NS	*	H < L *	NS	L < H *

3.1.2. Difference Isolation Point to Uniqueness Point

On average, HF-HND words were identified 92 ms before, HF-LND 96 ms before, LF-HND 13 ms after, and LF-LND 34 ms before the UP. The 2 × 2 LMER revealed an effect for *frequency* (t = 10.355, $p < 0.001$) and for the *frequency* by *density* (t = −3.024, $p < 0.01$) interaction but not for *density* (t = −0.498, $p = 0.619$). Additional LMER models were run separately for *frequency* (H vs. L) and for *density* (H vs. L), with participant as a random intercept. An additional LMER model was run for *density* (H vs. L) within high and low frequencies. There was an effect of frequency, with LF words being identified 80 ms later relative to the UP than HF words (t = 11.47 $p < 0.001$). There was also an overall effect for ND with LND words being identified 17 ms sooner than HND words (t = −2.304, $p < 0.05$). The effect was carried by differences at low frequency words, with no significant effect in high-frequency words (t = −0.413, $p < 0.68$) and a significant effect in low-frequency words with LND words being recognized 47 ms sooner relative to UP than HND words on average (t = −4.917, $p < 0.001$).

3.2. Gating by Time: Morphological

Difference—Isolation and Uniqueness Points

The results of the morphological experiments are summarized in Table 6. On average, RCP < UP words were identified 46 ms sooner, RCP = UP words were identified 15 ms sooner, and UP < RCP words were identified 79.23 ms later than the uniqueness point. The ANOVA revealed a significant effect for type overall (F = 83.906 value, $p < 0.001$). The first LMER model, with type as an independent variable and participant as a random intercept effect, revealed an effect for type: RCP < UP words were identified on average 31 ms before

RCP = UP words (t = −3.067, $p < 0.01$) and UP < RCP 95 ms later than the RCP = UP words (t = 9.363 $p < 0.001$). Results from the second LMER model comparing RCP < UP and UP < RCP words revealed a significant effect with UP < RCP identified 125 ms after RCP < UP relative to the uniqueness point (t = 11.93, $p < 0.001$).

Table 6. Morphological Results Summary (* indicates significance at $p < 0.05$).

	Overall	Diff from RCP = UP	RCP < UP vs. UP < RC
Gating by Time	*	* RCP < UP faster	* RCP < UP faster
Gating by Phoneme	*	* UP < RCP slower	* UP < RCP faster

3.3. *Gating by Phoneme: Lexical*

3.3.1. Isolation Points

The results for the lexical experiment using phoneme gating are summarized in Table 5. Average isolation points by type occurred at gate 4.125 for HF-HND words, at gate 3.896 for HF-LND words, at gate 4.536 for LF-HND words, and at gate 4.376 for LF-LND words. The 2 × 2 LMER revealed an effect for *frequency* (t = 5.613, $p < 0.001$) and for *density* (t = −3.327, $p = 0.001$) but not the *frequency* by *density* (t = 0.670, $p = 0.503$) interaction. Additional LMER models were run separately for *frequency* (H vs. L) and for *density* (H vs. L), with participant as a random intercept. An additional LMER model was run for *density* (H vs. L) within high and low frequencies. LF words were identified on average 0.435 gate later than HF words (t = 8.513, $p < 0.001$). Overall LND words were identified on average −0.17158 gate earlier than HND ones (t = −3.286, $p < 0.01$). The effect was carried by differences in low-frequency words, with no significant effect in high-frequency words (t = −0.153, $p < 0.878$) and a significant effect in low-frequency words, with LND words being recognized on average 0.160 gate sooner relative to UP than HND words (t = −2.25, $p < 0.05$).

3.3.2. Difference—Isolation and Uniqueness Points

On average, HF-HND words were identified −0.875 gate before, HF-LND −0.887/gate before, LF-HND 0.464 gate before, and LF-LND −0.624 gate before the UP. The 2 × 2 LMER revealed an effect for *frequency* (t = 5.477, $p < 0.001$) but not for *density* (t = −0.162, $p = 0.871$) or the *frequency* by *density* (t = −1.424, $p = 0.155$) interaction. Additional LMER models were run separately for *frequency* (H vs. L) and for *density* (H vs. L), with participant as a random intercept. An additional LMER model was run for *density* (H vs. L) within high and low frequencies. There was an effect for frequency with LF words being identified 0.330 gate later relative to the UP than HF words (t = 6.325, $p < 0.001$). There was no significant overall effect for ND (t value = −1.129, $p < 0.259$). There was an effect for ND low-frequency words, with LND words being recognized −0.160 gate sooner relative to UP than HND words on average (t value −2.25, $p < 0.05$) but no significant effect for ND for high-frequency words (t value −0.153, $p < 0.878$).

3.4. *Gating by Phoneme: Morphological*

Difference—Isolation and Uniqueness Points

The results for the morphological experiment are summarized in Table 6. On average, RCP < UP words were identified 0.887 gate before, RCP = UP words were identified 0.241 gate before, and UP < RCP words were identified 0.496 gate after the uniqueness point. An ANOVA with type as the independent variable and IP gate number as the dependent variable revealed a significant main effect (F value = 183.38, $p < 0.001$). The first LMER model, with type as an independent variable and participant as a random intercept effect, revealed an effect for type: RCP < UP words were identified on average 0.646 gate before RCP = UP relative to UP < RCP (t = −8.94, $p < 0.001$) and UP < RCP 0.737 gate after RCP = UP relative to uniqueness point (t = 10.121, $p < 0.001$). Results from the second LMER model comparing RCP < UP and UP < RCP words revealed a significant effect with

UP < RCP identified 1.3828 gate after RCP < UP relative to uniqueness point (t = 18.12, $p < 0.001$).

3.5. Recognition Point Results Summary

The recognition point (RP) results are summarized in Table 7 (lexical experiments) and in Table 8 (morphological experiments). In both lexical and morphological results in both traditional and phoneme gating paradigms, the results for RP and RP-UP did not differ in significance, magnitude, or direction of effect from those obtained with IP and IP-UP. There were two exceptions to this equivalence in the RP metric for neighborhood density comparisons (H vs. L). The direction of the effect remained the same as in the IP metric (LND words were identified more quickly than words with HND). In the gating by phoneme paradigm, the effect was significant for both high- and low-frequency words (as opposed to low frequency only), and in the traditional paradigm, it was significant for high- but not low-frequency words (as opposed to low-frequency only).

Table 7. Recognition point (RP) summaries for the lexical experiments. Results differing from those with IP and IP-UP, as dependent variables are bolded. The asterisk indicates statistical significance.

			Gating by Time		
	HF-HND	**HF-LND**	**LF-HND**	**LF-LND**	
RP	398 ms	368 ms	454 ms	439 ms	
RP-UP	−107 ms	−113 ms	−3 ms	−52 ms	
			Statistics		
	LMER 2 × 2	**Freq**	**ND**	**ND in HFreq**	**ND in LFreq**
RP	* Freqt = 5.642, $p < 0.001$ * ND t = −3.318, $p < 0.001$ Freq:ND t = 1.149, $p < 0.25$	* H 62 ms < L t = 9.055, $p < 0.001$	*** L 18 ms < H t = −2.577, $p < 0.05$**	*** L 30 ms < H t = −3.155, $p < 0.01$**	**L 15 ms < H t = −1.473, $p < 0.141$**
RP-UP	* Freq t = 10.035, $p < 0.001$ ND t = −0.667, $p < 0.505$ *** Freq:ND t = −2.918, $p < 0.01$**	* H 80 ms < L t = 11.12, $p < 0.001$	* L 18 ms < H t = −2.471, $p < 0.01$	t = −581, $p < 0.6$	* L 48 ms < H t = −4.87, $p < 0.001$
			Gating By Phoneme		
	HF-HND	HF-LND	LF-HND	LF-LND	
RP	4.074	3.852	4.49	4.302	
ID-UP	−0.926/gate	−0.931/gate	−0.504/gate	−0.698/gate	
			Statistics		
	LMER 2 × 2	Freq	ND	ND in HFreq	ND in LFreq
RP	* Freq t = 5.658, $p < 0.001$ * ND t = −3.179, $p < 0.01$ Freq:ND t = 0.270, $p = 0.787$	* H 0.424 < L t = 8.153, $p < 0.001$	*** L 0.184 < H t = −3.474, $p < 0.001$**	*** L 0.222 < H t = 3.091, $p < 0.01$**	*** L 0.194 < H t = −2.625, $p < 0.01$**
RP-UP	* Freq t = 5.509, $p < 0.001$ ND t = −0.070, $p = 0.95$ **Freq:ND t = −1.775, $p = 0.076$**	* H 0.319 < L t = 5.94, $p < 0.001$	t = −0.072, $p < 0.2$	t = −0.005, $p < 0.95$	* L 0.194 < H t = −2.625, $p < 0.01$

Table 8. Recognition point (RP) summaries for the morphological experiments. Results differing from those with IP and IP-UP, as dependent variables are bolded. The asterisk indicates statistical significance.

Gating by Time			
	RCP < UP	**RCP = UP**	**UP < RCP**
	−57 ms	−35 ms	67 ms
Statistics			
Overall		**RC < UP**	**UP < RC**
* F = 94.29, $p < 0.001$	**RC = UP**	* RC < UP 23 ms before t = −2.36, $p < 0.05$	* UP < RC 102 ms after t = 10.49, $p < 0.001$
	UP < RC	* RC < UP 124 ms before t = 12.149, $p < 0.001$	
Gating By Phoneme			
Summary			
	RCP < UP	**RCP = UP**	**UP < RCP**
	0.931	0.267	−0.469
Statistics			
Overall		**RC < UP**	**UP < RC**
* F = 178.06, $p < 0.001$	**RC = UP**	* RC < UP 0.663/gate before t= −8.939, $p < 0.001$	* UP < RC 0.736/gate after t = 9.866, $p < 0.001$
	UP < RC	* RC < UP 1.40/gate before t = 18.03, $p < 0.001$	

4. Discussion and Conclusions

This study represents the first time an auditory gating paradigm has been applied to spoken Hebrew to test lexical and morphological effects in word recognition. Using the gating paradigm allows us to observe how word information unfolds over time in the spoken signal and how lexical and morphological factors interact with auditory word recognition. Hebrew is an interesting test case because the Semitic templatic morphology has been shown, using other methods, to interact with word recognition in a way that is different from concatenative languages. Furthermore, we introduced and tested the phoneme-gating paradigm, which can greatly expand the number of stimuli and which has the potential to expand the kinds of questions a researcher can address using gating.

4.1. Lexical Results

Higher-frequency words were recognized at shorter gating times than lower-frequency words both in terms of IP and the IP-UP measures. This result is in line with previous findings in concatenative languages with the gating paradigm [19,20]. That is, less information is needed for a listener to recognize higher frequency words.

For higher-frequency words, there was no statistically reliable effect for neighborhood density. However, for lower-frequency words, words with lower neighborhood density were recognized at earlier gates than words with higher neighborhood density. These findings differ from those by Metsala's [20] results for English, where for higher-frequency words, low neighborhood density words were identified more quickly than high-density words, and for lower-frequency words, high-density words were identified more quickly than low-density words. It is always complicated to compare results such as these across languages because of the myriad ways in which any two languages may differ. Neighborhood density effects have been shown to differ between languages in previous studies. For example, while high neighborhood density has a facilitatory effect in Spanish [27], in English, high neighborhood density has an inhibitory effect (e.g., [28]). Furthermore, the inherent morphological complexity of Hebrew words may also contribute to the differing results. Therefore, our results should not be taken as a refutation of previous findings but perhaps as a further example of the complexity of comparing lexical effects across languages.

4.2. Morphological Results

Words in which the root completion point preceded the uniqueness point (RCP < UP) were identified with less signal information. In contrast, words in which the uniqueness point preceded the root completion point (UP < RCP) needed more signal information. That is, not having all root-phoneme information made it difficult to identify a word correctly. This result replicates, and extends to the online auditory domain, previous findings that the root is important for word recognition in templatic languages. In particular, during the process of recognition, having access to root information may narrow the scope of guesses not just acoustically but also morphologically, allowing for words to be identified with less information. This is an important extension of previous findings that root information plays a crucial role in word recognition in Hebrew, and it is novel in that the gating paradigm has allowed us to observe the time course of the process in comparing the effect of the UP to the RCP.

4.3. Paradigms

In both the lexical and the morphological experiments, the effect significance and effect direction did not differ between the two gating paradigms. This suggests that gating by phoneme is an appropriate methodology, at least for addressing certain types of research questions. Being able to gate by phoneme extends the types of research questions that could be addressed with gating, allowing for more careful control of information available to participants at each gate. Furthermore, this adaptation of the paradigm addresses the problem of having to control for the acoustic duration of phonemes across stimuli. We feel that this is a new and powerful research tool. While there are advantages to the phonemic gating paradigm, it much more difficult to apply than the traditional paradigm. Cutting stimuli such that there is access to only one additional phoneme at each gate requires precision and extensive acoustic phonetic training.

4.4. Differences between RP and IP Results

RP and IP results differed only in one aspect: the magnitude and statistical significance of neighborhood density effects in low vs. high frequencies with RP/IP as the dependent measures. While neighborhood density effects were only significant at low frequency with IP, with RP, they were only significant at high frequency or in both low and high frequency. Given that the difference between the RP and IP is whether a participant subsequently changed the answer from a correct guess to an incorrect one and then back again, differences with regards to neighborhood density (i.e., potential competitors) are not surprising. These differences may in fact be attributed to the frequency or more likely the location in the signal at which potential competitors (neighbors) appeared for the high- vs. low-frequency stimuli. That is, this difference may be the result of differences between high- and low-frequency words in the position in a word where changing a phoneme created a neighbor. If this position was later in the word for low-frequency words, this may cause more incorrect back-tracking after a correct guess. Thus, measuring from RP-UP, incorporating the uniqueness point at which no neighbors exist instead of just RP eliminates any differences in effects. In related work with the gating paradigm in English, Vitevich [26] found that neighborhood density effects were the result of neighborhood spread (the number of positions in the word at which potential neighbors could occur). The current stimuli were not designed to fully test this prediction, so it is left to future work to address this more rigorously.

4.5. Future Directions

In the morphological experiments of this paper, we focused on the role of the root overall and its importance in spoken word recognition. However, the role of roots and templates in recognition of words in Hebrew and other templatic languages is tied not only to the morphemes themselves but also to their productivity. For example, in Hebrew, Farhy, Verissimo, and Clahsen [29] found that morphological root priming occurred in words with a productive verbal template but not with a different, non-productive, verbal template. In Arabic, Boudelaa and Marslen-Wilson [30] found that morphological priming effects were only found in words with productive roots. Thus, taking into account factors, such as the predictability of a morpheme based on its context, could be applied to an investigation of the role of morphology in templatic spoken word recognition. In future work, we plan to extend our research to investigate spoken word recognition based on productivity of these root and template morphemes and their co-occurrence as well as context effects.

Author Contributions: M.O. and R.A.W. made equal contributions to conceptualization, methodology, and writing; M.O. was responsible for running the behavioral testing and for statistical analysis. All authors have read and agreed to the published version of the manuscript.

Funding: This research received no external funding.

Institutional Review Board Statement: The research procedures in this study were reviewed by the Human Subjects Division at the University of Washington and were deemed exempt from IRB review.

Informed Consent Statement: All subjects consented to participate in the experiment following exempt research guidelines.

Data Availability Statement: Data are available from the authors upon request.

Acknowledgments: The authors would like to acknowledge the assistance in stimulus preparation by Courtney Mansfield and in data preprocessing by Ajda Cokcen.

Conflicts of Interest: The authors declare no conflict of interest.

Appendix A. Stimuli Used in the Experiments

	Lexical Stimuli	
	HF-HND	
Word	**Root**	**Transliteration**
שחק	שחק	SaHak
שוטר	שטר	Soter
שיפור	שפר	Sipur
כרם	כרם	kerem
קשת	קשת	keSet
חורש	חרש	HoreS
חוקר	חקר	Hoker
פרט	פרט	peret
בשר	בשר	basar
סימון	סמן	simun

Cont.

Lexical Stimuli		
HF-LND		
Word	**Root**	**Transliteration**
חומרה	חמר	Humra
צמד	צמד	Temed
בדיחה	בדח	bdiHa
גפן	גפן	gefen
גרסה	גרס	girsa
רוטב	רטב	rotev
נוהג	נהג	nohag
תומך	תמך	tomeK
נציג	נצג	naTig
מקל	מקל	makel
LF-LND		
Word	**Root**	**Transliteration**
כחל	כחל	kaHal
מחט	מחט	maHat
קיטור	קטר	kitur
שחף	שחף	SaHaf
רתך	רתך	rataK
רחף	רחף	raHaf
חוטר	חטר	Hoter
שכירה	שׂכר	sKira
שחת	שחת	SaHat
גלף	גלף	galaf
LF-LND		
כפיס	כפס	kafis
מיסוך	מסך	misuK
גיהוץ	גהץ	gihuT
נחיל	נחל	neHil
נקז	נקז	nekez
נבג	נבג	neveg
ריגוש	רגש	riguS
רומח	רמח	romaH
תותח	תתח	totaH
סיבוך	סבך	sibuK

Cont.

Morphological Stimuli		
RC < UP		
Word	**Root**	**Transliteration**
גמלאי	גמל	gimlay
כורסה	כרס	kursa
קדמות	קדם	kadmut
קרנית	קרן	karnit
למדן	למד	lamdan
לכידה	לכד	leKida
שמלה	שׂמל	simla
ספרנות	ספר	safranut
חומצה	חמץ	HumTa
חרצית	חרץ	HarTit
RC = UP		
Word	**Root**	**Transliteration**
גלשן	גלשׁ	galSan
גיבוש	גבשׁ	gibuS
קרחון	קרח	karHon
כובש	כבשׁ	koveS
לפתן	לפת	liftan
לחישה	לחשׁ	leHiSa
סיפון	ספן	sipun
ספלון	ספל	siflon
חיריק	חרק	Hirik
חרוסת	חרס	Haroset
UP < RC		
Word	**Root**	**Transliteration**
גרדת	גרד	garedet
גוזל	גזל	gozal
כלבת	כלב	kalevet
כינור	כנר	kinor
לקט	לקט	leket
לבונה	לבן	levona
סבילות	סבל	svilut
סדיקה	סדק	sdika
חיגר	חגר	Higer
חכירה	חכר	HaKira

References

1. Taft, M. Morphological decomposition and the reverse base frequency effect. *Q. J. Exp. Psychol. Sect. A* **2004**, *57*, 745–765. [CrossRef]
2. Longtin, C.M.; Meunier, F. Morphological decomposition in early visual word processing. *J. Mem. Lang.* **2005**, *53*, 26–41. [CrossRef]
3. Leminen, A.; Leminen, M.; Kujala, T.; Shtyrov, Y. Neural dynamics of inflectional and derivational morphology processing in the human brain. *Cortex* **2013**, *49*, 2758–2771. [CrossRef]
4. Fiorentino, R.; Naito-Billen, Y.; Minai, U. Morphological decomposition in Japanese deadjectival nominals: Masked and overt priming evidence. *J. Psycholinguist. Res.* **2016**, *45*, 575–597. [CrossRef]
5. Frost, R.; Deutsch, A.; Gilboa, O.; Tannenbaum, M.; Marslen-Wilson, W. Morphological priming: Dissociation of phonological, semantic, and morphological factors. *Mem. Cogn.* **2000**, *28*, 1277–1288. [CrossRef]
6. Boudelaa, S.; Marslen-Wilson, W.D. Discontinuous morphology in time: Incremental masked priming in Arabic. *Lang. Cogn. Process.* **2005**, *20*, 207–260. [CrossRef]
7. Velan, H.; Frost, R. Letter-transposition effects are not universal: The impact of transposing letters in Hebrew. *J. Mem. Lang.* **2009**, *61*, 285–302. [CrossRef]
8. Perea, M.; Abu Mallouh, R.; Carreiras, M. The search for an input-coding scheme: Transposed-letter priming in Arabic. *Psychon. Bull. Rev.* **2010**, *17*, 375–380. [CrossRef]
9. Yablonski, M.; Ben-Shachar, M. The morpheme interference effect in hebrew: A generalization across the verbal and nominal domains. *Ment. Lex.* **2016**, *11*, 277–307. [CrossRef]
10. Geary, J.; Ussishkin, A. Morphological priming without semantic relationship in Hebrew spoken word recognition. *Proc. Linguist. Soc. Am.* **2019**, *4*, 9. [CrossRef]
11. Schriefers, H.; Zwitserlood, P.; Roelofs, A. The identification of morphologically complex spoken words: Continuous processing or decomposition? *J. Mem. Lang.* **1991**, *30*, 26–47. [CrossRef]
12. Balling, L.; Baayen, R.H. Morphological effects in auditory word recognition: Evidence from Danish. *Lang. Cogn. Process.* **2008**, *23*, 1159–1190. [CrossRef]
13. Frost, R.; Forster, K.I.; Deutsch, A. What can we learn from the morphology of Hebrew? A masked-priming investigation of morphological representation. *J. Exp. Psychol. Learn. Mem. Cogn.* **1997**, *23*, 829. [CrossRef]
14. Oganyan, M.; Wright, R.; Herschensohn, J. The role of the root in auditory word recognition of Hebrew. *Cortex* **2019**, *116*, 286–293. [CrossRef] [PubMed]
15. McClelland, J.L.; Elman, J.L. The TRACE model of speech perception. *Cogn. Psychol.* **1986**, *18*, 1–86. [CrossRef]
16. Marslen-Wilson, W.D.; Welsh, A. Processing interactions and lexical access during word recognition in continuous speech. *Cogn. Psychol.* **1978**, *10*, 29–63. [CrossRef]
17. Tyler, L.K. The structure of the initial cohort: Evidence from gating. *Percept. Psychophys.* **1984**, *36*, 417–427. [CrossRef]
18. Vitevitch, M.S.; Mullin, G.J. What Do Cognitive Networks Do? Simulations of Spoken Word Recognition Using the Cognitive Network Science Approach. *Brain Sci.* **2021**, *11*, 1628. [CrossRef]
19. Grosjean, F. Spoken word recognition processes and the gating paradigm. *Percept. Psychophys.* **1980**, *28*, 267–283. [CrossRef]
20. Metsala, J.L. An examination of word frequency and neighborhood density in the development of spoken-word recognition. *Mem. Cogn.* **1997**, *25*, 47–56. [CrossRef]
21. Frost, R.; Plaut, D. The Word-Frequency Database for Printed Hebrew. 2005. Available online: http://word-freq.huji.ac.il/index.html (accessed on 1 June 2020).
22. Charles-Luce, J.; Luce, P.A. Similarity neighbourhoods of words in young children's lexicons. *J. Child Lang.* **1990**, *17*, 205–215. [CrossRef] [PubMed]
23. Itai, A.; Wintner, S. Language Resources for Hebrew. *Lang. Resour. Eval.* **2008**, *42*, 75–98. [CrossRef]
24. Peirce, J.; Gray, J.R.; Simpson, S.; MacAskill, M.; Höchenberger, R.; Sogo, H.; Lindeløv, J.K. PsychoPy2: Experiments in behavior made easy. *Behav. Res. Methods* **2019**, *51*, 195–203. [CrossRef]
25. Sánchez-García, C.; Kandel, S.; Savariaux, C.; Soto-Faraco, S. The time course of audio-visual phoneme identification: A high temporal resolution study. *Multisens. Res.* **2018**, *31*, 57–78. [CrossRef]
26. Vitevitch, M.S. The spread of the phonological neighborhood influences spoken word recognition. *Mem. Cogn.* **2007**, *35*, 166–175. [CrossRef] [PubMed]
27. Vitevitch, M.S.; Rodríguez, E. Neighborhood density effects in spoken word recognition in Spanish. *J. Multiling. Commun. Disord.* **2005**, *3*, 64–73. [CrossRef]
28. Luce, P.A.; Pisoni, D.B. Recognizing spoken words: The neighborhood activation model. *Ear Hear.* **1998**, *19*, 1–36. [CrossRef]
29. Farhy, Y.; Veríssimo, J.; Clahsen, H. Universal and particular in morphological processing: Evidence from Hebrew. *Q. J. Exp. Psychol.* **2018**, *71*, 1125–1133. [CrossRef] [PubMed]
30. Boudelaa, S.; Marslen-Wilson, W.D. Productivity and priming: Morphemic decomposition in Arabic. *Lang. Cogn. Process.* **2011**, *26*, 624–652. [CrossRef]

brain sciences

MDPI

Article

Relating Suprathreshold Auditory Processing Abilities to Speech Understanding in Competition

Frederick J. Gallun [1,2,*], Laura Coco [1,2], Tess K. Koerner [1,2], E. Sebastian Lelo de Larrea-Mancera [3], Michelle R. Molis [2], David A. Eddins [4] and Aaron R. Seitz [3]

1 Oregon Hearing Research Center, Oregon Health & Science University, Portland, OR 97239, USA; coco@ohsu.edu (L.C.); koernert@ohsu.edu (T.K.K.)
2 VA RR&D National Center for Rehabilitative Auditory Research, VA Portland Health Care System, Portland, OR 97239, USA; michelle.molis@va.gov
3 Department of Psychology, University of California, Riverside, CA 92521, USA; elelo001@ucr.edu (E.S.L.d.L.-M.); aseitz@ucr.edu (A.R.S.)
4 Department of Communication Science & Disorders, University of South Florida, Tampa, FL 33620, USA; deddins@usf.edu
* Correspondence: gallunf@ohsu.edu; Tel.: +1-503-494-4331

Abstract: (1) Background: Difficulty hearing in noise is exacerbated in older adults. Older adults are more likely to have audiometric hearing loss, although some individuals with normal pure-tone audiograms also have difficulty perceiving speech in noise. Additional variables also likely account for speech understanding in noise. It has been suggested that one important class of variables is the ability to process auditory information once it has been detected. Here, we tested a set of these "suprathreshold" auditory processing abilities and related them to performance on a two-part test of speech understanding in competition with and without spatial separation of the target and masking speech. Testing was administered in the Portable Automated Rapid Testing (PART) application developed by our team; PART facilitates psychoacoustic assessments of auditory processing. (2) Methods: Forty-one individuals (average age 51 years), completed assessments of sensitivity to temporal fine structure (TFS) and spectrotemporal modulation (STM) detection via an iPad running the PART application. Statistical models were used to evaluate the strength of associations between performance on the auditory processing tasks and speech understanding in competition. Age and pure-tone-average (PTA) were also included as potential predictors. (3) Results: The model providing the best fit also included age and a measure of diotic frequency modulation (FM) detection but none of the other potential predictors. However, even the best fitting models accounted for 31% or less of the variance, supporting work suggesting that other variables (e.g., cognitive processing abilities) also contribute significantly to speech understanding in noise. (4) Conclusions: The results of the current study do not provide strong support for previous suggestions that suprathreshold processing abilities alone can be used to explain difficulties in speech understanding in competition among older adults. This discrepancy could be due to the speech tests used, the listeners tested, or the suprathreshold tests chosen. Future work with larger numbers of participants is warranted, including a range of cognitive tests and additional assessments of suprathreshold auditory processing abilities.

Keywords: auditory processing; hearing loss; speech perception; aging

Citation: Gallun, F.J.; Coco, L.; Koerner, T.K.; Larrea-Mancera, E.S.L.d.; Molis, M.R.; Eddins, D.A.; Seitz, A.R. Relating Suprathreshold Auditory Processing Abilities to Speech Understanding in Competition. *Brain Sci.* **2022**, *12*, 695. https://doi.org/10.3390/brainsci12060695

Academic Editor: Antoine Shahin

Received: 1 March 2022
Accepted: 25 May 2022
Published: 27 May 2022

1. Introduction

The ability to understand speech in the presence of competing sounds is a fundamental aspect of hearing that many older adults find challenging. Both age and degree of hearing loss have been shown to reliably predict a portion of the variability in performance on tasks of speech perception and have been studied extensively (for a review of earlier work, see Gordon-Salant [1]; for a more recent example, see Goossens et al. [2]). Hearing loss—customarily quantified on an audiogram as increased pure-tone detection thresholds—impacts speech perception through decreased overall audibility and reduced access to the temporal and spectral cues in speech. Similarly, normal aging can reduce the ability to use the cues needed for speech understanding in competition, although not all older listeners experience difficultly understanding speech in noise.

Previous experiments conducted in our laboratories and elsewhere have identified a set of psychophysical tasks of auditory processing (so-called "suprathreshold" auditory processing tests) that appear to predict performance on speech tasks with either noise or speech as the interference [2–7]. However, with a few notable exceptions—Rönnberg et al. and Marsja et al. [8,9]—most of these investigations have either explored no more than one or two auditory processing tasks at a time, or have lacked the statistical power to allow firm conclusions to be drawn (for an example, see Neher et al. [7]) making it difficult to determine which independent predictors contribute most to speech-in-competition performance. To address the limitations of traditional laboratory-based psychophysical assessments of auditory processing ability, we developed a new assessment platform—Portable Automated Rapid Testing (PART)—that allows direct testing of a diverse range of auditory processing abilities and facilitates the collection of large data sets with consistent procedures.

PART is an example of recent efforts to harness accessible technologies for hearing assessment in basic science and clinical telemedicine contexts, a current summary of which was recently created by the Acoustical Society of America's Task Force on Remote Testing [10]. The PART application has already been used successfully to collect data both in the laboratory [11–14] and remotely with participant-owned equipment in their own homes [15,16]. Currently in the United States, the limited availability of hearing healthcare professionals relative to the number of people with hearing loss limits access to hearing healthcare, particularly in rural areas—a gap that will be exacerbated by the projected growth of the aging population [17–19]. The use of portable technologies such as PART can improve access to hearing healthcare for patients who have difficulty traveling to the clinic and will allow research investigators to recruit and test a more diverse population of study participants [18,19].

Evidence for the Influence of Suprathreshold Auditory Abilities on Speech in Competition Performance

Behavioral and neurophysiological studies in humans and non-human primates have revealed substantial changes in brain structure, neurochemistry, and function associated with normal aging, even in the presence of normal or near-normal sensitivity or detection thresholds to pure tones (audibility) [20–29]. An additional set of auditory abilities is needed to discriminate different aspects of an audible signal. Since these processes operate on sounds that are above the audibility thresholds, they are referred to as suprathreshold auditory abilities. They include modulations of the amplitude of the acoustic signal over time (temporal modulation; TM) and modulation of the frequency spectrum of the acoustic signal relative to an unmodulated reference signal (spectral modulation; SM). Modulation of the spectrum that changes over time is called spectrotemporal modulation (STM). In addition, it is possible to modulate the phase of the signal, producing modulation in frequency of a pure tone or a narrowband noise (frequency modulation; FM). When the phase modulation is applied to both ears simultaneously it is called 'diotic', while applying FM to the signal at one ear, or different FM to the two ears is called 'dichotic'. See Palandrani et al. [30] for more details on these types of signals. Another way of thinking about the modulation of the signal is through what has been called modulation of the temporal fine structure (TFS). See Hoover et al. [31] for a discussion and investigation of a wide range of tests of TFS sensitivity.

It is not yet clear how speech understanding depends on the ability to detect and discriminate these types of acoustic signals. However, speech signals contain all of these types of modulations, and neurophysiological evidence shows that the auditory system is very sensitive to all of these modulations. For example, STM provides a robust signal for the characterization of auditory cortical receptive fields [32–37]. STM-based representations of acoustical stimuli have also successfully been applied to computational models of speech representation [38,39]. Furthermore, the ability to detect STM has been shown to be related to speech understanding in listeners with normal pure-tone detection thresholds [40–42], in listeners with cochlear damage [3,4,43], and in listeners who use cochlear implants [44–46].

Sensitivity to TFS also is related to speech understanding, especially in competition [47–49]. TFS sensitivity relies on precise neuronal firing in the time domain—the most precise of all neurosensory systems in healthy listeners [50]. Neurophysiological studies in animals and electrophysiological studies in humans show that aging is associated with degradation in precise temporal coding [21,51–53]. Introducing TFS distortion also disrupts speech understanding, which has been interpreted as evidence that aging leads to reduced neural synchrony and thus increased temporal jitter [54,55]. Füllgrabe et al. [6] related the perception of TFS cues to speech understanding in the presence of masking. Sensitivity to differences in time of arrival at the two ears ("interaural timing differences"; ITDs) is based on the same cues as is dichotic FM detection, which has been shown to be a reliable method of measuring TFS sensitivity [31]. Ellinger et al. [56] showed that when only presented with ITDs, older listeners experienced less spatial release from masking than did younger listeners while Eddins and Eddins [57] showed that when TFS and envelope modulation cues convey binaural information simultaneously, coding of TFS cues accounts for age-related deficits in binaural release from masking. These studies support a role for TFS cues in speech understanding, especially in the presence of competing sounds.

The goal of this study was to better understand the relationships among these suprathreshold auditory processing abilities and speech understanding in the presence of competing speech. To do so, we explored the relationships between a test of speech understanding in competition (with and without spatial separation among the talkers) and a battery of six suprathreshold auditory processing tests, all implemented in the PART application. The six tests in the battery were selected based on prior studies demonstrating their potential to predict the outcome of measures of speech understanding in competition [1,19,20]. SM, TM, and STM detection was measured using tasks similar to those in the literature [3,58–64]. TFS sensitivity was measured using diotic FM as well as a temporal gap detection task in which the gap was between two brief tone pulses [27,31,52,53,65].

The test of speech understanding in the presence of competing speech was chosen for two main reasons. Primarily, it involves a comparison of speech understanding with and without spatial separation between the target talker and two masking talkers. The calculated difference between the "colocated" and "separated" conditions is called spatial release from masking (SRM). SRM is expected to depend on binaural sensitivity and thus should be related to the ability to perform the dichotic FM task. In addition, performance on both component tasks and the derived SRM have known relationships to aging and pure-tone detection thresholds [24,66,67].

Data collection on the test battery, conducted with PART implemented on a portable tablet computer, took less than an hour for subjects to complete, satisfying the requirement for "rapid" testing. Note that this time included about ten minutes for two tests of tone in noise detection not reported here due to a programming error in the design of the measures. The data sample reported here, while not particularly large (41 listeners), is sufficient to demonstrate the utility PART could have in future explorations of similar relationships between these and other suprathreshold auditory assessments, cognitive assessments, or other measures determined in the future, and speech understanding.

In addition to answering the question of what happens when one attempts to use consumer-grade electronics to measure a battery of suprathreshold tests on a sample of participants varying in age and hearing loss, this study also seeks to explore the relationships among those tests and the speech in competition tasks, which have already been shown to be related to age and hearing thresholds [67]. While it is useful to know how much loss of spatial release should be expected for a listener with a particular set of hearing thresholds, after taking age into account, this does not explain why performance is reduced.

It is hoped that with the right set of additional test measures, it would be possible to say with some certainty which cues an individual listener is using and which they are not. This would open the door to a wide range of counseling and rehabilitative options that are currently quite difficult to pursue, given the uncertainty about why two people with similar audiograms often have different abilities to understand speech in complex environments.

2. Materials and Methods

2.1. Participants

Forty-one volunteers aged 23 to 80 years (mean: 51.1 yrs; standard deviation: 16.7 yrs), participated as listeners. Pure-tone hearing thresholds were obtained by an audiologist using traditional manual testing carried out in a sound-treated test room. All participants had audiometric hearing thresholds better than 85 dB HL between 0.25 and 8 kHz. Audiograms for all 41 participants are shown in Figure 1. Pure-tone average (PTA) thresholds based on an average of thresholds at 0.25, 0.5, 1, and 2 kHz ranged from −3.12 dB HL to 41.25 dB HL (mean: 15.58 dB HL; standard deviation: 11.19 dB HL). Hearing thresholds at the two ears differed by an average of 4 to 6 dB for frequencies below 2 kHz and an average of 8 to 10 dB for frequencies above 2 kHz. Two of the participants had interaural asymmetries greater than 10 dB at more than one audiometric frequency and so the regression analyses were conducted both with and without their data. Because the results of the regression analyses when those two participants were excluded were essentially unchanged (other than a reduction in statistical power) their data are included in all analyses reported below. The full data set, including supplementary material, is available at https://github.com/gallunf/SR2022 (accessed on 24 May 2022).

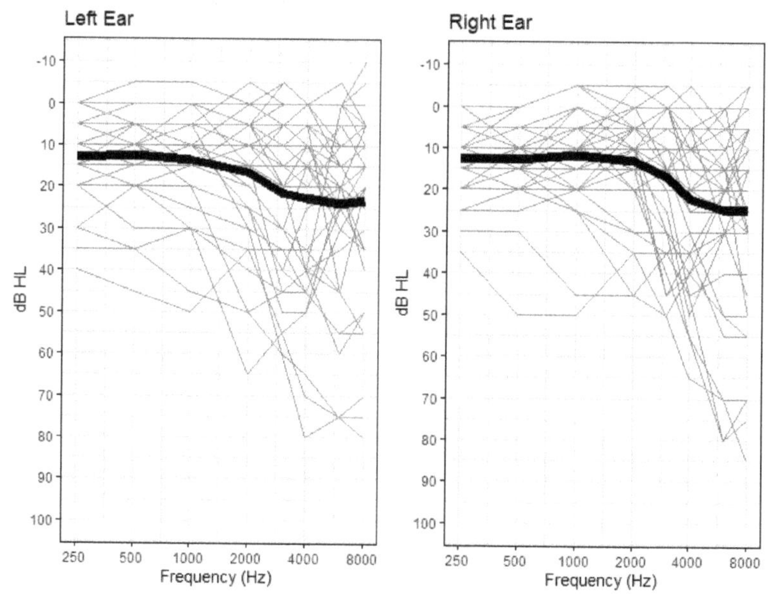

Figure 1. Audiograms for all 41 participants. Thin lines indicate individual listeners. Thick lines indicate mean values.

2.2. Stimuli and Procedures

Procedures and stimuli were a subset of those described in Diedesch et al. [14] and Larrea-Mancera et al. [12]. As for both of those studies, the tests described here (other than audiometric procedures) were conducted on an iPad running the PART application with calibrated Sennheiser HD 280 Pro headphones. These relatively inexpensive headphones were calibrated by the experimenters prior to use with a low-cost microphone and another iPad running a Sound Level Meter app, as in Gallun et al. [10]. Participants were seated in a comfortable chair in either a quiet room or a sound booth and were instructed to take breaks whenever necessary. Our previous work [12,16] has shown that test environment and even test equipment is not a significant predictor of performance for listeners using PART on a tablet or a smartphone, across a wide range of models of device and multiple headphones. For this reason, we felt comfortable using both a sound booth and a quiet room to increase the number of participants that could be tested with the time and resources available.

As described in more detail below, each participant completed a set of eight tasks, comprised of six two-cue two-alternative forced-choice (2C-2AFC) tasks assessing their ability to detect changes in various types of auditory stimuli and two speech tasks that assesses participants ability to correctly identity speech in the presence of competing talkers.

The 2C-2AFC task has been described in detail elsewhere [12,14,68], and involves four temporal intervals, the first and fourth of which contain the standard stimulus (unmodulated or otherwise lacking the target stimulus). These are the "cue" intervals. The second and third intervals are the "forced choice" portion, as one contains the target and one contains a standard stimulus and the listener is forced to choose which contains the target. In this case, correct answer feedback was given after every trial and the size of the target signal was adaptively adjusted based on the pattern of correct and incorrect responses.

2.2.1. Temporal Fine Structure

Three different tests of TFS sensitivity were used, each presented in a 2C-2AFC paradigm with adaptive tracking to estimate threshold: diotic FM (DioFM) and dichotic FM (DichFM) thresholds [31,52,65,69] and temporal gap (TGap) thresholds [27,31]. The FM tests used a pure tone with a frequency randomized between 460 and 550 Hz on each interval. Each interval contained a stimulus that was 400 ms in duration and was presented at a level of 75 dB SPL. The intervals were separated by 250 ms of silence. The standard stimulus was unmodulated and presented diotically (identically at the two ears). For both FM tasks, the target was a 2-Hz sinusoidal phase modulation that was either the same in both ears (DioFM) or that was out of phase in the two ears (DichFM). In the DioFM condition, the phase modulation created the percept of a change in the frequency of the tone at a rate of 2 Hz. The amount of this change (the "modulation depth") was then adjusted until the listener could just detect that the frequency was changing. The DichFM modulation, however, created a dynamic interaural time difference cue that resulted in a percept of a sound image that started at one side of the head and moved between the two ears at a rate of 2 Hz. In DichFM, the modulation depth was reduced until the depth was found at which the listeners could just detect that the sound image was moving. In both tasks, the modulation depth was adjusted on a logarithmic scale using the adaptive algorithm used by Larrea-Mancera et al. [12].

In the temporal gap detection task (TGap), stimuli were presented diotically, and each of the four intervals contained a pair of 4-ms 0.5-kHz tone bursts. In the standard intervals, the two bursts occurred with no temporal gap between them, while in the target interval there was a brief period of silence (a "gap") between when one burst ended and the other burst started. The gap started at 20 ms and was adaptively adjusted on a logarithmic scale using the methods described in Larrea-Mancera et al. [12].

2.2.2. Temporal, Spectral, and Spectrotemporal Modulation Sensitivity

The target stimuli in the temporal modulation task (TM) were 400-ms bursts of broadband noise that were amplitude-modulated at a rate of 4 Hz and that the listener was asked to discriminate from the three unmodulated broadband noise standards presented in the other three intervals [70]. For the spectral modulation task (SM), the target stimulus was a spectrally modulated noise (2 cycles/octave) with a random phase that the listener was asked to distinguish from unmodulated broadband noise [58,60]. For the spectrotemporal modulation task (STM), the target noise was both temporally modulated (4 Hz) and spectrally modulated (2 cycles/octave), compared to unmodulated broadband noise. All tasks employed adaptive-staircase procedures with step sizes scaled in dB units as described in Isarangura et al. [63] but otherwise the methods were comparable to those described by Bernstein et al. [3].

2.2.3. Speech in Competition

Speech understanding in competition was measured using sentence-level stimuli from the closed-set Coordinate Response Measure (CRM) corpus [71] and consisted of syntactically identical, time-synchronous sentences produced by three male talkers and presented in colocated and spatially separated listening conditions [72]. Participants were instructed to attend to a target talker located directly in front of them in a virtual acoustic spatial array while ignoring two masker talkers that either were also located directly in front of the participant (colocated condition; CO) or were located at +45° and −45° (separated condition, SEP). The virtual acoustic spatial array was implemented using a generic set of head-related transfer functions (HRTFs) following the methods developed and validated by Gallun and colleagues [24,67,73]. Each target and masker sentence took the form: Ready (CALL SIGN) go to (COLOR) (NUMBER) now; the target talker always used the callsign "CHARLIE". Each masker talker used one of seven other callsigns and different color and number combinations from those spoken by the target talker. Participants were instructed to indicate the color-number combination spoken by the target talker, using a 32-element color and number grid displayed on the iPad that contained buttons representing all possible combination of four colors and eight numbers.

Participants were familiarized with the response matrix during a short practice session in which the target "CHARLIE" sentences were presented from directly in front at 65 dB SPL without any distractor speakers. During testing, progressive tracking was used to reduce target-to-masker ratios (TMRs) from 10 dB to −10 dB in 2-dB steps, with two trials at each target-to-masker ratio. Participants were provided with feedback about correct or incorrect responses on each trial. Following the methods developed by Gallun et al. [24], the correct number of responses out of 22 trials was subtracted from the starting target-to-masker ratio of 10 dB to approximate target-to-masker thresholds (in dB). This measure estimates the point at which performance is 50%. In addition to reporting TMR thresholds, in dB, for the colocated and separated tasks, a derived measure of spatial release from masking (SRM), in dB, was calculated by taking the difference between thresholds in the colocated (CO) and spatially separated (SEP) conditions.

3. Statistical Analyses

Statistical analyses were centered on the question of the degree to which suprathreshold auditory processing abilities account for differences in individual performance that age and hearing loss cannot. Table 1 shows descriptive statistics for all predictors and outcome measures. First, correlations were calculated among the eight test measures, SRM (the difference between CO and SEP), age (operationalized as age in years at time of testing, referred to as Age), and hearing loss (operationalized as PTA). No corrections for multiple comparisons were applied, as the goal of the correlational analysis was primarily to give an idea of the strength of each variable when considered on its own. The main analysis involved backward linear regression, in which all of the predictors were entered into the model and then those not accounting for a significant proportion of the unexplained variance (at a criterion of $p < 0.100$) were eliminated. It should be noted, however, that as there were a total of 76 correlations calculated; if one wishes to consider each of the variables on its own it would be necessary to apply some form of correction for multiple comparisons. The most conservative approach would involve using the Bonferroni inequality and thus dividing the p-value for significance ($p < 0.05$) by 76, thus resulting in a critical value of $p < 0.00066$. All analyses were conducted in SPSS v27, which does not provide p-values less than 0.001, thus guaranteeing that all of the correlations would be regarded as non-significant.

Table 1. Descriptive Statistics.

	Units	Minimum	Maximum	Mean	Std. Deviation
CO	dB	0.00	4.50	2.23	1.21
SEP	dB	−9.00	5.55	−2.39	3.64
SRM	dB	−4.15	12.40	4.62	3.62
Age	Years	23.00	80.00	51.05	16.70
PTA	dB HL	−3.13	41.25	15.58	11.19
TGap	log2 (ms)	1.60	4.12	2.96	0.66
DioFM	log2 (Hz)	−2.17	3.29	0.68	1.43
DichFM	log2 (Hz)	−1.89	3.96	1.48	1.37
TM	dB	0.20	4.37	1.85	1.00
SM	dB	0.70	5.97	1.83	1.11
STM	dB	0.20	5.67	1.46	1.28

Note: dB = decibels; HL = hearing level; ms = milliseconds; Hz = Hertz; CO = speech with colocated target and maskers; SEP = speech with spatially separated target and maskers; SRM = difference between CO and SEP; PTA = 4 frequency pure-tone average; TGap = temporal gap; DioFM = Diotic FM; DichFM = Dichotic FM; TM = Temporal Modulation; SM = Spectral Modulation; STM = Spectrotemporal Modulation.

Rather than considering each correlation on its own, the approach taken here, backward regression, evaluates statistical significance based on the final model prediction. After each variable not accounting for a sufficient proportion of the unexplained variance has been removed, the final model is then evaluated for significance. The estimated effect size is based on the adjusted R^2 value of the final model, which is a measure of the variance in the dependent variable attributable to the linear regression model prediction, and includes an adjustment for the mathematical improvement in any model when an additional predictive variable is introduced.

To set the stage for the linear regression, the values in Table 2 are presented first for SEP and SRM, followed by correlations with Age and PTA. The significant correlations among the suprathreshold measures are then presented. None of the correlations between CO and any of the other measures reached a level of $p < 0.05$ and thus are not presented in Table 2.

Table 2. Correlations with *p*-values less than 0.05. Additional correlations available in the supplementary materials.

Variable 1	Variable 2	Pearson Correlation	Sig. (2-Tailed)
SEP	SRM	−0.944	<0.001
SEP	Age	0.471	0.002
SEP	PTA	0.467	0.002
SEP	DioFM	0.381	0.014
SEP	SM	0.318	0.043
SEP	DichFM	0.315	0.045
SRM	Age	−0.497	0.001
SRM	PTA	−0.476	0.002
SRM	DioFM	−0.377	0.015
SRM	SM	−0.358	0.022
Age	DichFM	0.500	0.001
Age	PTA	0.476	0.002
PTA	TGap	0.418	0.007
TGap	DioFM	0.592	<0.001
TGap	STM	0.451	0.003
TGap	SM	0.432	0.005
SM	STM	0.691	<0.001

Note: SEP = speech with spatially separated target and maskers; SRM = difference between CO and SEP; PTA = 4 frequency pure-tone average; DioFM = Diotic FM; DichFM = Dichotic FM; TGap = temporal gap; SM = Spectral Modulation; STM = Spectrotemporal Modulation.

Results: Linear Regression Modeling

Table 3 shows the performance of the final linear regression models for each speech measure (CO, SEP, SRM) after backward elimination of all variables not significantly accounting for variance at a level of $p < 0.100$. The final regression model for CO was unable to fit a model with an adjusted R^2 value that exceeded 0.000, and so no variables are listed. The linear regression model that provided the best prediction of thresholds in the SEP condition (adjusted $R^2 = 0.288$), included Age and Diotic FM detection. For SRM, the final regression model (adjusted $R^2 = 0.310$) also included Age and Diotic FM.

Table 3. Final linear regression models predicting CO, SEP, and SRM.

Condition	Predictors	Adjusted R^2	p	Error (dB)
CO	-	-	-	-
SEP	Age, DioFM	0.288	0.022	3.07
SRM	Age, DioFM	0.310	0.023	3.00

4. Discussion

In this analysis of 41 people of varying ages and hearing losses, with the exception of a diotic FM task, measures of suprathreshold auditory processing abilities were not strong predictors of variation in speech understanding in competing speech backgrounds. These results indicate that, while there are relationships between speech understanding and suprathreshold abilities, more work is needed to reconcile the results of the current study with the existing literature. The sections below compare these results to others in the literature and suggest a variety of potentially fruitful directions for future work in this area. Importantly, the current study demonstrates that any future investigations are likely to benefit from the further development of effective ways to test large numbers of participants on a wide range of tests.

4.1. The Relationships among Tests of Suprathreshold Processing and Speech Understanding

There were three speech measures examined in this study. The first, which did not include spatial separation of the talkers (colocated three talker speech; CO), showed no significant relationship with Age, PTA, or any of the other tests included in this analysis. The lack of correlation with Age and PTA is consistent with the data of Jakien and Gallun [67], who were able to explain only 5% of the variance among their listeners on the collocated condition using a model that included Age; PTA was not a significant predictor. It is possible that collecting a larger number of test runs on each participant would yield a relationship between Age and speech understanding, as stronger relationships were observed after more test runs in the Jakien and Gallun study. Nonetheless, the low proportion of variance explained for the colocated maskers by Age and PTA is consistent with the results of that study as well as others [24,56,66,74].

The second speech measure was the separated condition with the same three talker stimuli and task (SEP). For this condition, the Jakien and Gallun study reported that for the same amount of testing, a linear regression model based on Age and PTA explained 28% of the variance in thresholds estimated for their listeners. Here, PTA was not a significant predictor in the final regression model shown in Table 2. However, an alternative analysis included in the supplementary material did include PTA. In that analysis, the p-value to eliminate a variable was set at $p > 0.20$ rather than $p > 0.10$. In the model using the default criterion value for SPSS v27 (eliminate variables for $p > 0.10$), Age and DioFM explained 29% of the variance. This could be interpreted to mean that PTA and DioFM are tapping a similar aspect of the SEP measure, but another possibility is that the sample size tested here was simply insufficient to show an effect of PTA significant to allow to be retained in the final regression model. The two samples were very similar in age and PTA, but varied in the number of participants. Jakien and Gallun's 82 listeners had a mean age of 46.7 years and the 41 participants tested here had a mean age of 51.1 years. The mean

PTA of the participants in the Jakien and Gallun study (12.48 dB HL) was very similar to the mean of the listeners tested in this study (15.58 dB HL). However, the correlation between PTA and SEP was greater in the Jakien and Gallun study (r = 0.615) than in this study (r = 0.467), and the correlation between PTA and DioFM seen here was only 0.285 (see supplementary materials). However, the correlation between Age and PTA was 0.476, which probably explains why PTA was unable to account for a significant proportion of the variance once Age was included. The inclusion of DioFM is consistent with the notion that both performance on the DioFM and the SEP tasks reflect variations in suprathreshold abilities rather than audibility, but replication with a larger sample of participants is needed to before firm conclusions can be drawn.

The third speech measure, SRM, is derived from the other two and as such, might be expected to provide little additional information. Indeed, here the correlation between SEP and SRM was the strongest observed across the entire dataset ($\rho = -0.944$). Nonetheless, other studies have consistently reported differences in the variables that are associated with performance in SEP and SRM [24,56,67,73,74]. For example, in the Jakien and Gallun study, PTA was a significant predictor of SRM and SEP, while Age was a significant predictor of SEP but not SRM. In that study, PTA alone accounted for 23% of the variance in SRM. In the current study, Age was a significant predictor of SRM, also accounting for 23% of the variance. As with SEP, DioFM was able to account for another 8% of additional variance in the regression model. The results of the backward regression analysis in which variables were eliminated using the $p > 0.20$ criterion included PTA in the model along with Age and DioFM.

The role of the diotic FM detection task in predicting SEP and SRM are both consistent with the results of Strelcyk and Dau [75] who also demonstrated that TFS sensitivity was related to speech understanding in a binaural listening task. Such a relationship may be related to the need for very precise timing at the level of the cochlear nucleus for the extraction of binaural cues [76] that lead to improved speech understanding in the presence of spatially distributed sound sources. It is surprising, however, that the diotic FM task was a stronger predictor of spatial abilities than was the dichotic task, which actually involves a binaural judgment. Additional studies will be needed to better understand this unexpected result.

The poor predictive power of the STM detection task was surprising given the previous results of Bernstein and colleagues [3,4,43]. Similarly surprising is the contrasting results of Diedesch et al. [14], who reported stronger correlations with STM than were demonstrated here. One important difference between those studies and the current study is that both included a greater degree of hearing loss in their participants. Souza and her colleagues [68,77,78] have argued that listeners vary in their ability to use dynamic spectral cues to identify formant transitions in speech stimuli, and that listeners with hearing loss are more likely to have difficulty with this cue. Future work in this area would benefit from testing larger numbers of participants with an even wider range of hearing thresholds in order to capture listeners with a range of listening abilities and strategies. This will be facilitated by the increased availability of access to portable testing of the type employed in this study, thus providing easier access to large numbers of participants who are likely to vary in ways relevant to the hypotheses being tested.

4.2. The Importance of Cognitive Processing Abilities for Speech Understanding

Overall, the proportion of variance in the current data accounted for by even the best-fitting model was only 31%, lending support to a number of recent studies indicating that additional variables must be considered. Humes [79] and Nuesse et al. [80] both used similar statistical techniques to those reported here, but focused on cognitive variables instead of suprathreshold auditory processing, and both studies accounted for substantially more variance in the speech in noise tests they employed than is reported here. The results of Gallun and Jakien [81], who used measures of cognitive abilities to predict performance on the same speech tasks used here, are also consistent with the hypothesis that the variance left unaccounted for in the current data is likely related to differences in cognitive processing—specifically, differences in attention and working memory. In Gallun and Jakien [81], age and PTA were used as predictors along with performance on various auditory, visual, and auditory/visual working memory and attention tasks. In that case, 60% of the variance in the SEP condition was accounted for with a model that was based on PTA and auditory/visual working memory alone. In the CO condition, age and a measure of visual working memory span under conditions of uncertainty predicted 38% of the variance; and 45% of the variance in SRM was predicted by a model that included those same two variables (age and working memory span under response uncertainty).

The relationship between cognitive abilities and auditory processing performance among older adults is not yet fully understood. Loughrey et al. [82] used a meta-analysis of 36 studies and over 20,000 patients to show that age-related hearing loss is a significant predictor of a range of types of cognitive decline. However, even among older adults with normal hearing, cognitive performance predicts poor speech understanding; Marsja et al. [9] used structural equation modeling on data from 399 older listeners (199 with and 200 without hearing loss) which indicated that cognitive performance was a strong predictor of speech understanding.

Another important possibility is that the suprathreshold measures used here were not the ones that explain most of the variance, or that the measurement technique was too brief to provide sufficiently reliable threshold estimates. Establishing the relationship between cognitive abilities, age, and these specific auditory perceptual abilities will be an important target for future studies. As with hearing loss, the availability of portable testing will be helpful to researchers interested in testing the large numbers of heterogeneous participants necessary to test these hypotheses and will allow thresholds to be estimated with in a manner that allows the tradeoff between efficiency and accuracy to be controlled to a much greater extent than has been common with clinical research in the past. In addition to the measures described here, PART provides a robust signal processing environment that allows the researcher access to nearly every class of psychoacoustical tests that have been attempted over the past century. Furthermore, PART is constantly being upgraded and recently has been expanded to include several validated tests of cognitive function, such as working memory, divided and selective attention, response inhibition, and fluid intelligence. Multiple investigations are already underway and many more can and have been envisioned that will leverage the affordability and accessibility of PART to generate rich datasets.

5. Conclusions

The current study was designed to assess the extent to which suprathreshold auditory tests account for variation in speech understanding in competing speech backgrounds using a portable testing application—PART. The main result was that FM detection was the only suprathreshold ability that appeared to be strongly related to the ability to understand target speech in the presence of competing speech in which the talkers are repeating similar low-context closed-set sentences. Furthermore, this relationship was not present for speech in which there was not spatial separation between the target speech and the two competing talkers. Hypothesized relationships with detection of binaural differences and detection of spectrotemporal modulation were not observed in this data set, despite evidence in previous work supporting these hypotheses.

Future work carried out using an application such as PART will afford data collection on substantially larger numbers of participants. In addition, portable testing will allow greater examination of the relationships between speech understanding and a broad range of cognitive tests, not to mention additional assessments of suprathreshold measures, such as TFS processing and tone-in-noise perception. These studies will also address the possible influence of different types of hearing dysfunction that would be expressed as distinct non-linear relationships among the measures. The work presented here stands both as useful information about suprathreshold predictors of speech understanding and as an example of what is possible in future research using portable testing platforms.

Supplementary Materials: Supporting information can be downloaded at: https://github.com/gallunf/SR2022 (accessed on 24 May 2022).

Author Contributions: Conceptualization, F.J.G.; L.C.; D.A.E. and A.R.S.; Methodology, F.J.G.; L.C.; T.K.K.; E.S.L.d.L.-M.; M.R.M.; D.A.E. and A.R.S.; Software, F.J.G.; E.S.L.d.L.-M. and A.R.S.; Formal analysis, F.J.G.; Investigation, F.J.G.; Resources F.J.G.; Data curation, F.J.G.; Writing—original draft preparation, F.J.G.; L.C. and T.K.K.; Writing—review and editing, F.J.G.; L.C.; T.K.K.; E.S.L.d.L.-M.; M.R.M.; D.A.E. and A.R.S.; Visualization, F.J.G.; L.C. and T.K.K.; Supervision F.J.G.; Project administration, F.J.G.; Funding acquisition, F.J.G.; D.A.E. and A.R.S. All authors have read and agreed to the published version of the manuscript.

Funding: This work was supported, in part, by grant R01 DC 015051 (PIs: Gallun, Seitz, Eddins) and by grant R01 DC018166 (PIs: Gallun, Seitz, Stecker) both from NIH/NIDCD.

Institutional Review Board Statement: The study was conducted according to the guidelines of the Declaration of Helsinki and approved by the Joint Institutional Review Board of Oregon Health and Science University and the VA Portland Medical Center (study ID #15217, original approval 3 July 2016).

Informed Consent Statement: Informed consent was obtained from all participants involved in the study.

Data Availability Statement: The statistical analyses with accompanying figures for visualization are available at https://github.com/gallunf/SR2022 (accessed on 24 May 2022).

Conflicts of Interest: The authors declare no conflict of interest.

References

1. Gordon-Salant, S. Hearing loss and aging: New research findings and clinical implications. *J. Rehabil. Res. Dev.* **2005**, *42*, 9–24. [CrossRef] [PubMed]
2. Goossens, T.; Vercammen, C.; Wouters, J.; van Wieringen, A. Masked speech perception across the adult lifespan: Impact of age and hearing impairment. *Hear. Res.* **2017**, *344*, 109–124. [CrossRef] [PubMed]
3. Bernstein, J.G.W.; Mehraei, G.; Shamma, S.; Gallun, F.J.; Theodoroff, S.M.; Leek, M.R. Spectrotemporal Modulation Sensitivity as a Predictor of Speech Intelligibility for Hearing-Impaired Listeners. *J. Am. Acad. Audiol.* **2013**, *24*, 293–306. [CrossRef]
4. Bernstein, J.G.W.; Danielsson, H.; Hallgren, M.; Stenfelt, S.; Rönnberg, J.; Lunner, T. Spectrotemporal Modulation Sensitivity as a Predictor of Speech-Reception Performance in Noise With Hearing Aids. *Trends Hear.* **2016**, *20*, 2331216516670387. [CrossRef]
5. Dubno, J.R.; Dirks, D.D.; Morgan, D.E. Effects of age and mild hearing loss on speech recognition in noise. *J. Acoust. Soc. Am.* **1984**, *76*, 87–96. [CrossRef] [PubMed]
6. Füllgrabe, C.; Moore, B.C.J.; Stone, M.A. Age-group differences in speech identification despite matched audiometrically normal hearing: Contributions from auditory temporal processing and cognition. *Front. Aging Neurosci.* **2015**, *6*, 347. [CrossRef] [PubMed]
7. Neher, T.; Laugesen, S.; Jensen, N.S.; Kragelund, L. Can basic auditory and cognitive measures predict hearing-impaired listeners' localization and spatial speech recognition abilities? *J. Acoust. Soc. Am.* **2011**, *130*, 1542–1558. [CrossRef]
8. Rönnberg, J.; Lunner, T.; Ng, E.H.N.; Lidestam, B.; Zekveld, A.A.; Sörqvist, P.; Lyxell, B.; Träff, U.; Yumba, W.; Classon, E.; et al. Hearing impairment, cognition and speech understanding: Exploratory factor analyses of a comprehensive test battery for a group of hearing aid users, the n200 study. *Int. J. Audiol.* **2016**, *55*, 623–642. [CrossRef]
9. Marsja, E.; Stenbäck, V.; Moradi, S.; Danielsson, H.; Rönnberg, J. Is Having Hearing Loss Fundamentally Different? Multigroup Structural Equation Modeling of the Effect of Cognitive Functioning on Speech Identification. *Ear Hear.* **2022**. [CrossRef]
10. Peng, Z.E.; Buss, E.; Shen, Y.; Bharadwaj, H.; Stecker, G.C.; Beim, J.A.; Bosen, A.K.; Braza, M.; Diedesch, A.C.; Dorey, C.M.; et al. Remote testing for psychological and physiological acoustics: Initial report of the P&P Task Force on Remote Testing. *Proc. Meet. Acoust. Acoust. Soc. Am.* **2020**, *42*, 050009. [CrossRef]
11. Gallun, F.J.; Seitz, A.; Eddins, D.A.; Molis, M.R.; Stavropoulos, T.; Jakien, K.M.; Kampel, S.D.; Diedesch, A.C.; Hoover, E.C.; Bell, K.; et al. Development and validation of Portable Automated Rapid Testing (PART) measures for auditory research. *Proc. Meet. Acoust. Acoust. Soc. Am.* **2018**, *33*, 050002. [CrossRef]
12. Larrea-Mancera, E.S.L.; Stavropoulos, T.; Hoover, E.C.; Eddins, D.A.; Gallun, F.J.; Seitz, A.R. Portable Automated Rapid Testing (PART) for auditory assessment: Validation in a young adult normal-hearing population. *J. Acoust. Soc. Am.* **2020**, *148*, 1831–1851. [CrossRef] [PubMed]
13. Srinivasan, N.K.; Holtz, A.; Gallun, F.J. Comparing Spatial Release From Masking Using Traditional Methods and Portable Automated Rapid Testing iPad App. *Am. J. Audiol.* **2020**, *29*, 907–915. [CrossRef] [PubMed]

14. Diedesch, A.C.; Bock, S.J.A.; Gallun, F.J. Clinical Importance of Binaural Information: Extending Auditory Assessment in Clinical Populations Using a Portable Testing Platform. *Am. J. Audiol.* **2021**, *30*, 655–668. [CrossRef]
15. Larrea-Mancera, E.S.L.; Philipp, M.A.; Stavropoulos, T.; Carrillo, A.A.; Cheung, S.; Koerner, T.K.; Molis, M.R.; Gallun, F.J.; Seitz, A.R. Training with an auditory perceptual learning game transfers to speech in competition. *J. Cogn. Enhanc.* **2021**, *6*, 47–66. [CrossRef]
16. Larrea-Mancera, E.L.; de Stavropoulos, T.; Carrillo, A.A.; Cheung, S.; Eddins, D.A.; Molis, M.R.; Gallun, F.; Seitz, A. Portable Automated Rapid Testing (PART) of Auditory Processing Abilities in Young Normally Hearing Listeners: A Remotely Administered Replication with Participant-Owned Devices. 2021. Available online: https://psyarxiv.com/9u68p/ (accessed on 24 May 2022).
17. Coco, L.; Titlow, K.S.; Marrone, N. Geographic Distribution of the Hearing Aid Dispensing Workforce: A Teleaudiology Planning Assessment for Arizona. *Am. J. Audiol.* **2018**, *27*, 462–473. [CrossRef]
18. Planey, A.M. Audiologist availability and supply in the United States: A multi-scale spatial and political economic analysis. *Soc. Sci. Med.* **2019**, *222*, 216–224. [CrossRef]
19. Windmill, I.M.; Freeman, B.A. Demand for Audiology Services: 30-yr Projections and Impact on Academic Programs. *J. Am. Acad. Audiol.* **2013**, *24*, 407–416. [CrossRef]
20. Snell, K.B.; Frisina, D.R. Relationships among age-related differences in gap detection and word recognition. *J. Acoust. Soc. Am.* **2000**, *107*, 1615–1626. [CrossRef]
21. Walton, J.P. Timing is everything: Temporal processing deficits in the aged auditory brainstem. *Hear. Res.* **2010**, *264*, 63–69. [CrossRef]
22. Recanzone, G. The effects of aging on auditory cortical function. *Hear. Res.* **2018**, *366*, 99–105. [CrossRef] [PubMed]
23. Eddins, A.C.; Ozmeral, E.J.; Eddins, D.A. How aging impacts the encoding of binaural cues and the perception of auditory space. *Hear. Res.* **2018**, *369*, 79–89. [CrossRef] [PubMed]
24. Gallun, F.J.; Diedesch, A.C.; Kampel, S.D.; Jakien, K.M. Independent impacts of age and hearing loss on spatial release in a complex auditory environment. *Front. Neurosci.* **2013**, *7*, 252. [CrossRef] [PubMed]
25. Gallun, F.J.; Diedesch, A.C.; Beasley, R. Impacts of age on memory for auditory intensity. *J. Acoust. Soc. Am.* **2012**, *132*, 944–956. [CrossRef]
26. Shinn-Cunningham, B.; Ruggles, D.R.; Bharadwaj, H. How Early Aging and Environment Interact in Everyday Listening: From Brainstem to Behavior Through Modeling. In *Basic Aspects of Hearing*; Moore, B.C.J., Patterson, R.D., Winter, I.M., Carlyon, R.P., Gockel, H.E., Eds.; Springer: New York, NY, USA, 2013; pp. 501–510.
27. Gallun, F.J.; McMillan, G.P.; Molis, M.R.; Kampel, S.D.; Dann, S.M.; Konrad-Martin, D.L. Relating age and hearing loss to monaural, bilateral, and binaural temporal sensitivity. *Front. Neurosci.* **2014**, *8*, 172. [CrossRef]
28. Ozmeral, E.J.; Eddins, A.C.; Frisina, D.R.; Eddins, D.A. Large cross-sectional study of presbycusis reveals rapid progressive decline in auditory temporal acuity. *Neurobiol. Aging* **2016**, *43*, 72–78. [CrossRef]
29. Ozmeral, E.J.; Eddins, D.A.; Eddins, A.C. Reduced temporal processing in older, normal-hearing listeners evident from electrophysiological responses to shifts in interaural time difference. *J. Neurophysiol.* **2016**, *116*, 2720–2729. [CrossRef]
30. Palandrani, K.N.; Hoover, E.C.; Stavropoulos, T.; Seitz, A.R.; Isarangura, S.; Gallun, F.J.; Eddins, D.A. Temporal integration of monaural and dichotic frequency modulation. *J. Acoust. Soc. Am.* **2021**, *150*, 745–758. [CrossRef]
31. Hoover, E.C.; Kinney, B.N.; Bell, K.L.; Gallun, F.J.; Eddins, D.A. A Comparison of Behavioral Methods for Indexing the Auditory Processing of Temporal Fine Structure Cues. *J. Speech Lang. Hear. Res.* **2019**, *62*, 2018–2034. [CrossRef]
32. Kowalski, N.; Depireux, D.A.; Shamma, S.A. Analysis of dynamic spectra in ferret primary auditory cortex. I. Characteristics of single-unit responses to moving ripple spectra. *J. Neurophysiol.* **1996**, *76*, 3503–3523. [CrossRef]
33. Theunissen, F.E.; Sen, K.; Doupe, A.J. Spectral-Temporal Receptive Fields of Nonlinear Auditory Neurons Obtained Using Natural Sounds. *J. Neurosci.* **2000**, *20*, 2315–2331. [CrossRef] [PubMed]
34. Depireux, D.A.; Simon, J.Z.; Klein, D.J.; Shamma, S.A. Spectro-Temporal Response Field Characterization With Dynamic Ripples in Ferret Primary Auditory Cortex. *J. Neurophysiol.* **2001**, *85*, 1220–1234. [CrossRef] [PubMed]
35. David, S.V.; Mesgarani, N.; Shamma, S.A. Estimating sparse spectro-temporal receptive fields with natural stimuli. *Network Comput. Neural Syst.* **2007**, *18*, 191–212. [CrossRef] [PubMed]
36. Elliott, T.M.; Theunissen, F.E. The Modulation Transfer Function for Speech Intelligibility. *PLoS Comput. Biol.* **2009**, *5*, e1000302. [CrossRef]
37. Zatorre, R.J.; Belin, P. Spectral and Temporal Processing in Human Auditory Cortex. *Cereb. Cortex* **2001**, *11*, 946–953. [CrossRef]
38. Chi, T.; Gao, Y.; Guyton, M.C.; Ru, P.; Shamma, S. Spectro-temporal modulation transfer functions and speech intelligibility. *J. Acoust. Soc. Am.* **1999**, *106*, 2719–2732. [CrossRef]
39. Elhilali, M.; Chi, T.; Shamma, S.A. A spectro-temporal modulation index (STMI) for assessment of speech intelligibility. *Speech Commun.* **2003**, *41*, 331–348. [CrossRef]
40. Edraki, A.; Chan, W.-Y.; Jensen, J.; Fogerty, D. Speech Intelligibility Prediction Using Spectro-Temporal Modulation Analysis. *IEEE/ACM Trans. Audio Speech Lang. Process.* **2021**, *29*, 210–225. [CrossRef]
41. Spille, C.; Ewert, S.D.; Kollmeier, B.; Meyer, B.T. Predicting speech intelligibility with deep neural networks. *Comput. Speech Lang.* **2018**, *48*, 51–66. [CrossRef]
42. Chabot-Leclerc, A.; Jørgensen, S.; Dau, T. The role of auditory spectro-temporal modulation filtering and the decision metric for speech intelligibility prediction. *J. Acoust. Soc. Am.* **2014**, *135*, 3502–3512. [CrossRef]

43. Mehraei, G.; Gallun, F.J.; Leek, M.R.; Bernstein, J.G.W. Spectrotemporal modulation sensitivity for hearing-impaired listeners: Dependence on carrier center frequency and the relationship to speech intelligibility. *J. Acoust. Soc. Am.* **2014**, *136*, 301–316. [CrossRef] [PubMed]
44. Won, J.H.; Drennan, W.R.; Rubinstein, J.T. Spectral-Ripple Resolution Correlates with Speech Reception in Noise in Cochlear Implant Users. *J. Assoc. Res. Otolaryngol.* **2007**, *8*, 384–392. [CrossRef] [PubMed]
45. Aronoff, J.M.; Landsberger, D.M. The development of a modified spectral ripple test. *J. Acoust. Soc. Am.* **2013**, *134*, EL217–EL222. [CrossRef] [PubMed]
46. Saoji, A.A.; Litvak, L.; Spahr, A.J.; Eddins, D.A. Spectral modulation detection and vowel and consonant identifications in cochlear implant listeners. *J. Acoust. Soc. Am.* **2009**, *126*, 955–958. [CrossRef]
47. Hopkins, K.; Moore, B.C.J. The contribution of temporal fine structure to the intelligibility of speech in steady and modulated noise. *J. Acoust. Soc. Am.* **2009**, *125*, 442–446. [CrossRef] [PubMed]
48. Rosen, S.; Carlyon, R.P.; Darwin, C.J.; Russell, I.J. Temporal information in speech: Acoustic, auditory and linguistic aspects. *Philos. Trans. R. Soc. Lond. B Biol. Sci.* **1992**, *336*, 367–373. [CrossRef]
49. Viswanathan, V.; Shinn-Cunningham, B.G.; Heinz, M.G. Temporal fine structure influences voicing confusions for consonant identification in multi-talker babble. *J. Acoust. Soc. Am.* **2021**, *150*, 2664–2676. [CrossRef]
50. Frisina, R.D. Subcortical neural coding mechanisms for auditory temporal processing. *Hear. Res.* **2001**, *158*, 1–27. [CrossRef]
51. Tremblay, K.L.; Piskosz, M.; Souza, P. Effects of age and age-related hearing loss on the neural representation of speech cues. *Clin. Neurophysiol.* **2003**, *114*, 1332–1343. [CrossRef]
52. Grose, J.H.; Mamo, S.K. Frequency modulation detection as a measure of temporal processing: Age-related monaural and binaural effects. *Hear. Res.* **2012**, *294*, 49–54. [CrossRef]
53. Koerner, T.K.; Muralimanohar, R.K.; Gallun, F.J.; Billings, C.J. Age-Related Deficits in Electrophysiological and Behavioral Measures of Binaural Temporal Processing. *Front. Neurosci.* **2020**, *14*, 578566. [CrossRef] [PubMed]
54. Lorenzi, C.; Gilbert, G.; Carn, H.; Garnier, S.; Moore, B.C.J. Speech perception problems of the hearing impaired reflect inability to use temporal fine structure. *Proc. Natl. Acad. Sci. USA* **2006**, *103*, 18866–18869. [CrossRef] [PubMed]
55. Pichora-Fuller, M.K.; Schneider, B.A.; MacDonald, E.; Pass, H.E.; Brown, S. Temporal jitter disrupts speech intelligibility: A simulation of auditory aging. *Hear. Res.* **2007**, *223*, 114–121. [CrossRef] [PubMed]
56. Ellinger, R.L.; Jakien, K.M.; Gallun, F.J. The role of interaural differences on speech intelligibility in complex multi-talker environments. *J. Acoust. Soc. Am.* **2017**, *141*, EL170–EL176. [CrossRef] [PubMed]
57. Eddins, A.C.; Eddins, D.A. Cortical Correlates of Binaural Temporal Processing Deficits in Older Adults. *Ear Hear.* **2018**, *39*, 594–604. [CrossRef] [PubMed]
58. Saoji, A.A.; Eddins, D.A. Spectral modulation masking patterns reveal tuning to spectral envelope frequency. *J. Acoust. Soc. Am.* **2007**, *122*, 1004–1013. [CrossRef] [PubMed]
59. Ozmeral, E.J.; Eddins, A.C.; Eddins, D.A. How Do Age and Hearing Loss Impact Spectral Envelope Perception? *J. Speech Lang. Hear. Res.* **2018**, *61*, 2376–2385. [CrossRef]
60. Hoover, E.C.; Eddins, A.C.; Eddins, D.A. Distribution of spectral modulation transfer functions in a young, normal-hearing population. *J. Acoust. Soc. Am.* **2018**, *143*, 306–309. [CrossRef]
61. Dau, T.; Kollmeier, B.; Kohlrausch, A.A. Modeling auditory processing of amplitude modulation. I. Detection and masking with narrow-band carriers. *J. Acoust. Soc. Am.* **1997**, *102*, 2892–2905. [CrossRef]
62. Ewert, S.D.; Dau, T. Characterizing frequency selectivity for envelope fluctuations. *J. Acoust. Soc. Am.* **2000**, *108*, 1181–1196. [CrossRef]
63. Isarangura, S.; Eddins, A.C.; Ozmeral, E.J.; Eddins, D.A. The Effects of Duration and Level on Spectral Modulation Perception. *J. Speech Lang. Hear. Res.* **2019**, *62*, 3876–3886. [CrossRef]
64. Stavropoulos, T.A.; Isarangura, S.; Hoover, E.C.; Eddins, D.A.; Seitz, A.R.; Gallun, F.J. Exponential spectro-temporal modulation generation. *J. Acoust. Soc. Am.* **2021**, *149*, 1434–1443. [CrossRef] [PubMed]
65. Whiteford, K.L.; Oxenham, A.J. Using individual differences to test the role of temporal and place cues in coding frequency modulation. *J. Acoust. Soc. Am.* **2015**, *138*, 3093–3104. [CrossRef]
66. Marrone, N.; Mason, C.R.; Kidd, G. The effects of hearing loss and age on the benefit of spatial separation between multiple talkers in reverberant rooms. *J. Acoust. Soc. Am.* **2008**, *124*, 3064–3075. [CrossRef] [PubMed]
67. Jakien, K.M.; Gallun, F.J. Normative Data for a Rapid, Automated Test of Spatial Release From Masking. *Am. J. Audiol.* **2018**, *27*, 529–538. [CrossRef] [PubMed]
68. Souza, P.; Gallun, F.; Wright, R. Contributions to Speech-Cue Weighting in Older Adults With Impaired Hearing. *J. Speech Lang. Hear. Res.* **2020**, *63*, 334–344. [CrossRef] [PubMed]
69. Whiteford, K.L.; Kreft, H.A.; Oxenham, A.J. Assessing the Role of Place and Timing Cues in Coding Frequency and Amplitude Modulation as a Function of Age. *J. Assoc. Res. Otolaryngol.* **2017**, *18*, 619–633. [CrossRef]
70. Viemeister, N.F. Temporal modulation transfer functions based upon modulation thresholds. *J. Acoust. Soc. Am.* **1979**, *66*, 1364–1380. [CrossRef]
71. Bolia, R.S.; Nelson, W.T.; Ericson, M.A.; Simpson, B.D. A speech corpus for multitalker communications research. *J. Acoust. Soc. Am.* **2000**, *107*, 1065–1066. [CrossRef]

72. Marrone, N.; Mason, C.R.; Kidd, G. Tuning in the spatial dimension: Evidence from a masked speech identification task. *J. Acoust. Soc. Am.* **2008**, *124*, 1146–1158. [CrossRef]
73. Jakien, K.M.; Kampel, S.D.; Stansell, M.M.; Gallun, F.J. Validating a Rapid, Automated Test of Spatial Release From Masking. *Am. J. Audiol.* **2017**, *26*, 507–518. [CrossRef] [PubMed]
74. Jakien, K.M.; Kampel, S.D.; Gordon, S.Y.; Gallun, F.J. The Benefits of Increased Sensation Level and Bandwidth for Spatial Release From Masking. *Ear Hear.* **2017**, *38*, e13–e21. [CrossRef] [PubMed]
75. Strelcyk, O.; Dau, T. Relations between frequency selectivity, temporal fine-structure processing, and speech reception in impaired hearing. *J. Acoust. Soc. Am.* **2009**, *125*, 3328–3345. [CrossRef] [PubMed]
76. Gallun, F.J. Impaired Binaural Hearing in Adults: A Selected Review of the Literature. *Front. Neurosci.* **2021**, *15*, 610957. [CrossRef]
77. Souza, P.E.; Wright, R.A.; Blackburn, M.C.; Tatman, R.; Gallun, F.J. Individual Sensitivity to Spectral and Temporal Cues in Listeners With Hearing Impairment. *J. Speech Lang. Hear. Res.* **2015**, *58*, 520–534. [CrossRef]
78. Souza, P.; Wright, R.; Gallun, F.; Reinhart, P. Reliability and Repeatability of the Speech Cue Profile. *J. Speech Lang. Hear. Res.* **2018**, *61*, 2126–2137. [CrossRef]
79. Humes, L.E. Factors Underlying Individual Differences in Speech-Recognition Threshold (SRT) in Noise Among Older Adults. *Front. Aging Neurosci.* **2021**, *13*, 702739. [CrossRef]
80. Nuesse, T.; Steenken, R.; Neher, T.; Holube, I. Exploring the Link Between Cognitive Abilities and Speech Recognition in the Elderly Under Different Listening Conditions. *Front. Psychol.* **2018**, *9*, 678. [CrossRef]
81. Gallun, F.J.; Jakien, K.M. The Ability to Allocate Attentional Resources to a Memory Task Predicts Speech-on-Speech Masking for Older Listeners. In Proceedings of the International Congress on Acoustics, Aachen, Germany, 9–13 September 2019.
82. Loughrey, D.G.; Kelly, M.E.; Kelley, G.A.; Brennan, S.; Lawlor, B.A. Association of age-related hearing loss with cognitive function, cognitive impairment, and dementia: A systematic review and meta-analysis. *JAMA Otolaryngol. Neck Surg.* **2018**, *144*, 115–126. [CrossRef]

brain sciences

MDPI

Article

Social Priming in Speech Perception: Revisiting Kangaroo/Kiwi Priming in New Zealand English

Gia Hurring [1,2,*], Jennifer Hay [1,2], Katie Drager [3], Ryan Podlubny [4], Laura Manhire [2] and Alix Ellis [2]

1 New Zealand Institute of Language, Brain, and Behaviour, University of Canterbury, 20 Kirkwood Avenue, Upper Riccarton, Christchurch 8041, New Zealand; jen.hay@canterbury.ac.nz

2 Department of Linguistics, University of Canterbury, 20 Kirkwood Avenue, Upper Riccarton, Christchurch 8041, New Zealand; lauraman22@gmail.com (L.M.); rory.alix.ellis@gmail.com (A.E.)

3 Department of Linguistics, University of Hawai'i at Mānoa, 1890 East-West Rd, Honolulu, HI 96822, USA; kdrager@hawaii.edu

4 Department of Linguistics, University of Alberta, Edmonton, AB T6G 2G4, Canada; pudplace@gmail.com

* Correspondence: gia.hurring@pg.canterbury.ac.nz

Abstract: We investigate whether regionally-associated primes can affect speech perception in two lexical decision tasks in which New Zealand listeners were exposed to an Australian prime (a kangaroo), a New Zealand prime (a kiwi), and/or a control animal (a horse). The target stimuli involve ambiguous vowels, embedded in a frame that would result in a real word with a KIT or a DRESS vowel and a nonsense word with the alternative vowel; thus, lexical decision responses can reveal which vowel was heard. Our pre-registered design predicted that exposure to the kangaroo would elicit more KIT-consistent responses than exposure to the kiwi. Both experiments showed significant priming effects in which the kangaroo elicited more KIT-consistent responses than the kiwi. The particular locus and details of these effects differed across experiments and participants. Taken together, the experiments reinforce the finding that regionally-associated primes can affect speech perception, but also suggest that the effects are sensitive to experimental design, stimulus acoustics, and individuals' production and past experience.

Keywords: priming; speech perception; sociophonetics; lexical decision task; New Zealand English; Australian English

Citation: Hurring, G.; Hay, J.; Drager, K.; Podlubny, R.; Manhire, L.; Ellis, A. Social Priming in Speech Perception: Revisiting Kangaroo/Kiwi Priming in New Zealand English. *Brain Sci.* **2022**, *12*, 684. https://doi.org/10.3390/brainsci12060684

Academic Editor: Richard Wright

Received: 21 March 2022
Accepted: 20 May 2022
Published: 24 May 2022

1. Introduction

Can priming with regionally-associated images affect speech perception? Previous work suggests that it can [1]. New Zealanders primed with kangaroos appeared to shift their vowel perception to be more Australian-like. However, the task used in that experiment was not unambiguously about speech perception. Our paper adopts a more controlled, preregistered design, which more directly tests whether priming with regionally-associated images can lead to shifts in perception.

Across two experiments, participants are primed with images of kangaroos, kiwis, and horses, while conducting a lexical decision task with words containing a vowel that could be categorized differently in New Zealand English and Australian English. Both experiments find some evidence that the regional prime can affect what vowel is heard. Overall, the results reinforce the claim that regionally-associated primes can affect speech perception, but also provide some reason for caution—suggesting that such effects are sensitive to aspects of experimental design, stimulus acoustics, and individual differences.

2. Background

2.1. Linguistic Terminology

This paper, like the literature it builds on, uses lexical sets to refer to the vowels under investigation [2]. Lexical set terminology provides a dialect-neutral way of referring to classes of vowels, as opposed to, for example, using the international phonetic alphabet, which references the production of a specific token of a vowel but does not identify (in any dialect-neutral way) which class of vowel is being discussed. For example, in NZE and Australian English, the production of the vowel in the word 'fish' would receive quite different phonetic transcriptions, but in both dialects the vowel belongs to the KIT lexical set, and thus shares its production with a set of other KIT words (such as *bit*, *miss*, *him*, etc.). This paper is focused on vowels in the KIT lexical set, and also the DRESS lexical set—namely, words that share a vowel with 'DRESS' (such as *bet*, *mess*, and *hem*).

2.2. Social Priming in Speech Perception

An array of language external factors has been shown to influence speech perception. Among such factors is social information attributed to the talker, including gender [3,4], age [5,6], ethnicity [7], sexual orientation [8], socioeconomic status [5], region [9,10], attractiveness [11], and social persona [12]. Most of this work either uses photographs or videos paired with different talkers to manipulate perceived characteristics of the talker (e.g., [3,5]) or else manipulates listeners' expectations through explicit descriptions of the talkers and their characteristics (e.g., [9,12]).

Understanding social priming in listener perception is important because it influences the relevant scope of speech perception models. What is the extent to which non–acoustic information is integrated in the process of speech perception? And is the creation and interpretation of social meaning an integral part of the speech perception process, or is it a separate process which does not influence speech perception at all? These questions have been the topic of debate regarding the modularity of speech perception, leading to the emergence of models in which social factors play an important role in speech perception (see [13–15]). The reported social priming results have fed directly into that discussion (see [15–22]). We follow previous works, using the term 'speech perception' to describe how listeners perceive the acoustic signal. More precisely—in this paper we will be concerned with which phoneme/word an acoustic signal is mapped onto. A reviewer asks us to consider whether our results relate to 'perception' or a later stage of 'interpretation' (following [23]). We take the mapping of acoustic input to linguistic categories to be a fundamental part of 'speech perception', broadly construed. However, we note here that by using this term, we are not making any claim about the specific stages of processing that are involved in this mapping.

One key social priming study adopted a paradigm in which participants listen to a sentence and then match the realization of a target word from the sentence with one from a synthesized vowel continuum [9]. Niedzielski (1999) [9] showed that listeners from Detroit matched a target with a raised variant of a diphthong when they thought they were listening to a Canadian, but not when they thought the speaker was from Michigan. These findings were interpreted as showing that regional expectations about where a speaker was from could affect speech perception. The same task was subsequently employed in New Zealand, in a task in which half the listeners had 'Australian' written on the answer-sheet, and half had 'New Zealander' [10]. Those participants with Australian matched KIT vowels to more raised variants, typical of Australian English. The effect was present for women, but not men, although the gender ratio in the sample was not balanced. The participants were surveyed at the end of the experiment about where they thought the speaker was from, and nearly all participants responded that they thought the speaker was from New Zealand. This finding led the authors to speculate that the effect was not driven by expectations or beliefs about the speaker, but perhaps by a more automatic priming effect.

In Hay & Drager (2010) [1], we, therefore, tested this possibility explicitly through employing the same task as Hay, Nolan & Drager (2006) [10] but exposed participants to incidental social primes associated with the two regions. To achieve incidental priming, the experimenter pulled out from a cabinet one of two sets of stuffed animal toys prior to beginning the experiment, pretending like she did not know why they were there and setting them aside but within view of the participant. The first set of stuffed toys were kangaroos and koalas associated with Australia, and the second set were kiwis associated with New Zealand. The results from this second study were remarkably similar to those observed in the original study—including a gender difference. The presence of the regionally-associated stuffed toys—items that the participants were led to believe had nothing to do with the task—appeared to have influenced vowel perception in much the same way as regional labels.

Speculating that the observed difference between men and women may ultimately stem from attitudinal differences, Walker et al. (2018) [24] conducted the same task with an attitudinal prime, in which participants conducted a baseline condition, and then were exposed to one of three sets of facts about Australia (good, neutral, vs. bad facts). The manipulation shifted their performance in the subsequent task, and the authors concluded that perceptual adaptation towards a dialect can occur in the absence of a speaker of that dialect, and that these adaptations can be subject to a listener's affect towards the primed dialect region.

These experiments are all from New Zealand, and it is important to note that priming with regionally-associated images has not been replicated outside of New Zealand. In an experiment conducted in Australia, Walker, Szakay and Cox (2019) [25] found that Australian participants were not influenced by exposure to the animals. They suggest that the lack of an effect may be due to differences between Australians' and New Zealanders' metalinguistic awareness of the relevant variation (p. 21).

Attempts to replicate the general design from Niedzielski (1999) [9] in other dialect areas have also been mixed. For example, Jannedy et al. (2011) [26] showed that written dialect areas on an answer sheet (following [9]) significantly shifted perceptions of German fricatives. Whereas Lawrence (2015) [27] used a similar design and found limited evidence for social priming of BATH/STRUT vowels in speakers of Standard Southern British English.

While we do not necessarily expect these effects to replicate in places with different language experiences and stereotypes, the mixed results raise questions about the validity of the findings presented in Hay & Drager (2010) [1]. Did the results presented therein arise due to chance, or is the lack of a finding in other work due to, for example, differences in either exposure to or salience of sociolinguistic variation for the community of participants tested?

In addition to the failed replication attempts outside of New Zealand, the very task itself raises questions about how exactly to interpret the results. The above experiments all involve the same task (in which participants hear a vowel embedded within a sentence and are then asked to match it to the closest token on a synthesized continuum). This task is unnatural in that the process involves holding the target vowel in memory, while participants are exposed to other realizations of the same vowel and then perform a matching task. These effects might be attributed to memory then and are thus not unambiguously a meaningful part of speech perception. Because the task does not allow for an immediate response, we cannot be sure that the regionally-associated primes have actually influenced the perception of the target vowel. Such a task incorrectly presumes that memory does not degrade over time and that it is not influenced by the presentation of subsequent auditory input. Therefore, even if the primes were effective, and the observed effect was not due to chance, the primes may have influenced the selection of tokens from the vowel continua (i.e., a process downstream from the initial recognition and mapping) instead of influencing perception per se.

In the current paper we are primarily concerned with the priming of social information, whereby socially charged information that is deemed incidental to the speech signal influences listener-behaviour (see [28], to appear, for a discussion). This automatic social priming would thus exclude effects that stem from overtly manipulating expectations by describing characteristics of the talker, and it also excludes studies that manipulate social information attributed to a talker through the use of photographs or video because they may arise due to multisensory or multimodal integration (e.g., [29]) rather than priming per se.

Therefore, we ask: does exposure to social information influence the perception of sounds even when the social information is believed to be incidental to the talker, or the language forms they produce? One study in this literature [30] uses something incidental—listener location—with a task that is simpler and more directly related to speech perception. The incidental prime is not explicitly social, but rather—relates to previous experience in different locations. In a simple listening experiment, listeners listen to tokens on a HEAD-HAD continuum and identify what word they hear. Listeners who first conduct the listening experiment while sitting in a car, have a different threshold between DRESS and TRAP than those who first complete it in the lab. They claim that this result—more unambiguously about speech perception—is likely to arise from the same automatic mechanism that elicited the priming by the kangaroos and kiwis in Hay & Drager (2010) [1].

The current study addresses the methodological concerns by attempting to replicate the social priming reported in Hay & Drager (2010) [1] using a completely different experimental paradigm. Using a modified lexical decision task, we examine the extent to which drawings of kangaroos and kiwis may influence vowel perception. We do this by creating ambiguous vowels that are likely to be heard as 'KIT' in Australia and 'DRESS' in New Zealand. By embedding the vowels in different lexical frames, we can use responses in a lexical decision task to establish what was heard. If we embed an ambiguous vowel (X) in the frame fXzzy, for example, then a 'yes' response indicates that the KIT vowel was likely heard (i.e., real word *fizzy*), whereas a 'no' response would indicate that DRESS was heard (i.e., nonsense word *fezzy*).

2.3. *Australian and New Zealand English Vowels*

New Zealanders' and Australians' realizations of the front vowels are largely distinct. Figure 1 shows the vowel spaces for a number of dialects of English, recorded as stimuli for a perception experiment [31]. As shown in Figure 1, TRAP and DRESS vowels are both realized higher in F1-F2 space in New Zealand English (NZE) compared to Australian English (AusE), with DRESS overlapping with FLEECE for at least some NZE speakers. In AusE, KIT overlaps with FLEECE, whereas it is realized in a much more central position in NZE. Overlap of DRESS or KIT with FLEECE is not a feature that is present in the other dialects studied by Shaw et al. (2018) [31]. Our dialects of interest, then, are distinct from other varieties in having a very high front short vowel and can be differentiated from each other with respect to the specific identity of that vowel.

Due to the high front position of KIT in AusE and DRESS in NZE, it seems likely that a fairly high front short vowel in a lexically ambiguous context is likely to be perceived as KIT by Australians, and DRESS by New Zealanders. Indeed, by analyzing error patterns from the Shaw et al. (2018) [31] experiment, we can examine what happens when a New Zealander hears an Australian KIT vowel in isolation. The authors played recordings of words from NZE and AusE to listeners in each country. In some cases, these words were in isolation, and in some cases after a period of exposure to a speaker from one of the dialects reading a story. Listeners identified what vowel they heard in the word. We examined the patterns of response errors in the data from that experiment, to confirm that there is variation in New Zealand listeners' perceptions of KIT and DRESS. In particular, an Australian KIT vowel is more likely to be heard as DRESS than KIT, but the likelihood of identifying it accurately as KIT substantially increases for listeners who were exposed to an Australian voice prior to hearing the target word. New Zealanders' perception of DRESS, on the other hand, is more stable and more often heard as DRESS than any other phoneme, regardless of the dialect it is produced in.

The acoustic configuration of the DRESS and KIT vowels, together with this observed pattern of errors, suggests that it should be possible to create stimuli that are ambiguous between a NZE DRESS and an AusE KIT, and that it might be possible to influence the mapping of these signals during speech processing using regionally-associated primes.

For the two experiments presented herein, we created a set of words with synthesized vowels with F1 and F2 positioned between the NZE and AusE KIT acoustic spaces. For

New Zealand listeners, the vowel could therefore be heard as either a NZE KIT or an AusE KIT; and it could also be mapped to NZ DRESS, as in the Shaw results described above. We hypothesize that listeners are unlikely to classify the vowel as DRESS if they are primed towards Australia, and more likely to classify it as DRESS if they are primed toward NZ.

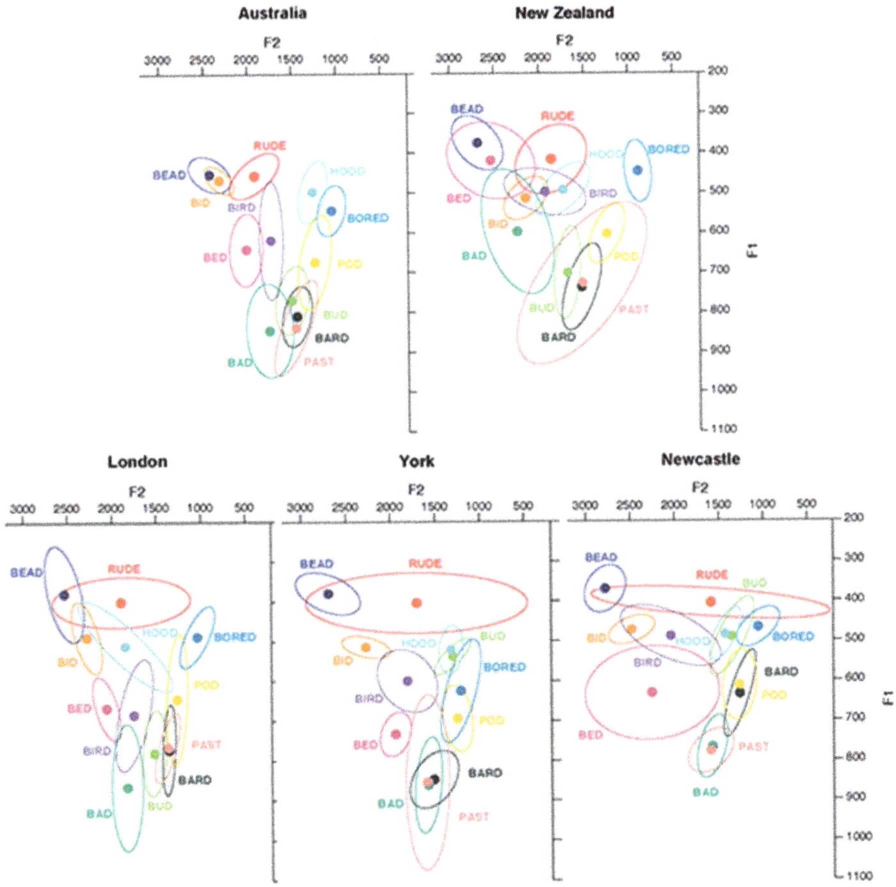

Figure 1. F1-F2 plot of vowels in Australian English (top left), NZE (top right), and three UK dialects (bottom) as reported in Shaw et al. (2018). F1 represents tongue height (high vs. low) while F2 represents tongue backness (front vs. back). For example, *BEAD* in the vowel plots depicts the tongue in the highest and furthest forward position in the mouth. (CC BY-4.0 https://creativecommons.org/licenses/by/4.0/, accessed on 12 February 2022).

The experiments use a lexical decision task so that we can infer from participants' responses which vowel they heard. The frames for the vowels form a real word context for only one of the vowels (i.e., either DRESS or KIT) and not the alternative vowel. For example, if a listener perceived the ambiguous vowel (X) as KIT, they would interpret stimuli such as tXpping and fXzzy as real words and scXptic and pXppered as nonsense words. Alternatively, the opposite would be true if they mapped the vowel to DRESS. We refer to contexts where a KIT vowel results in a real word as being in a 'KIT frame' and contexts where a DRESS vowel results in a real word as being in a 'DRESS frame'. Central to the present work, priming listeners with different, culturally charged animal images allows us to test whether use of a Kangaroo prime (associated with Australia) will lead

New Zealand listeners to hear the ambiguous vowel as KIT rather than DRESS (i.e., to respond in a more Australian-like way).

We preregistered a design, which is reported here as experiment one (Section 3). Experiment two (Section 4) adjusts various elements of the experimental design and repeats the experiment.

3. Experiment One

We used a lexical decision task to test our hypotheses that (1) exposure to regional primes would affect vowel perception in ways that are consistent with the relevant regional dialects and (2) that the effect of the primes is not solely due to listener expectations about where the talker is from.

3.1. Materials and Methods

The experiment was run on the internet. Participants completed a lexical decision task while images were presented on the screen. Responses to target words were to an ambiguous vowel embedded in one of two **frames** (a **KIT-frame** or a **DRESS-frame**).

We used line drawings of kangaroos, kiwis and horses as different **prime-types**. The prime-types were presented in one of two **presentation-types**—either as if they were the character talking the words (the **speaking** presentation-type), or just incidentally on the screen, presenting some instructions (the **incidental** presentation-type). The speaking presentation-type tests the effect of expectations or beliefs about the speaker (to the extent that a participant believes that the animal is 'speaking') in line with Hay, Nolan & Drager (2006) [10]. The incidental presentation-type tests the effect of incidental exposure to regional primes, in line with Hay & Drager 2010 [1].

Each participant was selected either into the baseline condition or the priming condition, and into either a speaking or incidental presentation-type. Each participant responded to words across three blocks, each of which used a different **voice** and used different images for each block. Participants in the **baseline condition** encountered pictures of three different coloured horses in the three blocks, presented as either speaking (for participants in the speaking presentation-type) or presented incidentally. Participants in the **priming condition** encountered pictures of a horse, a kangaroo and a kiwi, again either speaking or incidentally. More detail on all manipulations is given below.

3.1.1. Auditory Stimuli

Target items contained an ambiguous vowel embedded in one of two **frames:** a KIT-frame or a DRESS-frame. **KIT-frames** are items that are real words if they contain KIT but nonsense words if they contain DRESS, whereas the opposite is true for **DRESS-frames**. Filler items contained either LOT or STRUT. Example frames for target and filler items are shown in Table 1. A complete list of the frames used can be found in Appendix A.

Stimuli are all trochaic, with the target vowel in the first syllable. Half of the stimuli (90 items) contained the ambiguous vowel, half of them in a KIT frame (where only if the vowel was heard as KIT would the stimulus be a real word), and half of them in a DRESS frame. Fillers were designed so that half of the fillers would be heard as real words, and half would be heard as non-words. The fillers included words with STRUT and LOT, neither of which are likely to have led to major misunderstandings across dialects. Also, as fillers, we also included items which the listener would hear as real words regardless of which vowel was heard (**the Both-frame**—e.g., bXgger,), and which neither vowel would make the frame a real word (**the Neither-frame**—e.g., kXzzard).

The target stimuli (KIT-frame and DRESS-frame) have similar CELEX wordform frequencies [32]. The summary statistics are depicted in Table 2. Words with more than one CELEX entry were summed by their wordform and those with zero frequencies were included. All words have low frequency, and a Wilcoxon rank-sum test returned a p-value of 0.25, suggesting that the two stimuli frames are not significantly different from each other. The highest frequency words in each frame (KIT-frame and DRESS-frame, respectively) were *giving* and *spending*, while the lowest frequency words (with zero frequency wordforms in CELEX) were *stingers* and *sketchers* (KIT-frame and DRESS-frame, respectively).

Table 1. Distribution of stimuli for experiment one across word-type for target items and fillers over three speakers.

Frame Type	Example	Count
KIT-frame	dXgging	45
DRESS-frame	sXnding	45
Filler stimuli	**Example**	**Count**
Real LOT	bothers	15
Fake LOT	fomments	15
Real STRUT	custom	15
Fake STRUT	duppet	15
Both-frame	bXgger	15
Neither-frame	kXzzard	15

Table 2. Distribution of CELEX wordform frequencies (counts per 17.9 million) for KIT-frame and DRESS-frame stimuli.

	Minimum	Median	Max	Mean
KIT-frame	0.0	29.0	2270.0	161.7
DRESS-frame	0.0	26.0	827.0	81.78

3.1.2. Stimuli Recording and Vowel Resynthesis

The ambiguous vowels were created by taking recordings of two voices producing the target KIT word—a New Zealand voice and an Australian voice—and using these voices to synthesize a stimulus that was intermediate between the two vowels.

In stimuli generation, our goal was to synthesize a realistic sounding, stepwise progression that spans true New Zealand and Australian KIT vowels at either extreme. For example, formant values associated with the vocalic portion of a given New Zealand syllable would be manipulated systematically, and incrementally throughout the intermediary steps to become more like that vowel's Australian counterpart until the actual AusE vowel is incorporated as the final step. To create stimuli in different voices, to span our different blocks, we created three continua for each stimulus item, by recording three NZE speakers, and mixing them each together with a single AusE speaker.

Four female speakers were recruited with both age and height in mind, aiming to reduce inter-stimulus differences that might be rooted in physiology. Thus, three speakers of NZE (mean age = 23 years, mean height = 174.6 cm) and one speaker of AusE (age = 32 years, height = 177 cm) were recorded producing wordlists that included both the target words (KIT-variant) and filler items. For labelling purposes, the three speakers from New Zealand are differentiated as NZ1, NZ2, and NZ3 and the Australian speaker is identified as AU. Recordings were captured in a sound-attenuated booth on the University of Canterbury campus using a Beyerdynamic Opus 55.18 MK II head-mounted condenser microphone, and a digital VU meter to identify and compensate for differences in speaker loudness. Signals were routed through a Sound Devices USBPre 2 audio interface and recorded as WAV files on a late-2013 Macbook Pro laptop computer via Praat [33] at a sampling rate of 44.1 kHz and bit-depth of 16. In order to use the *SpeechInNoise* tool to run the experiment (see Section 3.1.3), which best suited our needs for an online presentation of the experiment, it was necessary to later downsample these source stimuli to 22,050 Hz and convert them to MP3. The MP3 format was required by the platform, and downsampling the stimuli dramatically reduced stimuli load times for participants; piLOT tests incorporating the 44.1 kHz stimuli had problematically long wait times that participants sometimes misinterpreted as crashes/errors.

The creation of continua for each item, for each speaker, was automated using a Praat script authored by Winn (2014) [34] which involves a hybrid of parametric and concatenative synthesis (this script can be accessed at: http://www.mattwinn.com/praat/Make_Formant_Continuum_v30.txt, accessed on 7 April 2017). The script allows users to independently isolate a target time range in each of two sound files, which serve as the bases for the parametric synthesis; the script also excises segments immediately preceding and following each target for appropriate concatenation following the generation of each synthetic target. Original intensity contours are retained. All steps in a given continuum are framed by the same preceding and following segments, although the user specifies from which file those segments are taken. For example, a target removed from the middle of hypothetical "SoundNZ" could serve as one extreme of the continuum, whereas another target removed from the middle of hypothetical "SoundAus" would serve as the basis for the other extreme. The script provides a form that allows the user to set pertinent parameters to values that suit their needs before creating the designated number of steps, interpolating formant values for F1-F4 every 40 ms. The script then outputs sound files for each step where all steps are framed by the preceding and following segments from "SoundNZ" or "SoundAus" as directed. The following parameter settings resulted in satisfactory outputs for our needs: we elected to have five steps per continuum; specified Praat should recognise 4 formants within the specified pitch range (with the exception of targets with adjacent nasals, in which case we specified 5); maximum formant frequency was set to 5000 Hz (with the exception of targets with adjacent nasals, in which case we specified 6000); and the output file intensity was set to normalise all segments and steps to an amplitude of 73 dB. All other parameters retained their default settings. All told, three continua were generated for each stimulus item, pairing each of NZ1, NZ2, and NZ3 with AU.

Preliminary informal testing indicated the third continuum point was perceived by listeners to be ambiguous between an Australian KIT and New Zealand DRESS, and this step was selected for use in experiment one (see Figure 2 and Table 3). Thus, the vowel in experiment one is not exactly like any vowel from NZE or AusE. It is halfway between a NZE KIT and an Australian KIT (and thus also approximately equidistant between a NZE DRESS and an NZE KIT). It is not near an Australian DRESS. Furthermore, the analysis revealed that these vowels were not dissimilar in length and remain around the same length in both these recorded word lists and in a corpus of NZE natural conversation speech (for NZE KIT and DRESS). Having the vowel lengths near equal minimizes the likelihood for a listener to be able to distinguish these vowels by their length. Thus, if primed with an Australian image, we would expect fewer DRESS-like responses to the stimulus, and more KIT-like responses whereas a New Zealand image might be expected to produce a more balanced distribution between KIT/DRESS.

We attempted to synthesize a larger number of stimuli than shown in Table 1, and the final number of words was reduced to words whose synthesis was successful across all three speakers. The range of formant values for an example continuum are depicted in Table 3.

Table 3. Continua formant values (Hz) for the word 'fixture' at T-step 15 (median).

	F1	F2	F3
NZ3 original *fixture*	568	1983	2879
***Fixture* synthesis 1**	568	1983	2879
***Fixture* synthesis 2**	531	2099	2871
***Fixture* synthesis 3**	495	2222	2864
***Fixture* synthesis 4**	460	2352	2856
***Fixture* synthesis 5**	426	2492	2849
AU original *fixture*	426	2492	2849

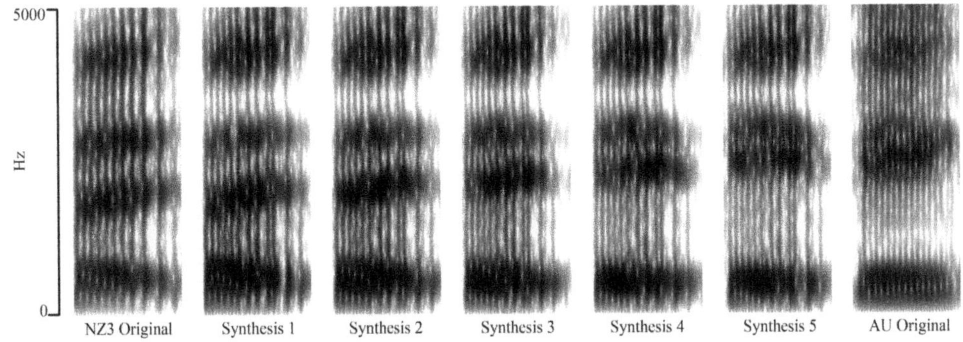

Figure 2. Five-step vowel continuum synthesis of the word 'fixture' proceeding from most New Zealand-like (NZ3 original recording) to most Australian-like (AU original recording).

3.1.3. Conditions and Prime Type

Three cartoon images were commissioned for use as visual primes in this study, all composed by a single artist (Andrew Kepple, see an overview of Kepple's work here: https://en-academic.com/dic.nsf/enwiki/2553002, accessed on 31 March 2018). These different animals will be referred to as **prime-types**. The kiwi is a prevalent national emblem of New Zealand, so was included to prime listeners for NZE. Similarly, the kangaroo is a prevalent national emblem of Australia and was selected for priming AusE. An image of a horse was included as a 'neutral' or baseline prime (see Figure 3). Thus, these animals were selected for their cultural significance in Australasia, or in the context of the horse for its relative neutrality. Additionally, the use of the kiwi and kangaroo make for a more ready comparison to the primes used by Hay & Drager (2010) [1]. We made two additional versions of the horse by manipulating the colour and the aspect ratio of the image.

Figure 3. Commissioned artwork of prime-types (horse, kiwi, and kangaroo).

The four conditions were presented as follows in Figure 4a–d (note the black arrow indicates participants clicking 'next' to start listening to the stimuli). Participants were in a baseline-speaking, baseline-incidental, priming-speaking or priming-incidental condition. Each block had an initial instruction screen, and then an experiment screen that remained visible during the block. Each screen contained an animal and a stick figure. In the speaking condition, the animal is speaking, and the stick figure delivers instructions. In the incidental condition, the stick figure is speaking, and the animal delivers instructions. The instruction-giver remains present on the screen throughout the block.

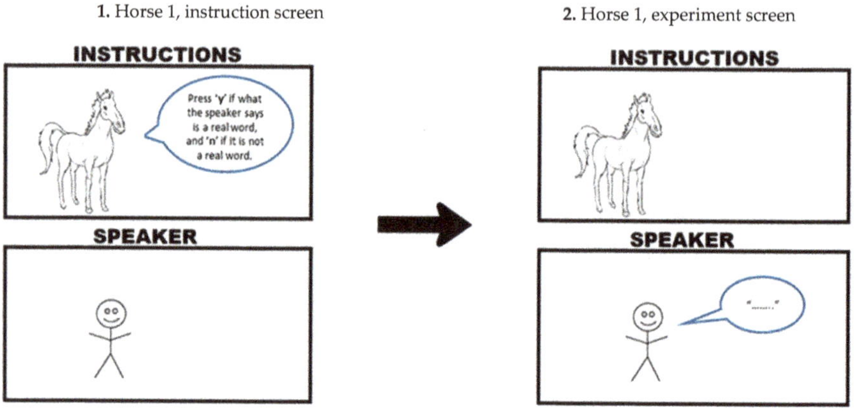

(a)

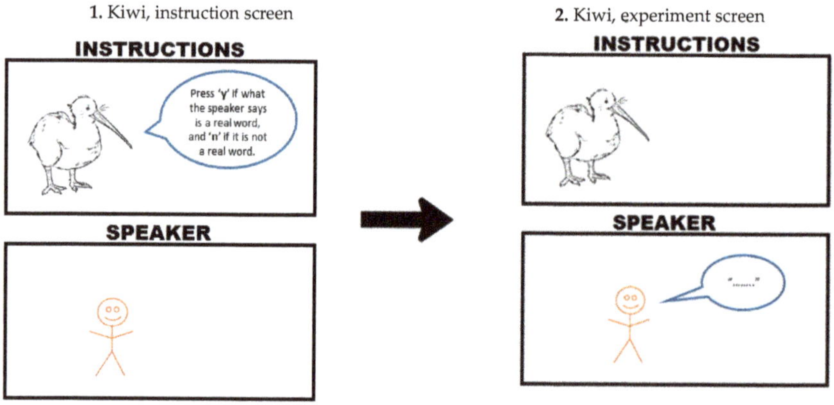

(b)

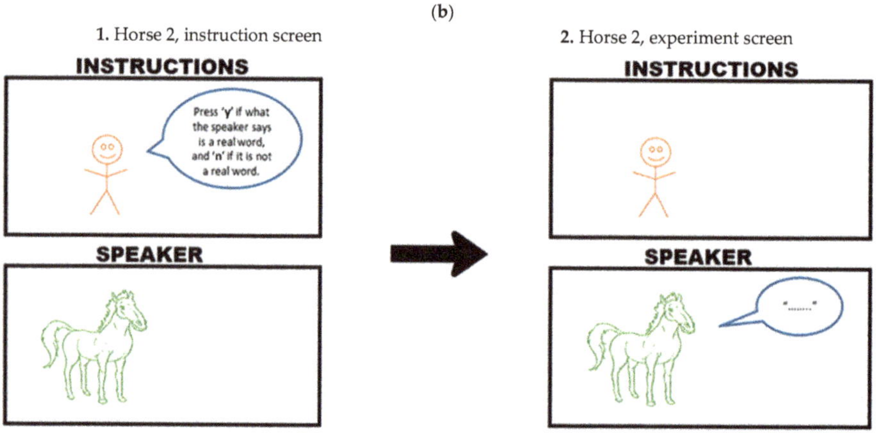

(c)

Figure 4. *Cont.*

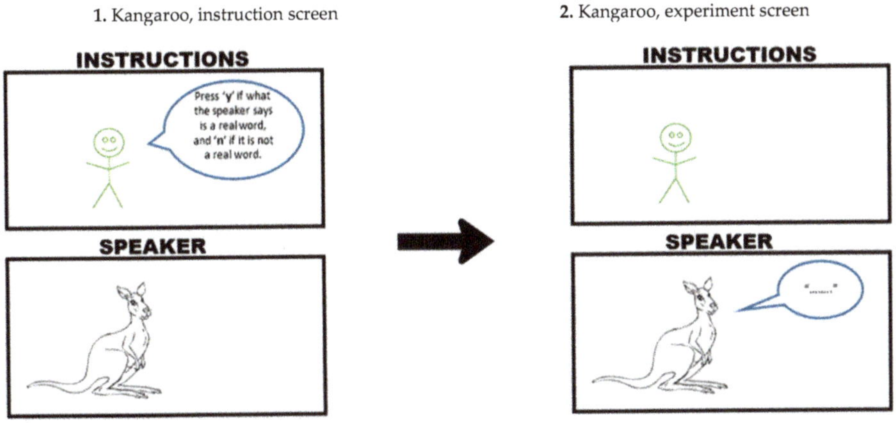

(**d**)

Figure 4. (**a**) Example block from **baseline-incidental** condition. (**b**) Example block from **priming-incidental** condition. (**c**) Example block from **baseline-speaking** condition. (**d**) Example block from **priming-speaking** condition.

3.1.4. Online Word Recognition Task

The experiment was designed using a *Speech In Noise 2* platform, developed by Chan (2015) [35] (this documentation can be accessed at: https://northwestern.app.box.com/s/9g2rigz1iqh4ymfkgunq6t31u3iycbpr, accessed on 1 June 2020). Data is immediately uploaded to a web-linked *Firebase* console, a Google-owned app development platform (Firebase is accessible here: firebase.google.com, accessed on 22 July 2020).

Individual playlists were generated in R. These playlists were counterbalanced and involved assigning randomized voice-to-prime type pairings. Each playlist included three voice-to-prime type blocks (e.g., black horse blocked with NZ1, kiwi blocked with NZ2, kangaroo blocked with NZ3). Block order was randomised within and between playlists. Participants only heard each word once, and every playlist incorporated a quasi-random word-to-block order (i.e., no playlist had the same 60 words blocked to a voice compared to another playlist). Participants encountered one block at a time, where each block included 60 unique words with one prime-type before proceeding to the next block. No playlist was ever used twice. The experiment randomly assigned each participant a unique playlist that allocated them to one of four conditions above (see Figure 4a–d). The experiment was designed for cross-participant comparison, so each participant only ever encountered one condition. Participants heard both stimuli and filler words over the course of a session. Blocking voices to prime-type ensured participants would make no crossing associations between a particular prime-type and a particular voice if we were to randomise prime-type and stimuli.

Experiment one briefed participants to listen to English words that may be real or not real and asked them to press key 'n' for 'no' (not a real word) or key 'y' for 'yes' (real word). The instructions also suggested that participants use two hands for this experiment, keeping one index finger on key 'y' and the other on key 'n'. Participants were able to take a break between each block and were prompted to rest before they continued. In turn, allowing us to minimize listener fatigue. The full experiment took an average of 20 minutes to complete. A questionnaire followed the listening portion of the experiment. It concerned participant background and attitudes towards Australia and New Zealand. Last, we gave participants a debrief to read and a secondary 'yes' or 'no' consent option.

3.1.5. Recruitment and Participants

The experiment was available to participants on the internet, and we recruited participants through personal connections and paid Facebook advertising. The advertisement described the experiment as a word recognition task. Participants were told that we were interested in seeing how New Zealanders hear words. It asked that participants wear earphones/headphones when doing the experiment. The advertisement also included an incentive of NZD$10 e-voucher which eligible participants could optionally claim. We limited the experiment to NZE speakers who have no hearing impairments, learned English as one of their first languages, and have lived in New Zealand since the age of seven with no extensive gaps (over 1 year).

Following our preregistered criteria, we excluded data which was 2.5 sd outside the mean; if a participant answered *yes* to having a hearing impairment; if the participant did not learn English from birth and/or if participants were not living in New Zealand since before the age of seven; or have lived outside New Zealand for more than a year. Furthermore, we excluded unreliable data if the filler words were answered below 68% accuracy (2.5 sd from the mean). After excluding outliers, 119 participants were eligible for analysis (one short of our pre-registered minimum of 120). Their distribution across conditions is shown in Table 4.

Table 4. Distribution of participants across conditions (number of men shown in parentheses).

	Speaking	**Incidental**	**Total**
Baseline	24 (6)	30 (12)	54
Priming	27 (5)	38 (12)	65
Total	51	68	119

3.2. Preregistered Predictions

These are the predictions that appeared in our preregistration (we have reworded these slightly to align with the terminology we have adopted in this paper, but we have not changed the predictions. See https://aspredicted.org/rw4kq.pdf (accessed on 18 May 2022)):

1. We expect the kangaroo to increase 'yes' responses to KIT-frames and decrease them to DRESS-frames.
2. We expect the differences between the animal primes to be greater in the priming-speaking than in the prime-incidental condition.
3. We do not expect differences between the all-horse baseline presentation-types or the different horses within these conditions.
4. There may be block and trial effects.
5. Predictions (1) and (2) may be mediated by listener gender or Australian English experience or Attitudes toward Australians.

3.3. Statistical Approach

The effect of the prime was not retained as significant in our preregistered model:

(a) Grouping Prime-Speaking and Prime-Incidental Conditions together in one model (with PrimeType having 3 levels—horse, kiwi, kangaroo), and grouping AllHorse-Speaking and AllHorse-Incidental Conditions together in another (with PrimeType having 3 levels—horse1, horse2, horse3) we will test: YES ~ PrimeType × StimulusType × PresentationType + Block × Trial-within-Block + (1 + PrimeType + StimulusType + PresentationType | Speaker) + (1 + PrimeType + StimulusType | Listener) + (1 + PrimeType + PresentationType | Word).

However, while our preregistration indicated a possibility that there would be block effects, our preregistered model did not allow for the possibility that the effect of the animal might differ strongly across blocks. Exploration of the data indicated a strong effect of prime in the first block only, which then seemed to persist through the remaining two blocks. As such, the effect of the animal on the screen in the later blocks appeared counter to hypothesis (e.g., if the kiwi was shown first, the effect of the kiwi was still evident in the data when the kangaroo was later shown), cancelling out any effect in any model that did not take this persistence into account.

In order to assess the significance of these apparent effects, we modified our planned modelling procedure to take into account two non-planned factors—a binary category indicating whether the block is the participant's first block, and a category indicating which animal was the animal displayed to the participant during the first block.

Two separate generalised linear mixed-effects regression (glmer) models were implemented for the baseline condition and the priming conditions data, initially starting with the same effects and interactions. We used a backwards stepwise procedure involving ANOVA comparisons find the best model per dataset.

The dependent variable in the model is our estimate of what vowel our participants appeared to hear in the experiment—either a KIT or DRESS vowel (we note this dependent variable also departs from our preregistration, which planned to model whether the participant answered). This estimate was coded as KIT-consistent if participants answered 'yes' to a vowel in a KIT frame (such as dXgging), or 'no' to a DRESS frame (such as sXnding). The opposite set of answers (no to a word like dXgging and yes to a word like sXnding) were coded as DRESS-consistent. It should be noted that this process involves some assumptions and likely over-simplifications. When someone answers 'no' to dXgging, they may have heard it as 'degging', but it is also possible that they heard an alternative vowel, such as 'deeging'. The results from Shaw et al. (2018) [31] (outlined above), show that the most common mishearing of an Australian KIT vowel is DRESS, and vice-versa, but it is not the only mishearing. We also conducted preliminary exploratory modelling that treats the DRESS-frame and KIT-frame words separately and found the same key results as reported in this paper. Grouping the results together into the same model leads to greater clarity and fewer models. In all cases *frame* × *prime* is tested, to allow for the possibility that the responses to the two frame types should be treated separately.

Two sets of interactions were tested in the modeling procedure. The first set involved the prime present on the screen. However, given that experiment one was a within-participant blocked design, and following exploratory analysis, we also wanted to allow for the possibility that the prime presented in the very first block would have a pervasive effect. We therefore also included a set of interactions involving the identity of the first prime.

We fit down from:

KIT-consistent-response ~ primetime × presentationtype × (firstblock + frametype + orderwithinblock) + firstprime × presentationtype × (firstblock + frametype + orderwithinblock)

Random effects of *id* (individual participants) and *word* (stimuli) were included. Slopes were explored but led to non-convergent models, and so were dropped following the preregistered procedure, which was retained for the baseline model, and retained through most of the modeling for the priming model, but then dropped to obtain convergence. Early modeling included speaker intercepts representing the 3 different speakers, but these led to multiple convergence issues and explained little variance, so they were dropped (dropping slopes to fix convergence problems was anticipated in the preregistration).

The fitting procedure involved iteratively removing first interactions then main effects and comparing minimally different models via ANOVA comparisons. If an interaction or main effect did not lead to a significantly improved model, it was excluded.

3.4. Results

The baseline model revealed no significant interactions, and two main effects. The main effects were first block—with responses in the first block leading to more KIT-consistent responses, and frame-type, with KIT frames eliciting more KIT-consistent responses. Importantly, which of the three horses was presented had no significant effect, consistent with prediction (3).

The prime model is shown in Table 5. Like the baseline model, it also includes the increased KIT-consistent response in the first block, and a bias toward more KIT-consistent responses for KIT-frames. It contains an additional order effect, in which the proportion of KIT responses reduces slightly over the course of each block. Presentation-type is not retained as significant. The non-significance of presentation-type contradicts prediction (2) that a prime presented 'speaking' the stimuli should invoke greater perceptual shift in listeners.

Table 5. Selected model for experiment one (dependent variable = KIT-consistent response. Intercepts = participant and word).

	Estimate	Std. Error	z Value	Pr (>|z|)
(Intercept)	−0.469	0.20714	−2.264	0.03
firstblock = yes	0.34007	0.07018	4.845	<0.0001
scaled order within block	−0.0691	0.03283	−2.105	0.04
frame = KIT	2.07562	0.21986	9.441	<0.0001
firstanimal = kanga	−0.2732	0.24416	−1.119	0.26
firstanimal = kiwi	−0.8441	0.23989	−3.518	<0.0001
frame = KIT: firstanimal = kanga	0.06594	0.16007	0.412	0.68
frame = KIT: firstanimal = kiwi	0.88564	0.16411	5.397	<0.0001

Counter to prediction (1), there is no significant effect of the prime-type. However, while the on-screen prime had no significant effect upon participant responses, we did find a significant interaction between the prime presented in the first block and the frame-type. The first prime listeners encountered in the experiment shifted their perception boundary for DRESS words and this persisted in their responses for the remaining two blocks. Once listeners began hearing the stimuli a certain way, it remained this way for the rest of the experiment.

This interaction is shown in Figure 5. With 'percentage KIT-consistent responses' on the y-axis, if our prediction that the kangaroo prime will induce greater KIT responses (cf. prediction (1)) is borne out in the data, then the blue kangaroo point would be the highest out of three prime points on the y-axis. Likewise, we expect the green kiwi point to be the lowest out of the three prime points because we predicted that the kiwi prime would induce fewer KIT responses (and inversely, greater DRESS responses). The graph is divided by DRESS frames and KIT frames on the x-axis to show the significant interaction. Looking first at the DRESS frame, Figure 5 suggests that the kiwi prime shifted listener perception in the predicted direction. In other words, listeners who saw the kiwi prime in the first block were more likely to give DRESS-consistent responses.

We note that the horse is not positioned between the kangaroo and the kiwi. Rather, the horse elicits the most KIT-consistent responses, and is not significantly different from the kangaroo. The kiwi elicits the fewest KIT-consistent responses, and relevelling the model confirms that it is significantly different both from the kangaroo and from the horse baseline (with kiwi as intercept: horse est = 0.84, $p < 0.001$; kanga est = 0.57, $p < 0.05$; KIT × horse est = −0.89, $p < 0.0001$; KIT × kanga est = −0.82, $p < 0.001$). Thus, while the difference between the kangaroo and the kiwi is as predicted, the inclusion of the horse reveals that the effect is being driven by the kiwi.

Turning now to the KIT frame, Figure 5 shows a likely ceiling effect for all three primes. This effect suggests that when presented in a KIT frame, listeners hear the ambiguous stimuli as KIT regardless of what prime they saw in the first block.

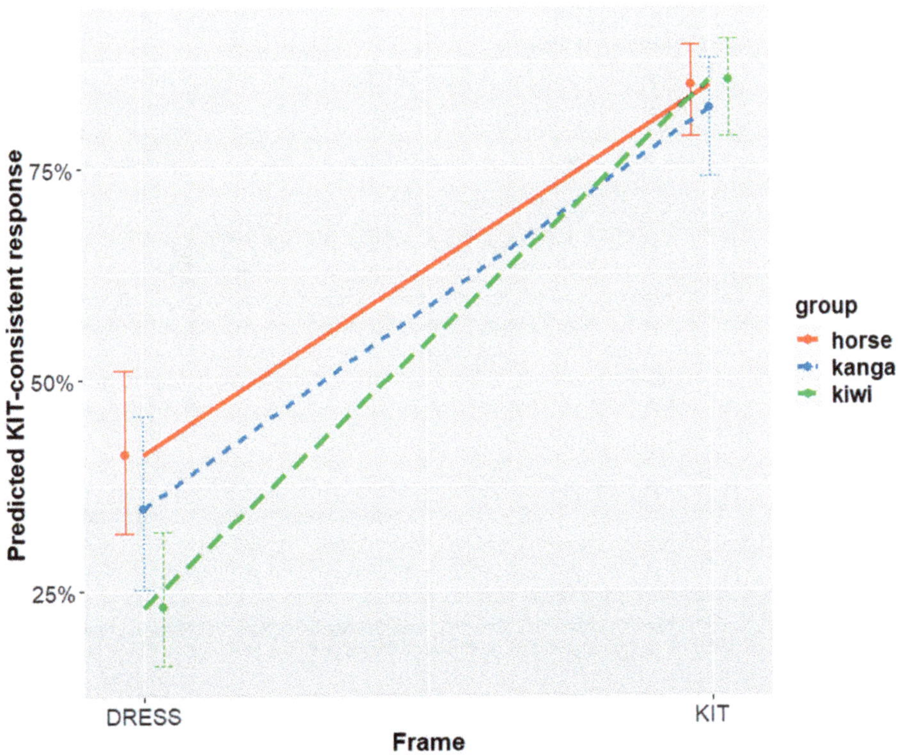

Figure 5. Model prediction plot from the model of experiment one (see Table 5).

3.5. Interim Summary

This experiment supported our overall hypothesis that participants would report hearing more KIT-consistent vowels when primed with a kangaroo than a kiwi. However, it had several limitations. First, the within-participant design did not work. Once exposed to a prime, participants maintained their behaviour throughout the rest of the experiment. Second, the stimulus was clearly not ambiguous enough. Responses were overwhelmingly associated with 'KIT', even when for the DRESS-frames which are expected to induce a perception of DRESS due to the Ganong effect [36]. Likewise, the KIT frames were responded to positively at a rate close to ceiling. A third limitation is that the significant difference between the kangaroo and the kiwi is unlikely to be caused by a strong priming effect of the kangaroo (as predicted). Rather, the greatest departure from the baseline horse responses is observed in the kiwi condition; the difference between the horse and the kangaroo is not significant and is in the opposite direction as predicted. One interpretation of the greater effect of the kiwi prime may be that listeners did not begin the experiment in a 'New Zealand English' listening mode. This response could be caused by the online environment, in which many accents are heard, together with the acoustic properties of the target vowel which (by design) did not actually match either NZE or AusE completely. Thus, the default in this context may have been to expect an 'other' accent, and it is only the presentation of the kiwi that triggers a more NZ-like listening model and accepts the ambiguous vowel as a viable 'DRESS'. We will return to this suggestion in the discussion.

While this experiment provides some overall support for the hypothesis that kangaroos and kiwis can elicit different speech perception behaviors, it also has the above limitations. We attempt to rectify these issues and replicate the key result in experiment two.

4. Experiment Two

4.1. Methods and Materials

A modified second experiment was conducted as a follow up to the first experiment. In particular, given that the blocked design was not successful, we switched to a cross-participant design in which each participant was only exposed to one prime. The experiment remained a lexical decision task with some changes explained below.

4.1.1. Conditions and Prime Types

We changed the experiment to be a cross-participant design, where listeners would only see one prime-type. Furthermore, we excluded the baseline condition (three horse primes) given that we know the condition is performing as expected and is having no influence on listener perception. We maintained the speaking/incidental condition to ensure its influence, or lack thereof, in listener perception. Participants thus saw one of six images (a horse, kiwi, or kangaroo, in either speaking or incidental condition). Participants all responded to the same stimuli, in the same voice, but in different random orders.

4.1.2. Stimuli

Given the apparent ceiling effect of the perceived KIT vowel, we decided to use stimuli that were at step 4 of the vowel resynthesis (refer to Figure 2 to see synthesis step 4 in the continuum). This stimulus is a step closer to an Australian KIT (and thus New Zealand DRESS) vowel; it is more likely to be heard as DRESS by our NZE participants and should therefore reduce the KIT ceiling effect. Speaker NZ1 was the voice used for all stimuli. The total number of word-types were dropped from 180 in experiment one to 160 in experiment two (80 target stimuli and 80 filler stimuli).

4.1.3. Online Word Recognition Task

Experiment two followed the same procedure as experiment one: that is, using the *Speech In Noise 2* program to run the experiment online, collating data to a new Firebase console. Like experiment one, participants were asked to listen to English words which may be real or not real and asked them to press key 'n' for 'no' (not real word) or key 'y' for 'yes' (real word). Following the listening task, we added six post-listening questions regarding the prime pictures—"What animal is this?" (where participants would type an answer), and "Does this animal suggest any particular country to you?" (where participants could select one from multiple answers). The addition of these questions reinforced that the prime animals were being accurately identified, and associated with the target country (e.g., the kangaroo prime could potentially be confused as a wallaby, which lives in both Australia and New Zealand). The results of this task showed high identifiability of the animals, and reliable associations with the target country for the kiwi and the kangaroo. The same personal and attitudinal/exposure questionnaire used in experiment one was used to finish this experiment.

4.1.4. Recruitment and Participants

We used the same recruitment, payment and inclusion criteria described in experiment one. After filtering out responses from participants who did not meet the criteria, and those whose data were 2.5 sd outside the mean, data from 136 participants were eligible for analysis. The distribution of speakers over conditions can be seen in Table 6.

Table 6. Distribution of participants across conditions. Number of men indicated in parentheses.

	Speaking	Incidental	Total (by Prime)
Kangaroo	18 (6)	28 (9)	46
Kiwi	20 (8)	21 (7)	41
Horse	26 (7)	23 (5)	49
Total (by condition)	64	72	136

4.2. Statistical Approach

We did not separately preregister experiment two. However, in our modeling, we undertook a two-step procedure that followed the spirit of our original preregistration. In a first step, we modeled overall effects, without regard to social factors, following (the spirit of) the model in our original preregistration. In a second step, we attempted to investigate any mediating effects of social factors. The modeling procedure for this step is described in detail in the Supplementary Materials (available here: https://github.com/jenniferhay/kangaroo-kiwi (accessed on 1 June 2020)). The data contained significant four-way interactions, which ultimately led us to a modeling procedure which dealt with the men and the women in separate models. We considered effects of *prime-type*, *order* and *presentation-type*, in addition to social characteristics of *age*, *gender*, *experience* (amount of time spent in Australia), and *attitude* (numerical scale based on post-questionnaire attitude questions). Random effects of *id* (individual participants) and *word* (stimuli) were included. Like experiment one, model slopes lead to convergence issues and are not pursued.

As described below, both the men's and women's models included effects of prime-type and order within the experiment. The women's data also included significant interactions involving social characteristics.

4.3. Results

4.3.1. Primary Results (No Social Factors)

We did not update our preregistration between experiments one and two, but in this analysis, we largely followed the model preregistered for experiment one.

The preregistered fixed effects for the experiment one model were:

primetype × frametype × presentationtype + block × trial-within-block.

Experiment two did not contain blocks, so block was not included in the experiment two model. Informed by the results of experiment one, we also wanted to allow for the priming effect to evolve over the course of the experiment. To this end, we included trial order within the interaction rather than as a separate effect, testing a four-way interaction between *presentation-type, order, frame-type and prime-type*. We simplified the random effects to obtain convergence, following the procedure outlined in the preregistration. The effect of order was scaled and centred.

Pruning this model, we observed a significant effect of prime-type, in interaction with frame-type and order (shown in Table 7). Interactions involving presentation-type resulted in significant improvement to the model, but these models failed to converge. Checking separate models of the different frame-types yielded convergent models with no significant effect of presentation-type. Presentation-type was then dropped from the model.

Table 7. Selected model for experiment two (dependent variable is KIT-consistent response. Intercepts are participant and word).

	Estimate	Std. Error	z Value	Pr (>|z|)
(Intercept)	−1.486276	0.249029	−5.968	<0.0001
primetype = kanga	0.132854	0.241932	0.549	0.58
Primetype = kiwi	0.2711	0.249015	1.089	0.28
scaled order	−0.33445	0.061424	−5.445	<0.0001
frame = KIT	2.836929	0.274642	10.33	<0.0001
prime = kanga: scaled order	0.290842	0.086805	3.351	<0.0001
prime = kiwi: scaled order	0.286136	0.089319	3.204	0.002
prime = kanga: frame = KIT	0.174694	0.130499	1.339	0.18
prime = kiwi: frame = KIT	−0.002354	0.132787	−0.018	0.99
scaled order x frame = KIT	0.028814	0.087952	0.328	0.74
prime = kanga: scaled order: frame = KIT	−0.075548	0.125966	−0.6	0.55
prime = kiwi: scaled order: frame = KIT	−0.388737	0.129347	−3.005	0.003

The significant three-way interaction can be seen in Figure 6. Contrary to the result from experiment one, the primary difference between the kangaroo prime and the kiwi prime is now found in the KIT-frame rather than the DRESS-frame. In the DRESS-frame, while there appear to be some differences involving the horse, the kangaroo and kiwi are not different from each other. In the KIT-frame, however, we see a general decline through the experiment in the KIT-consistent responses—except in the case of the kangaroo, which maintains a consistently high rate of KIT-consistent responses.

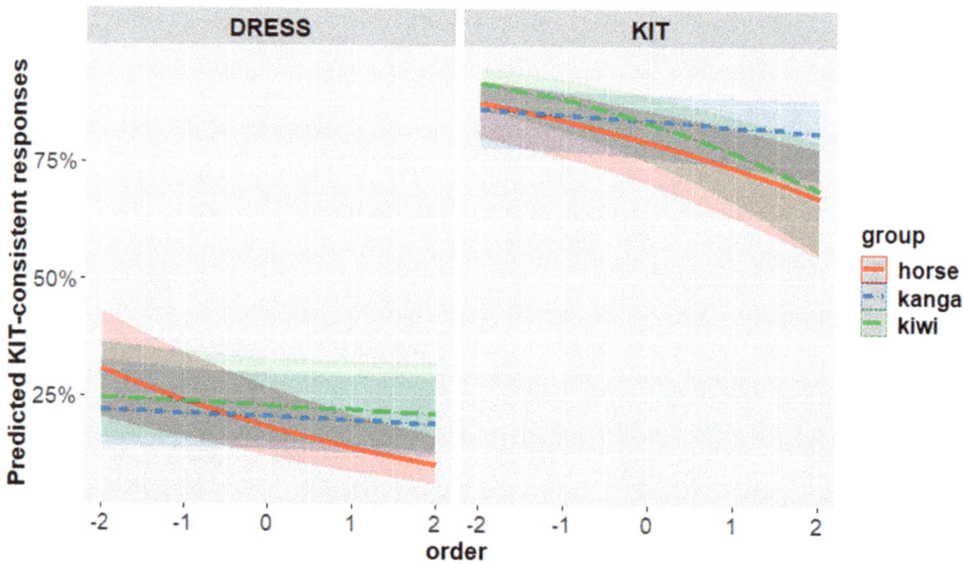

Figure 6. Model prediction plot from model of experiment two (see Table 7). Three-way interaction of order (x-axis), primetime (lines), and frame (panels), predicting percentage of KIT-consistent responses.

Thus, this model provides further support for predictions (1) (the effect of the animal prime on perception), and (3) (the possibility of order effects). It does not, however, support prediction (2) (the stronger effect in the speaking rather than incidental conditions).

4.3.2. Social Factors

Our preregistration also flagged social factors as an interest, and prediction (4) noted that we expect certain social factors may be playing an important role in this work. We explore the roles of these factors in the analysis below. The models we pursued involved fitting multiple subparts of the data to reveal sub-regularities without striking convergence issues. While the general social factors explored echoed those anticipated in the preregistration, the specific models were more complex; the preregistration planned simple interactions between individual social factors and the prime, without consideration of a mediating effect of order or the interactions between social factors themselves. Because we deviated significantly from the planned structure, and found some unpredicted effects for social characteristics, the model fitting reported in this section should be regarded as strictly exploratory.

Our attempt to explore social factors revealed some differences in the men versus the women in the experiments (two nonbinary participants were omitted from this secondary analysis—see models and model fitting procedure in Supplementary Materials), resulting in separate models for the men and women. Due to gender imbalance in the participants, the model for the women contains more than twice as many participants as that for the men.

In addition to the overall KIT-frame result found in Section 4.3.1, the men also showed an effect in the DRESS-frame that emerged over the course of the experiment. In this model, we observed increased KIT-consistent responses to the kangaroo and decreasing KIT-consistent responses to the kiwi (Figure 7—consistent with prediction (1)).

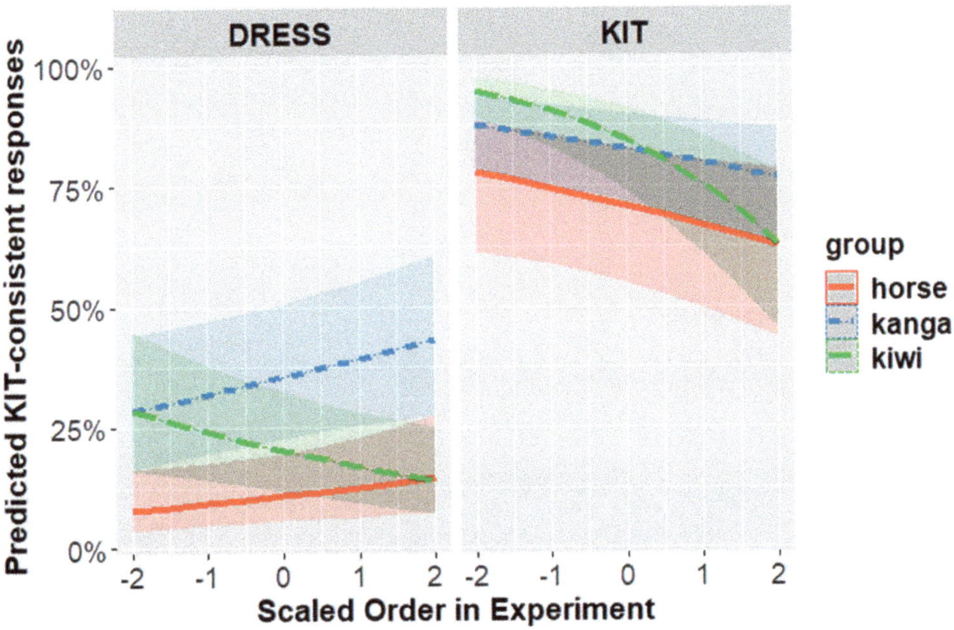

Figure 7. Model prediction plot from the model fit to men only (model details in Supplementary Materials). The figure plots a set of 3 two-way interactions involving order (x-axis), prime-type (lines), and frame (panels).

For women, the priming effect appears to be mediated by time spent in Australia. Listeners who have spent more than a month in Australia show the predicted effects for KIT and DRESS. Listeners who have spent less than a month in Australia, however, show no effect for KIT and an effect in the non-predicted direction for DRESS (Figure 8). A shift in the non-hypothesized direction by any group of participants was not predicted, but it is consistent with some results in the wider literature that will be discussed below.

Prediction (4) suggested that any observed priming effects may be "mediated by listener gender or AusE experience or Attitudes towards Australians". Exploration of these social factors supports that these effects do not behave the same way for all participants, consistent with earlier work (cf. Hay, Drager and Nolan 2006; [1]). However, as these results reflect modeling that incorporates a number of social factors, which carved data up in a number of ways, they should be taken as tentative. It is certainly not guaranteed that another sample would yield these same interactions. What the interactions suggest, however, is that different participants or groups of participants respond to the stimulus and primes somewhat differently. It is likely that the locus and even direction of the effect may be mediated by the relationship between the stimulus acoustics, the participant's own production, and the participant's previous experiences and stereotypes.

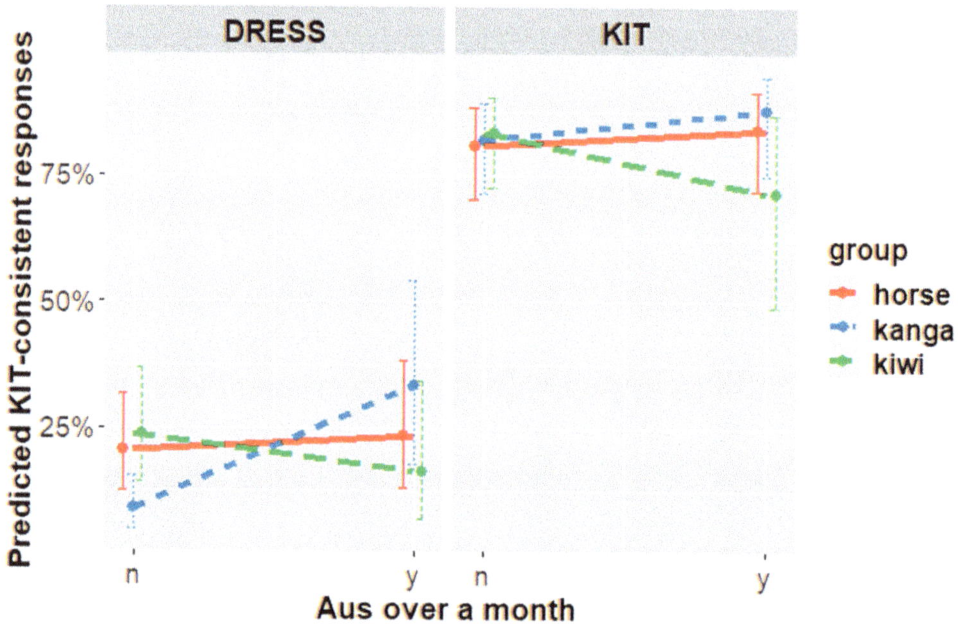

Figure 8. Experiment two model prediction plot from model fit to women only (model details in Supplementary Materials). The figure shows a three-way interaction involving time spent in Australia (binary—on x-axis), prime-type (lines), and frame (panels).

5. Summary

Experiment one showed a strong priming effect, in which the kiwi (relative to the kangaroo) elicited fewer KIT-consistent responses (i.e., answered 'yes' more) to vowels in a DRESS-frame. This effect persisted through blocks and was not overridden by subsequent presentations of other animals. In this experiment, the kangaroo appeared to act more similar to the horse, and the kiwi looked to be the image that produced different behaviors in the participants.

Experiment two changed to a cross-participant design, with a different stimulus that aimed for a greater level of ambiguity between DRESS and KIT. In this experiment, our overall analysis found an effect in the KIT-frame where the kiwi aligned with the horse and the kangaroo elicited a different response, emerging over the course of the experiment.

The exploration of social factors in experiment two also indicated some differences across participant subgroups. Men showed priming in the predicted direction for both frame-types in an effect that emerged over the course of the experiment. Women with some exposure to AusE showed the expected effect for both KIT and DRESS frames. However, women with very little exposure showed no effect for KIT frames and an effect in the wrong direction for DRESS frames.

In terms of the preregistered predictions, we can summarize the results as follows.

1. We expect the kangaroo to increase 'yes' responses to KIT-frames and decrease them to DRESS-frames.

Overall, this prediction was supported. In both experiments, we did find a difference between the animals in the predicted overall direction, with the kangaroo eliciting more KIT-consistent responses than the kiwi. In experiment one, the initial presentation of the kiwi (relative to the kangaroo AND the horse) elicited more 'yes' responses to the DRESS-frame in an effect that persisted across blocks. In experiment two, the kangaroo elicited more 'yes' responses (relative to the kiwi and the horse) in the KIT-frame, a result that emerged over the course of the experiment.

2. We expect the differences between the animal primes to be greater in the priming-speaking than in the prime-incidental condition.

   We found no evidence in support of this prediction.

3. We do not expect differences between the all-horse baseline presentation-types, or the different horses within these conditions.

   This prediction was supported.

4. There may be block and trial effects.

This prediction was supported. The block and trial effects were significant, and incorporating them into our analysis required deviation from our preregistered modeling procedure.

5. Predictions (1) and (2) may be mediated by listener gender or Australian English experience or attitudes toward Australians.

This prediction was tentatively supported. Separate models fit to data from male vs. female participants show different results. The men's data were consistent with our overall prediction (1), whereas the women's data were mediated by experience with Australia. Results from this social analysis should be regarded as tentative; however, because they were found through extensive post hoc exploration of the data.

6. Discussion

When we consider our two experiments together, we find good evidence that priming with regionally-associated animal images can affect patterns of speech perception. However, the location and nature of such effects seem to rely on alignment between participants' own productions, the nature of the acoustic stimuli, as well as experimental design. In this section, we explore several issues arising from the results reported above.

One thing to consider here is the social and physical differences between our primetype cartoon drawings and that of physical stuffed toys used in Hay & Drager (2010) [1]. The design of our experiment meant listeners were participating virtually while looking at a virtual animal, unlike Hay & Drager (2010) [1] who had in-person participants and handleable stuffed toys. These physical toys could potentially have carried greater social weight for the participants, particularly because these types of toys are often bought as souvenirs. Therefore, it would not be hard to imagine that participants realise some sort of social significance towards the toys and any inherent history that may belong to them. Our line-drawing prime-types are not necessarily associated with exactly the same social meaning. However, despite this, we still see some evidence of social priming with the online line drawings.

6.1. Lack of an Effect of Incidental vs. Speaking

A number of different projects have reported effects of expectations or beliefs about the speaker. If a listener is led to believe that a speaker is male vs. female [3], Canadian vs. American [9], older vs. younger [6], or working-class vs. middle class [5], these beliefs can affect their reported speech perception patterns. Munson et al. (2017) [37] show that the effect of gendered expectations on the perception of fricatives is weaker when the gendered expectation is implied (via what was said), as opposed to explicit (via photos).

The specific pair of papers we set out to pseudo-replicate found similar patterns of priming by writing 'Australian' vs. 'New Zealander' on the response form [10], and by priming with stuffed toy animals in the room [1]. Even though the strength of the effects in Hay, Nolan, Drager (2006) [10] and Hay & Drager (2010) [1] did not appear markedly different, we nonetheless expected that if the animal was represented as speaking, it would have a stronger effect than if it was incidentally presented on the screen. The intuition behind this prediction was linked to the fact that there is an extensive literature (cited above) showing that beliefs about a speaker can influence speech perception, whereas there are few studies that have looked at incidental social priming. While a greater number of studies need not necessarily imply a stronger effect, Munson et al. (2017) [37] finding that somewhat more explicit priming of gender reveals a stronger effect is consistent with our prediction.

This is the first study that used both types of priming in the same task to overtly test whether the effects are of the same type and magnitude. With regard to speech processing, the literature holds more work explicitly exploring the role of the speaker than studies about incidental priming patterns, or the differential effects reported by Munson and colleagues. Considering the available information, we predicted stronger effects might be observed when our animals appeared to be talking; this prediction was preregistered (as prediction (2)). Indeed, notwithstanding the literature, we would also presume that there is a significantly increased utility of adjusting one's speech perception strategy in response to the speaker, as opposed to responding to seemingly unrelated images or objects. This informal assessment gave rise to the expectation that we would see a difference between the speaking and the incidental conditions.

However, across both experiments our final models did not retain presentation-type as a viable predictor. The non-significance of presentation-type reinforces the interpretation that what we are seeing is a relatively automatic priming effect in which animal images are sufficient to activate the social category of 'Australian' or 'NZ'. This activation, in turn, likely progresses jointly with speech memories and/or models that are indexed to these social categories.

Of course, there are limits to our 'speaking' condition that must be considered in interpreting this result. While film, media and television have acclimatized us all to the idea that cartoon animals come from particular places and have particular accents, it is also true that most of us know that the voices produced by cartoon animals are actually voiced by a third-party voice-actor. That is, no one really believes that these animals are actually producing the speech that we hear. Even in our 'speaking' condition, then, our participants would not truly believe that they are listening to a voice actually produced by the animal. In that sense, perhaps the use of animal images makes both of our conditions somewhat 'incidental'—that is, the listener knows (at least at a subconscious level) that the voice has been produced by someone not seen on the screen. For a truer 'speaker' condition, we would need photos of actual people, and to lead the listener to believe the voice is truly produced by those people.

We thus have good evidence for the contribution of incidental priming, and that it does not matter whether an animal is represented as speaking or not. A stronger comparison of incidental vs. speaking effects would require an altered methodology, which must include "speakers" that are not so easily ruled out or dismissed.

6.2. The Acoustics of the Stimulus and the Locus of the Effect

There is a clear Ganong effect in the results, wherein listeners are more likely to interpret our ambiguous vowel as a phoneme that would complete a real word [36]. The average by-item 'yes' response for words containing the ambiguous vowel was 71%. Some word frames elicited this effect much more strongly than others. The frame most often identified as a real word was 'tXxture' (97%), whereas the least was 'thXnnest' (17%). A large amount of variation is no doubt linked to coarticulatory effects, word frequency, and the distinctiveness of the phonological frame. Some are also linked to specifics of the wider frame. We noted, for example, low rates of 'yes' responses to some filler items that should have been heard as words for both DRESS and KIT, such as 'bXdder'. Double-checking of these recordings revealed that the second syllable was produced with a degree of stress and a NURSE vowel rather than a schwa vowel, which would be more common in NZE, likely leading listeners to hear it as a nonce compound. We did not exclude any items from our analysis but rather relied on the random intercept for word to have controlled for this variation. It is very possible that priming occurred to greater or lesser degrees in some words depending on the strength of bias of other factors.

One respect in which our experiments differ from Hay, Nolan & Drager (2006) [10] and Hay & Drager (2010) [1] is that we have included a baseline condition. Adding this condition enables us to tell not only whether there is a difference between the kangaroo and kiwi conditions, but also whether one or both differ from a baseline in which there is no strong regional prime. While we found differences between the kangaroo and the kiwi in both of our experiments, the relationship with the baseline differed. In experiment one, the kangaroo and the horse elicited similar responses, and the kiwi elicited more DRESS-like responses. The direction of this prime is in the predicted direction, but the fact that it is the kangaroo (rather than the kiwi) that patterns with the baseline was not predicted.

Our interpretation of this result is that listeners may not have initially engaged with the task expecting that they were listening to a New Zealander. The task was conducted online, where the majority of voices one hears are probably not NZE voices. The voice encountered in our stimuli presents an ambiguous vowel which is not exactly typical of NZE or AusE, but instead exists in a perceptual space utilized by some more remote dialects (cf. [31] data, above). In this interpretation, listeners do not automatically/necessarily engage with the voice as if it is a New Zealander, except in the case where they are presented with a kiwi prime. The kiwi prime elicits a more NZE like listening mode and, thus, more 'DRESS' responses. This interpretation is also consistent with the high KIT ceiling-effect observed in experiment one: many dialects produce KIT somewhere in the range between the NZE and the AusE variant, but none of the other dialects in Figure 1 contain a DRESS vowel in that vicinity. The consequence with respect to priming is that a KIT-consistent response appears to be the default. Priming mainly occurs when the DRESS-frame and the kiwi prime are combined, two factors that together can increase a DRESS-consistent response.

Experiment two shifted the ambiguous stimulus to be higher and fronter in the vowel space. The stimulus is now not so close to a NZE KIT vowel, but close to an AusE KIT vowel and a NZE DRESS. This change had the overall effect of slightly increasing DRESS-consistent responses. Overall, the responses shifted from 57.5% KIT consistent in experiment one (78.6 for KIT frames and 36.5% for DRESS frames) to 50.4% KIT consistent in experiment two (73% to KIT frames and 27.9% for DRESS frames). In the main analysis of experiment 2 we see the locus of priming occurring when we have a combination of the KIT-frame and the kangaroo. It seems likely, then, that shifting toward a stimulus that is closer to something a New Zealander produced increased the degree to which listeners engaged with the speaker as a New Zealander; the priming effect which shifts perception away from this mapping is driven by the kangaroo. In each experiment, the prime that differs from the horse is the one that is aligned with the frame (kiwi for DRESS-frame and kangaroo for KIT-frame). A categorization that is consistent with the frame (i.e., resulting in a word) is further facilitated by priming with regional primes.

6.3. Effects of Blocks and Order

In experiment one we attempted a within-participant design, expecting that participants' responses to our stimuli would change when we altered the visual prime. Across three blocks we changed the voice of the speaker, and the image that was being looked at. However, while the perceptual behavior exhibited in the first block showed evidence of priming across participants, this influence upon perception appeared to persist for participants across subsequent conditions for the remainder of the experiment.

While we used three different voices in experiment one, we would like to draw attention to the fact that these voices were not radically different from one another. They were all synthesized combinations of the same Australian female, together with each of three young New Zealand female speakers. While our impression was that the speakers were distinguishable, the speaker change across blocks may not have been particularly apparent to our listeners, who were simply working quickly and focused on finishing the task at hand. Switching to radically different voices between blocks may have better facilitated a 'reset' of the listening behavior for participants, and thus allowed for increased influence through a new prime. We included some informational screens between blocks, but the pause between blocks will have been minimal for most participants. A between-blocks distractor task that was completely unrelated to the experiment may also have facilitated a more successful within-participant design.

This type of effect, involving initial exposure conditions persisting through later conditions, have been well-documented. Bordens and Abbot (2002) [38] describe such carryover effects as one of the most serious disadvantages of a within-participants design. Recent examples involving speech perception include Hay et al. (2017) [30], who examined whether the vowel boundary on a DRESS-TRAP continuum was affected by the location of the listener. The authors found a significant difference between the locations of the first completion of the task, but this perception persisted to the second location. Hashimoto (2019) [39] looks at social (topic-based) priming in a speech production experiment and shows persistence effects across blocks.

When we shift to an across participant design in experiment two, we see an effect emerging over the course of the experiment. A general trend towards decreasing KIT responses is absent when the kangaroo is the prime-type, leading to a significant separation at the end of the experiment between the kangaroo condition and the kiwi/horse conditions. This order effect may seem to contrast with experiment one in which we reported a persistent effect, with no effect of order. However, in posthoc analysis, we can see a trend toward a non-significant order effect in experiment one as well. The difference between the two experiments is likely the length of the analysis period and the exposure—just 30 target items in block one of experiment one, vs. 80 for experiment two, lending more granularity to an analysis of order.

6.4. The Role of Experience versus Stereotypes

While the priming generally shifted responses in the predicted direction, there is one pocket of data where effects were observed in an unanticipated direction. Women with little prior experience of AusE (i.e., have spent less than a month in Australia), showed no priming of KIT word-types, and an effect in the unpredicted direction for DRESS word-types. That is, when the frame was DRESS, they were more likely to answer 'yes' when looking at the kangaroo than the kiwi or the horse. While this result was not predicted and should be regarded as tentative, it is nonetheless consistent with some previous results in the literature that seem to reflect some New Zealanders holding a false stereotype of a high Australian DRESS vowel.

Ludwig (2007) [40] shows that, when presented with words in isolation and asked to rate them as produced by a New Zealand or Australian speaker, New Zealanders are very good at rating them when it comes to KIT but not when it comes to DRESS. In her study, Ludwig suggests that New Zealanders are prepared to accept both the New Zealand and the Australian variant as a NZ DRESS, unless it was nasalized, in which both nasalized variants were perceived as Australian. This pocket of the data, then, connects to a thread of literature suggesting that primes can sometimes activate stereotypes in the absence of extensive experience with a dialect.

False stereotypes about AusE DRESS are likely to be less common among individuals who have more direct experience with AusE. There is some evidence that differences in experience influence behavior in production tasks. Sanchez, Hay & Nilson (2014) [41] explored the effect of conceptual activation of Australia on New Zealanders' vowel productions, using both corpus analysis and a word-list elicitation task. In the corpus analysis, they found that talk about Australia shifted KIT and TRAP in a more Australian direction. Both the corpus and the wordlist reading showed an unexpected interaction for DRESS. Speakers who have experience with AusE produced more Australian-like DRESS in contexts that were primed with Australia. However, speakers with less Australian experience actually produced less Australian-like vowels in the Australian-like context. Sanchez, Hay

& Nilson (2014) [41] argue that the non-experienced New Zealanders are influenced by a false stereotype that Australians produce high DRESS vowels.

We have three possible explanations for why this effect was seen within the women but not the men. One is simply that we do not have many men in our study and might observe a more congruent effect through further sampling. A second explanation is that there may be some genuine difference in stereotypes, experience or attitudes that align with gender. This explanation would be consistent with the gender differences also observed in Hay, Nolan and Drager (2006) [10] and Hay & Drager (2010) [1], and the exploration of attitude by Walker et al. (2018) [24]. Finally, a third possibility is that there may be a difference in the participants' own production that leads to different behaviors in the task. There is certainly evidence that individual variation in production of various sounds can be related to their perception [42–44]. We have seen between experiment one and experiment two that a subtle shift in the acoustics of the stimulus matters. It seems possible, then, that subtle differences in the participants' own production could influence their perceptual processes. Based on the literature exploring DRESS raising within New Zealand, we might expect the men to have lower DRESS vowels than the women (e.g., [45]), thus making the stimulus a potentially close match to their own DRESS vowel than the women.

Indeed, for the participants in the experiment 2 'neutral' horse condition, the men responded with 41% KIT-consistent responses as opposed to the women with 51%. It is thus likely the men identified the speaker as a New Zealander, rather than an overseas accent with a non–local KIT vowel, causing the baseline and the kiwi to group together. This slightly greater bias toward hearing DRESS may have increased the potential for priming towards KIT with the kangaroo, even in the DRESS frame. In sum, when there is already a KIT bias (experiment one) the kiwi can push the listener more towards hearing DRESS. When there is a DRESS bias (experiment two, for men) the kangaroo can push the listener more towards hearing KIT. In the most balanced of our subsets (women in experiment two), we see some evidence of priming by both animals, however, when the women have little prior experience of Australia this effect appears to be influenced by stereotype rather than experience.

Drilling too far into the shifting loci of the effect would drift into conjecture, though it becomes clear as we shift between different acoustic stimuli and different subpopulations who themselves no doubt have different productions (men vs. women), stereotypes and experiences, that the ability to manipulate perceptions in different directions is inconsistent. The distance of the stimuli from a local production may also lead to differing assumptions about where a speaker is from, which can also influence the baseline in different ways. A study which explores individual participant's productions, and the relevant vowels' distances from our stimuli might help shed light on the dynamics at play here.

6.5. Limitations

In the present work we set out to address the limitations in Hay, Nolan & Drager (2006) [10] and Hay & Drager (2010) [1]. We addressed these limitations by designing a new task that is more clearly linked to speech perception, by incorporating larger participant samples, by preregistering our predictions, and by repeating the task across two different versions of the experiment. In doing so, we found evidence to support the claim that priming with regionally-associated animals can affect speech perception patterns.

However, we departed from our preregistered model in various ways, and in both experiments our reported results involved aspects of exploratory analysis. In experiment one, we needed to isolate the prime presented in block one to explain response patterns later in the experiment. In experiment two, our analysis took into account interactions involving order, which weren't anticipated in the preregistration. In our analysis of social factors (experiment two), we followed a complex model fitting procedure to find models that both converged and represented the patterns we find in the data (see Supplementary Materials, available here: https://github.com/jenniferhay/kangaroo-kiwi (accessed on 1

June 2020)). For strict proponents of preregistration who are not supportive of exploratory modeling, our results will leave something to be desired.

An ongoing challenge for replication is that these types of results are population-specific, including the particulars of participants' own production patterns and previous exposure to AusE. Even if we were to repeat the experiment in Christchurch, sound change involving the DRESS and KIT vowels might lead listeners to have a different relationship with our stimuli. Indeed, there is also an ongoing regional variation with DRESS [46] that may also lead to variable behavior in the experiment, and which we have not accounted for here.

Embedded within our results are many reasons why priming effects may be hard to find, and why it might be hard to see them extended across a range of contexts. These effects can be dependent on the specific stimulus and the individual and are sensitive to the experimental design. However, with that said, the above results support claims that such priming effects do in fact exist. We hope that further work can build upon this experimental methodology to explore relationships between speaker production, speech perception, and social primes.

7. Conclusions

To address a variety of limitations noted in Hay & Drager (2010) [1], we employed an alternate experimental design aiming to replicate their main reported effect. We wanted to shift to a design in which the results could more unambiguously be interpreted as reflecting priming-based differences in speech perception.

The design detailed above shows promise as a means to investigate how contextual factors can influence speech perception. Placing an ambiguous vowel in a Ganong context facilitated the perception of that vowel in a word-consistent manner. A categorization that was consistent with the frame (i.e., resulting in a word) was facilitated by priming with the appropriate regional prime. The overall result across two experiments was that exposure to a kiwi facilitates more DRESS-consistent responses to our stimulus, as compared to a kangaroo which facilitated more KIT-consistent responses.

Exploratory analysis of social factors indicated variability across different groups of listeners. One unexpected example that we noted was that women listeners with little previous experience of Australia tended to show an opposite effect for DRESS. This finding is consistent with reports in the literature that New Zealanders with little exposure to AusE may be influenced by false stereotypes regarding their DRESS vowel.

Overall, our attempt to replicate kangaroos/kiwis as affecting speech perception for New Zealanders was a qualified success. On the one hand, we preregistered a new design and found significant results across two experiments consistent with this interpretation. On the other hand, the modeling at every step involved some aspect that was not fully anticipated in the data, and both experiments contain a subgroup within the data where the priming did *not* occur. The success of the research program, therefore, comes with caveats and caution.

Our own overall interpretation is that these experiments reinforce the idea that such priming effects exist, but also very clearly illustrate the degree to which these effects are mediated by many factors. Such factors are likely to include the experimental design and order effects, the lexical frame, the acoustics of the stimulus, as well as the listener's own production, previous experience, and stereotypes. We would therefore not anticipate that this exact design would necessarily elicit results in Australia, for example, where listeners have different associations with the animals, and experience with the accents. It would not even necessarily elicit the same results if rerun in New Zealand, given the ongoing changes in KIT and DRESS both in NZE and AusE. Exploring the specifics of the boundaries and limitations on this type of priming will provide an interesting avenue for further research.

Supplementary Materials: Data and code to build the reported models and plots are available at: https://github.com/jenniferhay/kangaroo-kiwi (accessed on 1 June 2020).

Author Contributions: Conceptualization, J.H., K.D. and R.P.; data curation, G.H., J.H., R.P., L.M. and A.E.; formal analysis, G.H. and J.H.; funding acquisition, J.H.; investigation, G.H., J.H., L.M. and A.E.; methodology, G.H., J.H., K.D., R.P., L.M. and A.E.; project administration, G.H. and J.H.; resources, J.H.; supervision, J.H.; validation, G.H., J.H., K.D., L.M. and A.E.; visualization, G.H. and J.H.; writing—original draft, G.H.; writing—review and editing, G.H., J.H., and R.P. All authors have read and agreed to the published version of the manuscript.

Funding: This research received no external funding.

Institutional Review Board Statement: The study was approved by the Human Research Ethics Committee of University of Canterbury (HEC 2017/91).

Informed Consent Statement: Informed consent was obtained from all subjects involved in the study.

Acknowledgments: This research used the SpeechinNoise experiment software, developed by Chun-Liang Chan at Northwestern University. The researchers would like to thank Wakayo Mattingley for her assistance and knowledge in setting up XAMPP and Firebase. Robert Fromont's technical help was also instrumental in deployment. Many thanks to the student Research Assistants Vicky Watson, Nick Hight, and Sidney Wong for conducting the synthesis process and making the methodology as easy as possible for the researchers. Bob Haywood contributed code that produced all the JSON playlists and some data extraction, and Michael Field contributed python expertise. We would also like to thank Andrew Kepple for designing the cartoons on which this priming experiment depended. We thank the New Zealand Institute of Language, Brain and Behaviour, and the College of Arts at the University of Canterbury for their support of this project.

Conflicts of Interest: The authors declare no conflict of interest.

Appendix A

Full stimuli list for experiment one.

DRESS-Frame	KIT-Frame	Both-Frame	Neither-Frame	Fake LOT	Real LOT	Fake STRUT	Real STRUT
bXckoned	chXckens	bXdded	bXnims	choggy	bothers	busky	budget
bXnchers	dXffers	bXdder	bXnking	cothers	boxers	duppet	cupping
chXcker	dXgger	bXdding	chXdgets	doppers	chomping	fusting	custom
chXckered	dXgging	bXgger	cXdges	fomments	choppers	hudget	dusting
chXckers	dXgit	dXnted	dXgments	fopping	coffee	hupper	fudging
dXbit	dXpper	dXnting	fXctin	gonsored	coffin	funding	hupping
dXnceness	dXshes	mXsses	fXtchen	gopied	comets	munding	fungus
dXnntists	dXtching	mXssy	gXcking	honvicts	comments	pummy	gummy
fXncer	fXdget	pXcker	gXppered	mosmos	convicts	pussocks	hunted
fXnces	fXgment	pXckers	gXtching	noctors	copied	pustom	husky
fXnnder	fXshes	pXnnies	gXxton	poffee	cosmos	smuppies	puppet
fXstive	fXtness	pXnny	kXvers	shoxers	doctors	studging	puppies
fXtching	fXxes	sXxes	kXzzard	snoffin	foggy	suffered	sunted
hXctick	fXxture	tXnder	pXbbons	stomets	popping	tuffered	supper
hXxing	fXzzy	tXnter	vXpit	tomping	sponsored	tungus	tussocks
jXstures	gXddy						
kXchup	gXmmick						
mXnnding	gXving						
pXndent	hXccupped						
pXndents	hXnging						
pXppered	hXtches						
pXssky	kXcker						
pXstered	kXdney						
scXptic	mXnute						
sXckters	nXpping						
sXnding	pXgment						
skXcher	pXnkish						
skXches	pXstons						
spXckter	pXtcher						
spXckters	pXvots						
spXnder	sXfter						

DRESS-Frame	KIT-Frame	Both-Frame	Neither-Frame	Fake LOT	Real LOT	Fake STRUT	Real STRUT
spXnders	skXmpy						
pXnnding	spXnsters						
tXmpered	stXcker						
tXmpter	stXcking						
tXnnding	stXngers						
tXssted	stXnted						
tXsster	thXcker						
tXssters	thXckets						
tXssting	thXnnest						
tXxture	tXckets						
tXxtures	tXmid						
vXnnom	tXpping						
vXstment	vXcar						
vXxing	vXctims						

References

1. Hay, J.; Drager, K. Stuffed toys and speech perception. *Linguistics* **2010**, *48*, 865–892. [CrossRef]
2. Wells, J.C.; Wells, J.C. *Accents of English: Volume 1*; Cambridge University Press: Cambridge, UK, 1982.
3. Strand, E.; Johnson, K. Gradient and visual speaker normalization in the perception of fricatives. In *Natural Language Processing and Speech Technology*; Gibbon, D., Ed.; Mouton de Gruyter: Berlin, Germany, 1996; pp. 14–26. [CrossRef]
4. Strand, E. Uncovering the role of gender stereotypes in speech perception. *J. Lang. Soc. Psychol.* **1999**, *18*, 86–100. [CrossRef]
5. Hay, J.; Warren, P.; Drager, K. Factors influencing speech perception in the context of a merger-in-progress. *J. Phon.* **2006**, *34*, 458–484. [CrossRef]
6. Drager, K. Speaker Age and Vowel Perception. *Lang. Speech* **2011**, *54*, 99–121. [CrossRef]
7. Casasanto, L.S. Does social information influence sentence processing? In Proceedings of the Annual Meeting of the Cognitive Science Society, Washington, DC, USA, 23–26 July 2008; Volume 30.
8. Munson, B.; Jefferson, S.V.; McDonald, E.C. The influence of perceived sexual orientation on fricative identification. *J. Acoust. Soc. Am.* **2006**, *119*, 2427–2437. [CrossRef]
9. Niedzielski, N. The Effect of Social Information on the Perception of Sociolinguistic Variables. *J. Lang. Soc. Psychol.* **1999**, *18*, 62–85. [CrossRef]
10. Hay, J.; Nolan, A.; Drager, K. From fush to feesh: Exemplar priming in speech perception. *Linguist. Rev.* **2006**, *23*, 351–379. [CrossRef]
11. Babel, M.; McGuire, G.; King, J. Towards a more nuanced view of vocal attractiveness. *PLoS ONE* **2014**, *9*, e88616. [CrossRef]
12. D'Onofrio, A. Personae and phonetic detail in sociolinguistic signs. *Lang. Soc.* **2018**, *47*, 513–539. [CrossRef]
13. Johnson, K. Resonance in an exemplar-based lexicon: The emergence of social identity and phonology. *J. Phon.* **2006**, *34*, 485–499. [CrossRef]
14. Foulkes, P.; Docherty, G. The social life of phonetics and phonology. *J. Phon.* **2006**, *34*, 409–438. [CrossRef]
15. Sumner, M.; Kim, S.K.; King, E.; McGowan, K.B. The socially weighted encoding of spoken words: A dual-route approach to speech perception. *Front. Psychol.* **2014**, *4*, 1015. [CrossRef] [PubMed]
16. Drager, K. Sociophonetic variation in speech perception. *Lang. Linguist. Compass* **2010**, *4*, 473–480. [CrossRef]
17. Pierrehumbert, J.B. Phonological representation: Beyond abstract versus episodic. *Annu. Rev. Linguist.* **2016**, *2*, 33–52. [CrossRef]
18. Foulkes, P.; Hay, J. 13 The Emergence of Sociophonetic Structure. *Handb. Lang. Emerg.* **2015**, *87*, 292.
19. Weatherholtz, K.; Jaeger, T.F. Speech perception and generalization across talkers and accents. In *Oxford Research Encyclopedia of Linguistics*; Oxford University Press: Oxford, UK, 2016.
20. Campbell-Kibler, K. Towards a cognitively realistic model of meaningful sociolinguistic variation. In *Awareness and Control in Sociolinguistic Research*; Babel, A.M., Ed.; Cambridge University Press: Cambridge, UK, 2016; pp. 123–151.
21. Kleinschmidt, D.F. Structure in talker variability: How much is there and how much can it help? *Lang. Cogn. Neurosci.* **2019**, *34*, 43–68. [CrossRef]
22. Nygaard, L.C.; Tzeng, C.Y. Perceptual integration of linguistic and non-linguistic properties of speech. In *The Handbook of Speech Perception*; Blackwell Publishing Limited: Hoboken, NJ, USA, 2021; pp. 398–427.
23. Zheng, Y.; Samuel, A.G. Does seeing an Asian face make speech sound more accented? *Atten. Percept. Psychophys.* **2017**, *79*, 1841–1859. [CrossRef]
24. Walker, A.; Hay, J.; Drager, K.; Sanchez, K. Divergence in speech perception. *Linguistics* **2018**, *56*, 257–278. [CrossRef]
25. Walker, M.; Szakay, A.; Cox, F. Can kiwis and koalas as cultural primes induce perceptual bias in Australian English speaking listeners? *Lab. Phonol.* **2019**, *10*, 7. [CrossRef]
26. Jannedy, S.; Weirich, M.; Brunner, J. The Effect of Inferences on the Perceptual Categorization of Berlin German Fricatives. In Proceedings of the ICPhS, Hong Kong, China, 17–21 August 2011; pp. 962–965.
27. Lawrence, D. Limited evidence for social priming in the perception of the BATH and STRUT vowels. In Proceedings of the ICPhS, Glasgow, UK, 10–14 August 2015.

28. Walker, A.; Drager, K.; Hay, J. The use of priming in language attitudes research. In *Research Methods in Language Attitudes*; Kircher, R., Zipp, L., Eds.; Cambridge University Press: Cambridge, UK, in press.
29. Rosenblum, L.D. Primacy of multimodal speech perception. In *The Handbook of Speech Perception*; Pisoni, D.B., Remez, R.E., Eds.; Blackwell Publishing Ltd.: Hoboken, NJ, USA, 2008; pp. 51–78. [CrossRef]
30. Hay, J.; Podlubny, R.; Drager, K.; McAuliffe, M. Car talk: Location-specific production and perception. *J. Phon.* **2017**, *65*, 94–109. [CrossRef]
31. Shaw, J.; Best, C.T.; Docherty, G.; Evans, B.G.; Foulkes, P.; Hay, J.; Mulak, K.E. Resilience of English vowel perception across regional accent variation. *Lab. Phonol.* **2018**, *9*, 11. [CrossRef]
32. Baayen, R.H.; Piepenbrock, R.; Gulikers, L. *The CELEX Lexical Database (CD-ROM), LDC*; University of Pennsylvania: Philadelphia, PA, USA, 1995.
33. Boersma, P.; Weenink, D. *Praat: Doing Phonetics by Computer* [Computer Program]. Version 6.1.21 2017, retrieved 21 September 2017. Available online: http://www.praat.org/ (accessed on 12 July 2018).
34. Winn, M.B. *GUI-Based Wizard for Creating Realistic Vowel Formant Continua from Modified Natural Speech*. Version 30. 2014. Available online: https://www.mattwinn.com/praat/Make_Formant_Continuum_v30.txt (accessed on 7 April 2017).
35. Chan, C. *Speech in Noise 2*; Northwestern University: Evanston, IL, USA, 2015. Available online: https://northwestern.app.box.com/s/9g2rigz1iqh4ymfkgunq6t31u3iycbpr/folder/5925856385 (accessed on 1 June 2020).
36. Ganong, W.F. Phonetic categorization in auditory word perception. *J. Exp. Psychol. Hum. Percept. Perform.* **1980**, *6*, 110–125. [CrossRef] [PubMed]
37. Munson, B.; Ryherd, K.; Kemper, S. Implicit and explicit gender priming in English lingual sibilant fricative perception. *Linguistics* **2017**, *55*, 1073–1107. [CrossRef]
38. Bordens, K.S.; Abbott, B.B. *Research Design and Methods: A Process Approach*; McGraw-Hill: New York, NY, USA, 2002.
39. Hashimoto, D. Sociolinguistic effects on loanword phonology: Topic in speech and cultural image. *Lab. Phonol.* **2019**, *10*, 11. [CrossRef]
40. Ludwig, I. Identification of New Zealand English and Australian English Based on Stereotypical Accent Markers. Master's Thesis, University of Canterbury, Christchurch, New Zealand, 2017. Available online: https://ir.canterbury.ac.nz/handle/10092/985 (accessed on 1 June 2020).
41. Sanchez, K.; Hay, J.; Nilson, E. Contextual Activation of Australia can affect New Zealanders' vowel productions. *J. Phon.* **2014**, *48*, 76–95. [CrossRef]
42. Fridland, V.; Kendall, T. Exploring the relationship between production and perception in the mid front vowels of US English. *Lingua* **2012**, *122*, 779–793. [CrossRef]
43. Hay, J.; Drager, K.; Gibson, A. Hearing r-sandhi: The role of past experience. *Language* **2018**, *94*, 360–404. [CrossRef]
44. Yu, A. On the nature of the perception-production link: Individual variability in English sibilant-vowel coarticulation. *Lab. Phonol.* **2019**, *10*, 2. [CrossRef]
45. Maclagan, M.; Hay, J. The rise and rise of New Zealand English DRESS. In Proceedings of the Australian International Conference on Speech Science and Technology, Sydney, Australia, 8–10 December 2004; pp. 183–188.
46. Ross, B.; Ballard, E.; Watson, C. New Zealand English in Auckland: A Papatoetoe snapshot. *Asia-Pac. Lang. Var.* **2021**, *7*, 62–81. [CrossRef]

brain sciences

MDPI

Article

DIANA, a Process-Oriented Model of Human Auditory Word Recognition

Louis ten Bosch *, Lou Boves and Mirjam Ernestus

Center for Language Studies, Radboud University, 6525 HT Nijmegen, The Netherlands; lou.boves@ru.nl (L.B.); mirjam.ernestus@ru.nl (M.E.)
* Correspondence: louis.tenbosch@ru.nl

Abstract: This article presents DIANA, a new, process-oriented model of human auditory word recognition, which takes as its input the acoustic signal and can produce as its output word identifications and lexicality decisions, as well as reaction times. This makes it possible to compare its output with human listeners' behavior in psycholinguistic experiments. DIANA differs from existing models in that it takes more available neuro-physiological evidence on speech processing into account. For instance, DIANA accounts for the effect of ambiguity in the acoustic signal on reaction times following the Hick–Hyman law and it interprets the acoustic signal in the form of spectro-temporal receptive fields, which are attested in the human superior temporal gyrus, instead of in the form of abstract phonological units. The model consists of three components: activation, decision and execution. The activation and decision components are described in detail, both at the conceptual level (in the running text) and at the computational level (in the Appendices). While the activation component is independent of the listener's task, the functioning of the decision component depends on this task. The article also describes how DIANA could be improved in the future in order to even better resemble the behavior of human listeners.

Keywords: speech comprehension; computational model; process-oriented model

Citation: ten Bosch, L.; Boves, L.; Ernestus, M. DIANA, a Process-Oriented Model of Human Auditory Word Recognition. *Brain Sci.* **2022**, *12*, 681. https://doi.org/10.3390/brainsci12050681

Academic Editors: Richard Wright and Benjamin V. Tucker

Received: 17 March 2022
Accepted: 10 May 2022
Published: 23 May 2022

1. Introduction

This paper presents DIANA, a new, computational model of human speech processing. This model has been developed over a number of years. Implementation details of the model and specific simulations have been described in [1–7]. The current paper presents DIANA at the conceptual level and explains how its features are inspired by psycholinguistic and neurophysiological data. In addition, it makes explicit how and why DIANA differs from existing models of speech comprehension. Computational details that are relevant for the operation of DIANA are described in the Appendices.

In the following subsections, a number of existing computational models of human speech processing and their characteristics are described. There are more models, such as those based on episodes, but those mentioned here provide a framework for discussion about DIANA's position. In Section 2, we introduce DIANA and describe how this model differs from existing models at the conceptual level. In Sections 3 and 4, we describe and illustrate the operation of two of DIANA's components, while in Section 5 a number of future research directions are discussed.

1.1. Computational Models of Speech Processing

A substantial part of psycholinguistic research focuses on the cognitive processes that take place when listeners perceive speech. Based on a vast body of empirical psycholinguistic results obtained since the nineteen-eighties, a number of influential models of human speech comprehension have been developed. These models are based on three basic principles that are assumed to underly human speech processing. These principles are: (1) during the unfolding of the acoustic signal, multiple word candidates are activated in parallel; their activation is based on the degree of match between the input speech signal and their representations in the mental lexicon, (2) this mental lexicon contains information about the pronunciations and meanings of words, (3) the comprehension process is incremental; listeners do not wait until the end of a word before they start interpreting the input.

Most current theories of spoken-word recognition are computationally implemented. Computational models have the advantage that they may be able to simulate the conditions of experiments. They thereby allow a direct comparison between model predictions and behavioral results obtained from human listeners using the same stimuli. An unavoidable potential drawback of any computational model is that various implementational assumptions need to be made that are possibly unsupported by empirical data or are left unspecified by psycho-linguistic theories [8].

1.1.1. Cohort Model

The Cohort model [9–11] was one of the first models of spoken word recognition. It used phonemic transcriptions as input and accounted for incremental processing. In this model, spoken-word recognition is modeled as a three-stage process, involving access, selection, and integration. The input is dealt with phone-by-phone. Only words for which the beginnings match with the phonemic transcription of the input speech, aligned from a specific onset, are activated and make up a cohort (access). During processing of the next phone in the input, candidate words that no longer match are removed from this cohort. In the end, only one candidate remains (selection). At that moment, the semantic and syntactic properties of the winning word become available (integration).

A challenge for the Cohort model is that it cannot recover from early local mismatches: for instance, a /k/ instead of /g/ in the input blocks the activation of 'garden', no matter the support for this word after the /k/. Because the properties of the winning word only become available after selection, the cohort model also cannot use word frequency information during the recognition process. This behaviour is not in agreement with empirical data: many speech comprehension experiments have shown that recovery from errors is possible, and that word frequency has a substantial impact on accuracy and speed (see, e.g., [12] for an overview). Its successor version Cohort II [13,14] addressed these issues, but a major challenge for the cohort models remained the impossibility of defining activation based on the later parts in the word [15].

The Cohort model, like most models (see below), explains specific aspects of the speech comprehension process at Marr's computational level [16]. The model assumes that the acoustic signal is converted into a prelexical representation. It is this prelexical representation that is then matched with the words presented in the mental lexicon. In addition, the Cohort model assumes that this prelexical representation consists of phones (or phonemes). The advantage of a prelexical level consisting of categorical units is that the matching of the prelexical representation with the lexical representations is unproblematic. For example, different realizations of /a/ as produced by a male and female speaker, while acoustically very different, can be mapped on the same prelexical unit /a/, which then maps on any lexical /a/. It is unclear, however, how these categorical units are extracted from the acoustic signal because individual sounds are often highly ambiguous. As phone annotation tasks show, listeners can often only solve these ambiguities after they have recognized the word, based on other acoustic properties of the word or based on the linguistic context. The same is suggested by recent neurophysiological studies which indicate that how a phone sequence is recognized is influenced by the patterns in the lexicon from the very start [17–19]. It is therefore not likely that, just on the basis of the acoustic input, categorical decisions on the identity of units are made before lexical access takes place, and it is doubtful whether categorical units are instrumental in the comprehension process proper, e.g., [20].

1.1.2. TRACE

The TRACE model [21] has an entirely different design. It is a connectionist interactive-activation model that consists of three layers: a feature, a phoneme, and a word layer. The input to TRACE consists of a sequence of multidimensional (manually crafted) feature vectors, and each word's pronunciation in the TRACE lexicon is represented as a phoneme sequence. TRACE activates multiple word candidates that match any part of the speech input in proportion to their degree of fit with the complete input. As a result, partially overlapping words are considered in parallel. After nodes are activated, their activation spreads through the layers (feature nodes spread activation to matching phoneme nodes, phoneme nodes spread to word nodes).

In the TRACE model, inhibition takes place within the phoneme layer and within the word layer; the phoneme with the highest activation suppresses candidate phonemes with lower activations, and idem for words. Finally, the candidate word that matches the input best is 'recognized'. The activation of a word does not decrease in the presence of mismatching input. In its original version, word frequency was not taken into account, but later versions of TRACE do (see, e.g., [22]).

The model includes a 'lexical feedback loop', which makes it possible to revise the phonemic interpretation of feature vectors to make these comply with the phonemic representation of words. The use of such a feedback loop was criticized by [23] on the basis of the argument that such a loop would not be necessary and was theoretically unjustifiable. This argument continues to play a role in recent models (see, e.g., [24], and commentaries). Another aspect that received criticism was the implausible architecture of the network—each time the next phoneme in the input is to be processed, the search network has to be entirely duplicated.

1.1.3. Shortlist and Shortlist B

The Shortlist model [23] can be considered a response to the TRACE model. A major aim of Shortlist [23] was to show that the lexical feedback loop in TRACE is unnecessary. Its input consists of a phoneme string (again, handcrafted on the basis of an acoustic signal). It consists of two stages. Shortlist's first stage consists of an exhaustive serial lexical search, which results in a shortlist of maximally 30 candidate words that match the input processed so far (other candidates are not considered). In the competition stage, these candidate words compete in an interactive-activation network in which the word candidates that receive support from the same sequence of input phonemes are connected via inhibitory links. Mismatches with the acoustic signal do not completely block the recognition of a word but lead to decreasing word activation. The word with the highest activation inhibits candidate words with lower activations, and finally the candidate word that best matches the input is recognized. Shortlist's interactive activation network is equivalent to the word layer of TRACE. Instead of adapting the existing shortlist, the entire process is repeated with each new phoneme symbol in the input, which necessitates a new shortlist for each input phoneme.

Shortlist B [25] is an updated version of the Shortlist model. The theoretical assumptions underlying Shortlist B are identical to Shortlist, but it implements the word competition as a Bayesian update process. Its input is created as follows: first a phonemic transcription is created (by hand) of the speech signal, after which this transcription is transformed into a sequence of phone–phone confusion probabilities. These phone–phone confusion probabilities (defined over three time slices per phoneme) are derived from a large-scale perception study using gated diphones [26,27]. By using these probabilities as input, instead of categorical descriptions, Shortlist B addresses listeners' capability to process ambiguous speech signals. Shortlist B incorporates word frequencies as prior probabilities, and deals with matches and mismatches using the framework of likelihoods. There is no inhibition, and there is no feedback in the sense of higher layers modulating computations in lower layers. A drawback of Shortlist B is that it does not specify how it would extract information about phone-phone confusion probabilities from the acoustic signal and instead produces them from combining a phone transcription of the acoustic signal with data from perception experiments. In addition, the strict use of the Bayesian framework leads to a rather particular interpretation of how listeners process novel words: listeners can only process an unknown word after they have produced a prior for the acoustic realisation of that new word.

1.1.4. Fine-Tracker

The Fine-Tracker model [28,29] is based on the principles underlying Shortlist B. This model is specifically developed to account for the role of fine phonetic detail in speech comprehension. It is one of the first models that takes acoustic speech signals as input, rather than some kind of segment-level symbolic transcription. Fine-Tracker is a two-stage model. The first stage uses an artificial neural network (ANN) to convert the acoustic signal into a sequence of articulatory-phonetic feature vectors. In Fine-Tracker's lexicon, words are represented as sequences of such feature vectors, instead of phone labels. In the lexical representations the phonetic features have values 0 (absent) or 1 (present), or NA (not applicable, for example for the component plosive in a lexical feature vector representing a vowel). Phonetically longer segments are lexically represented by duplication of the vectors of those segments. For instance, the first syllable of the English words 'ham' and 'hamster' differ from each other in their lexical representations in that the vowel æ of 'ham', which is reportedly longer than that of "hamster" [30], is duplicated. The bottom-up ANN outputs real-valued feature vectors for which each component can take any value between 0 and 1. The use of the ANN vectors and the lexicon's vectors allows feature values to 'spread' into neighboring feature vectors through assimilation and co-articulation. Fine-Tracker's word recognition stage uses a probabilistic word search based on classical dynamic programming to find the most likely word sequence.

Fine-Tracker has the advantage of using a flexible signal representation in the form of feature vectors. TRACE also uses feature vectors, but these are essentially recoded phonemic symbols. Another advantage is Fine-Tracker's ability to use real speech as input. The model has two disadvantages. The performance of Fine-Tracker crucially depends on the ANN: If the ANN makes an error, Fine-Tracker cannot recover. Finally, the exact definition of the match between full-dimensional estimated feature vectors (by the ANN) and the (possibly partially defined) canonical lexical feature vectors is an unsolved issue, since it is unclear how to faithfully compare distances between fully specified vectors and distances between partially specified vectors in the definition of the match between the input signal and lexical representation.

1.1.5. EARSHOT and LDL-AURIS

Recently, computational models have been proposed that avoid pre-lexical levels consisting of explicit abstract units or phonetic/articulatory features. EARSHOT [24] and LDL-AURIS [31] do so by mapping the acoustic signal directly to vectors in a distributed semantic vector space, instead of to words, as in 'localist' models, by using neural networks: a two-layer long short-term memory (LSTM) neural network [32] (which models non-linear mappings) in EARSHOT, and a linear discriminative learner (with a linear mapping) in LDL-AURIS. These models are end-to-end in the sense that they circumvent explicit pre-lexical and lexical representations during the processing of the input; instead, these representations may be implicitly present in the layers of these networks. The semantic target vectors can be defined in different ways, e.g., chosen randomly or based on the outcome of a word-to-vector algorithm (e.g., word2vec [33]).

EARSHOT and LDL-AURIS do not claim to explain all putative cognitive processes involved in speech comprehension. Instead, they aim to serve as a cognitive model of human speech recognition without explicit phonetic training and by replacing words by distributed semantic representations, thereby leaving a word's articulation entirely unspecified.

2. Towards DIANA, A Novel Process-Oriented Model

In the past, the absence of empirical evidence about processes in the brain involved in speech comprehension was a valid argument for limiting models to the computational level. The rapid advancement of brain imaging techniques, and especially the availability of a growing corpus of knowledge derived from electrocortocography (ECoG) recordings, e.g., [34,35], make it possible to develop models that are also realistic at the neurophysiological level. DIANA takes into account the limitations that the 'wetware' of the human brain imposes on the type of computational processes than can be implemented [36–38]. In addition, it is based on psycho-linguistically motivated principles underlying the group of 'localist' computational models (including the Cohort model, Shortlist, Shortlist B and Fine-Tracker). From these 'localist' models, DIANA adopts the use of a lexicon, the concept of word activations and the unfolding of word hypotheses in parallel (i.e., the activation of words and competition among words as a function of time). DIANA does not assume a prelexical layer in which hard decisions have to be made about abstract prelexical units before lexical access. Instead, the acoustic signal is converted into representations that are neurophysiologically attested. These representations have a statistical relation with the representations in the mental lexicon.

In contrast to nearly all other models, DIANA is process-oriented by including activation and decision processes about word candidates in line with what we know about the neurophysiological basis of perception (via spectro-temporal receptive fields) and human decision making (ambiguity resolution). This will be elaborated upon in Sections 3 and 4. DIANA's behavioral adequacy can be tested as it takes as its input the acoustic signal and produces as its output decisions (e.g., on the identity of a word or on whether the word is a real word) and reaction times. It can therefore simulate a literate adult listener who takes part in a psycholinguistic experiment.

Figure 1 shows the architecture of DIANA. The model contains three interrelated components: an activation component, a decision component and an execution component. The activation component implements acoustic processing and activation of words; the decision component implements the word competition and the decision about the winning hypothesis. The activation and the decision components operate in parallel: the decision component receives a full set of activation scores at each time step from stimulus onset to stimulus offset. The execution component simulates the externalization of the decision, mimicking the time it takes for traveling neural signals to be effectuated eventually as an overt decision. This component adds a constant time (in the current implementation: 200 ms) to DIANA's RT prediction, and we will not discuss this component further in this article.

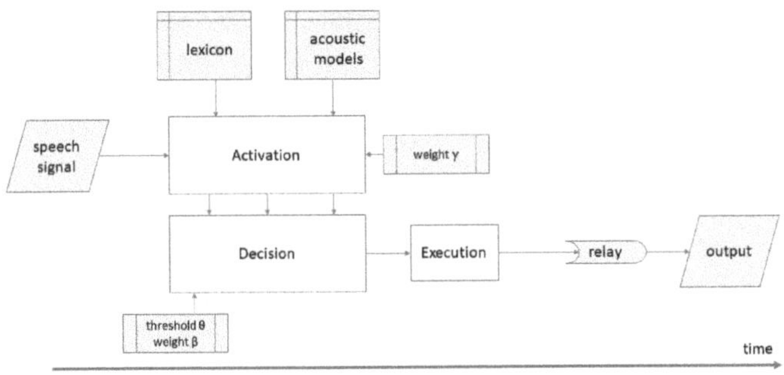

Figure 1. Overall architecture of the DIANA model. The acoustic signal is input for the activation component. During the unfolding of the input, the activation component computes activations and hypotheses which are input for the decision component. The output of DIANA is an overt decision (e.g., word identification or word evaluation) and corresponding reaction time. The two activation and decision components operate in parallel, while the decision and execution components operate serially.

3. The Activation Component

Given the input speech signal, the activation component computes activations of words in the lexicon, on the basis of which the decision component decides how the input is evaluated (e.g., the identity of the word is established). Before word activations can be computed, the acoustic signal has to be interpreted and represented in such a way that it can connect with the mental lexicon. This section first describes this process, then the assumptions about the mental lexicon, and finally the details of the activation process via a number of examples.

3.1. *From the Input Signal to Spectro-Temporal Receptive Fields*

Experiments producing electrocorticography data (ECoGs, e.g., [34]) with speech input suggest that the neural responses in the primary auditory cortex can be described in the form of so called spectro-temporal receptive fields (STRFs, [39]). STRFs describe the spectro-temporal processing in the human superior temporal gyrus (STG) during natural speech processing (see, e.g., [34,40,41]), and form a neural representation for time-varying sounds, reminiscent of conventional sonagrams [42]. One STRF contains information from both the static spectral (stable portions) and the dynamic spectro-temporal properties (transients) of a short stretch (approximately 20–30 ms) of the speech signal. STRFs also obey the 'tonotopic' frequency-locus relation, known from cochlear processing [43].

Approximations of the 'cortical' STRFs can be computed directly from the audio signal (see, e.g., [40]). This property is used in DIANA to map the input speech signal into a computational approximation of an STRF sequence in two steps. The first step is the mapping of the input speech to a sequence of feature vectors. Each feature vector represents the static and dynamic part of a 25 ms short stretch of the audio signal. This choice is based on knowledge about temporal alternation of stable regions and transients in speech [44–46]. The stable part is coded by 13 Mel-frequency cepstral coefficients, MFCC, [47]. These coefficients take into account the tonotopic properties of cochlear representations and the frequency and loudness sensitivity of the human auditory system (see, e.g., [48]). The dynamic changes of the spectrum are coded by the first and second time derivatives of the MFCCs, cf. [48]. The feature vectors (of dimension 39) are updated every 10ms. Taken together, each audio input is represented by a trajectory of (39-dimensional) feature vectors in the MFCC space, with a sampling rate of 100 per second. Such a trajectory captures the acoustic fine structure of the audio input to a degree that is sufficient for nearly all types of speech analyses [49].

The second step converts the MFCC feature vectors into the STRFs as used in DIANA. These 'audio-based' STRFs are very similar to STRFs based on ECogG data (e.g., [34], see also [50]), and they distinguish phones and broad phonetic classes as the cortical STRFs do (see Appendix A.1 for more details on how STRFs are computed). Figure 2 shows the relation between frequent phones (vertical axis) and DIANA's STRFs (indexed along the horizontal axis). The off-diagonal cells indicate patterns that are shared among related phones. Importantly, they are very similar to the relation between ECoGs and phones found in neurophysiological studies [34].

STRFs form the link between the pronunciation representations in the lexicon, on the one hand, and the MFCC feature vectors that encode acoustic signals on the other. The match between audio input and a word is computed via the statistical match between the MFCC vectors from the audio input and the STRFs associated with the lexical representation of that word (see Appendix A.1).

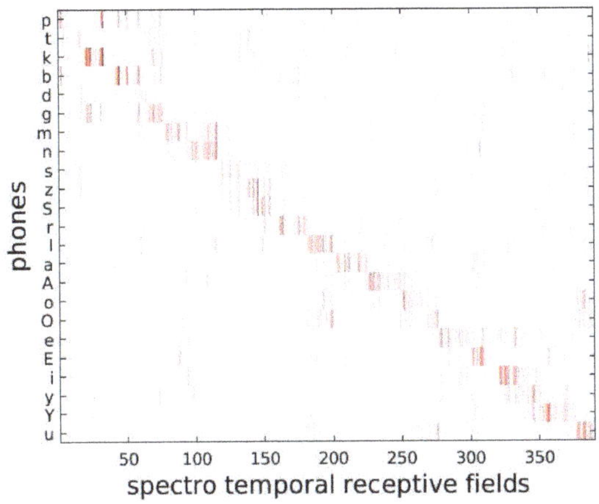

Figure 2. Correspondence between DIANA's STRFs and frequent phones, organized along broad phonetic classes. Phones are represented using SAMPA.

3.2. The Lexicon in DIANA

DIANA uses an internal lexicon, in which words with their pronunciations are stored. The pronunciations are described in the form of phone sequences. A phone-based representation helps to explain how listeners may divide a word in speech sounds, and it enables DIANA to differentiate between word candidates during the word competition on a phonetically-linguistically relevant level. Another advantage of lexical phone sequences relates to sufficiency; listeners are usually not aware of subtle phonetic differences between different instances of a phone that may arise during speech production.

The lexical representation of words determines how they are modeled in DIANA's computations. When any two words share a phone with the same pre- and post-context, that phone is modeled by the same articulatory model. For example, since the words 'speech' and 'speed' share the same word-initial /s p/ in their lexical description, their pre-context is the same (word start) and the post-context is the same (/i/), they share the same /s p/ model. In contrast, 'spell', 'speed' and 'speech' only share the /s/ model, but not the /p/ model, because the post-context of the /p/ is different in 'spell'. In the same vein, the words 'ham' and 'hamster' share the same /h æ/ model, but not the /h æ m/ model. If word stress is not expressed in DIANA's lexicon, it is not taken into account. That is, words such as 'household' and 'leasehold' (with stress on the first syllable) share their word-final three-phone model with words such as 'withhold', 'behold' and 'uphold' (with have stress on the second syllable), because the phone representation for the final syllable is the same. Due to the context-dependency, DIANA can process coarticulation effects within a limited scope.

For each (context dependent) phone in the lexical representation, the corresponding articulatory model is a three-state Markov model, in which each state is associated with an STRF. Via self-loop probabilities, the Markov model can deal with duration variation in the input, while the use of three states reflects the head-body-tail structure of the acoustic-phonetic realisation of that unit.

Two observations must be made. First, even though words may share parts of their lexical representations, they can still be in competition with each other. This will be clear from the examples in Section 3.4. Second, the fact that the pronunciation of a word is represented by a sequence of symbols does not imply that these symbols must be (completely) present in the audio input. This flexibility is based on the probabilistic relation between feature vectors (MFCCs) and lexical representations (STRFs) (see Appendix A.1).

3.3. Obtaining Activation Scores from Bottom-Up and Top-Down Information

Neurophysiological research using the phonetic mismatch negativity (a measure of mismatch between expected and actual phonetic input) in EEG traces has shown that, from word onset onwards, listeners develop expectations about which word is uttered, based on both the bottom-up information from the acoustic signal, and the top-down expectations from the (linguistic) context [51,52]. In DIANA, the words' activations are also based on a combination of both types of evidence. The bottom-up support for a word is formed by the match between the MFCC vectors from the audio input with the STRFs associated to the lexical representation of that word (see Section 3.1, and Appendix A.2 for details). The top-down support for a word depends on the task. When a word has to be recognized out of context (e.g., in a psycholinguistic experiment), the bottom-up supports boils down to the word's frequency of occurrence. In a meaningful context, instead, the top-down information is approximated by the probability of the word given the preceding words, which is computed with a statistical language model (in terms of, e.g., conventional word N-grams).

Since DIANA is a model for spoken word comprehension with as input the speech signal unfolding over time, the activation component does not only assign activations to complete words, but also to cohorts of those words. Longer word candidates match a longer stretch of the acoustic input than short word candidates and, therefore, longer word candidates receive more bottom support. Nevertheless, the input stretch of speech may consist of a series of short words rather than of a long one. In order to compare activations of word candidates with different durations, word activations are normalized by dividing by the word candidate's duration.

Activations can be computed for words, pseudo-words and parts of words via essentially the same combination of bottom-up and top-down support. Pseudo-words do not appear in the lexicon but obey the phonotactic patterns in the lexicon. They can be neologisms the listener has not heard before, or they can form the pseudo-words in a lexical decision experiment. During the search, DIANA can create pseudo-words as hypotheses on the fly, on the basis of a phone network in which phones are represented as nodes such that only those phone combinations that are phonotactically licensed appear as possible paths through the network. The top-down support for pseudo-words may be very low (e.g., for neologisms in a conversation), but in simulations of experimental outcomes they can be adjusted, e.g., to model the listener's updated estimation of the proportion of pseudo-words in a lexical decision experiment. Details about the involved computations can be found in Appendix A.3.

3.4. Examples of Word Activations

This section presents a number of concrete examples of activations, with emphasis on their evolution during the unfolding of the input signal. The first example, shown in Figure 3, shows the activations of the words 'housing' and 'houses' and parts thereof, while the speech input is 'housing'.

In the figure, the vertical and horizontal axes show the frame-normalised word activation and time, respectively. The black traces show the activation of individual cohorts, the phonetic transcription of which (using SAMPA symbols [53]) are shown at the right hand side of the figure. For the sake of clarity, the figure only shows the activations of the words 'housing' and 'houses' and their cohorts (instead of all words in DIANA's lexicon). The activation of the word 'housing', shown by the red trace, starts to 'win' over all other hypotheses at about 500 ms after stimulus onset, and it remains on top until the end of the input. Note that hypotheses that have activations at stimulus offset do not necessarily correspond to existing words, since partial word forms that are part of longer existing words may still be activated on the basis of the complete input signal.

Another example is presented in Figure 4, in which the audio input is the word 'hamster'. At $t = 380$ ms after onset, the competing word 'ham' branches of from the winning hypothesis, indicating that the acoustic information disfavours 'ham' in the competition with 'hams' and other longer cohorts of 'hamster'. The figure also shows the effect of shared representations of 'ham' and the first syllable of 'hamster'; both activation plots overlap, until $t = 380$ ms.

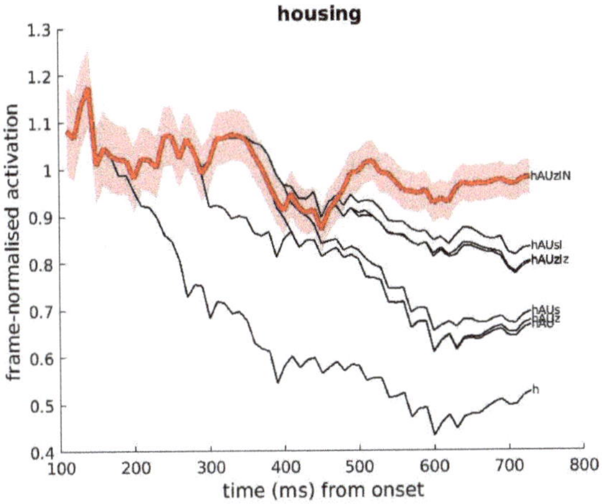

Figure 3. Example of DIANA's activation over time corresponding to the input English word 'housing'. The figure shows the activations of the competing words 'housing' and 'houses', and their word starts (cohorts). The red line shows the evolution of the winning candidate over time. The pink band around the red line indicates the $p = 0.05$ confidence interval. The competing forms are denoted (using SAMPA) at the right-hand side of each plot. A few competitors almost overlap with each other until stimulus offset.

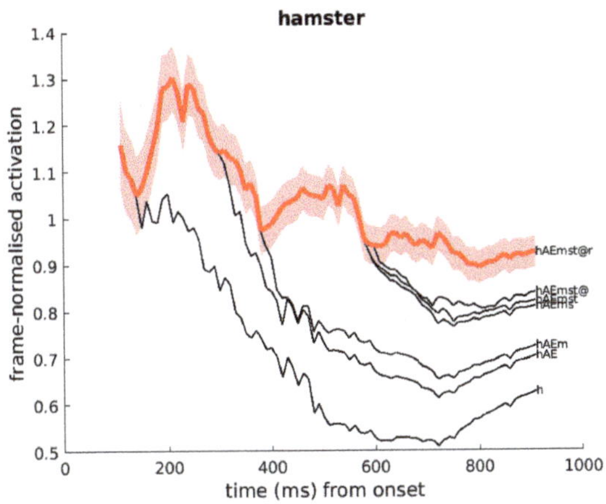

Figure 4. Example of DIANA's activation over time corresponding to the input English word 'hamster'. The figure shows the activations of the competing word forms 'ham' and 'hamster' and their cohorts. The competing forms are denoted (using SAMPA) at the right-hand side. A few competitors overlap with each other until stimulus offset.

The following example is in Dutch. Figure 5 presents the activations of the Dutch noun-noun compound '`pindakaas`' (SAMPA /pIndakas/, Eng. *'peanut butter'*). Certain cohorts of this word are real words themselves, such as the Dutch semantically unrelated word 'pin' (/pIn/), which can be a noun and a verb form (as its English equivalent 'pin'), and the first constituent of the compound '`pinda`' (/pInda/, Eng. *'peanut'*). The figure shows that the full word 'pindakaas' receives its activation from its cohort /pIn/ until about t = 250 ms, while later in the signal, 'pindakaas' receives it activation from /pInda/. In general, each full form adopts its activation from its shorter cohorts underway, representing the idea that these shorter cohorts are considered as part of the full form under development.

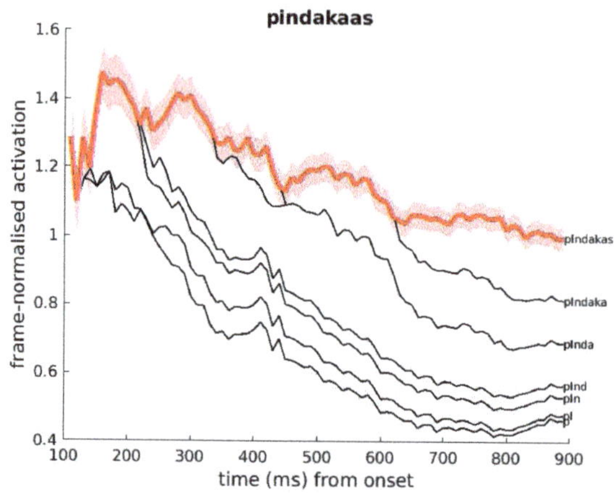

Figure 5. Example of DIANA's activation over time corresponding to the Dutch word 'pindakaas' (SAMPA /pIndakas/; Eng. *'peanut butter'*). The pink band around the red line indicates the $p = 0.05$ confidence interval.

3.5. Presence of Noise in the Input

DIANA behaves like humans in that it can recognize words that are partly produced in noise. This can be seen by comparing Figures 6 and 7. Figure 6 (clean condition) shows the competition between the Dutch derived words `begroting` (/bəxrotɪŋ/, Eng. *'budget'*) and `begroeting` (/bəxrʊtɪŋ/, Eng. *'greeting'*), which only differ in the vowel in the syllable that carries word stress. As soon as this vowel is processed, the hypotheses bəxro, and bəxru, and their longer counterparts, are clearly distinct from each other, showing that activations can differentiate hypotheses on the basis of their final segment.

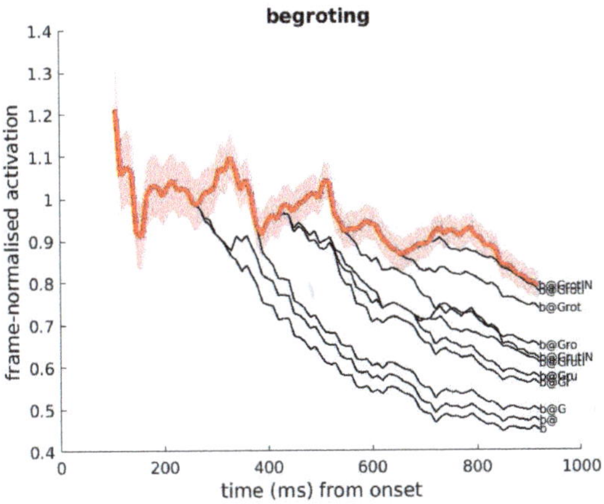

Figure 6. Activation plot of two competing words which form a minimal pair (clean recording condition): the Dutch real word 'begroting' (SAMPA /bəGrotIN/, IPA /bəχrotıŋ/, Eng. *'budget'*) with 'begroeting' (SAMPA /bəGrutIN/, IPA /bəχrutıŋ/, Eng. *'greeting'*). The audio is the real word 'begroting'. The competitor word 'begroeting' looses directly after the /o/, at time 450 ms from onset. Clearly, many competitors overlap until the stimulus offset.

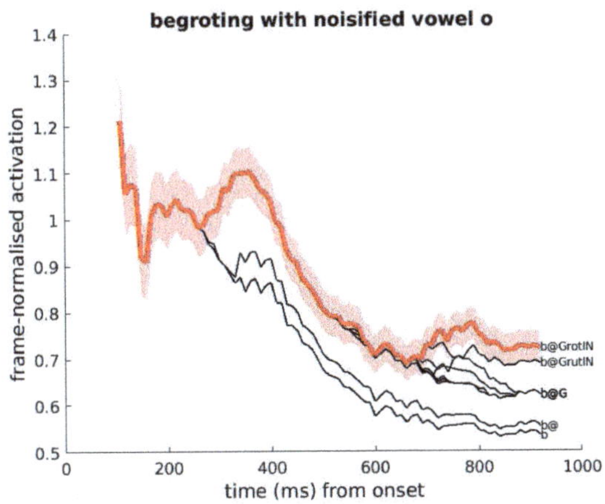

Figure 7. Same as Figure 6, but the noise on the vowel /o/ in the second syllable now yields a reduced and delayed differentiation in the activations of competitors, manifest from 450 ms after onset. Many competitors overlap with each other until stimulus offset.

Figure 7 (noisy condition) shows the activation as a function of time when the word `begroting` is distorted by superimposing background noise (white noise) on the stressed vowel /o/ in the second syllable with a signal-to-noise ratio of −5 dB. Comparison with the clean condition in Figure 6 shows that the activations are identical between stimulus onset and the noise onset, while soon after the noise onset differences emerge. Compared to the clean condition, the distortion has two substantial effects: first and foremost, the activation score of the 'correct' word in the noisy condition shows a steep drop that is completely absent in the clean condition. Second, the divergence between competing cohorts is much smaller in size and occurs later in the noisy condition, compared to the clean condition. (The smaller difference in activation in the case of noise slows down DIANA's decision, as will become clear in Section 4). In the end, the word is still recognized correctly.

These effects on activation scoress are observed in all cases of segment noisification. Quantitative effects appear substantially stronger in case of distortion of those segments that differentiate between real words, such as the /o/ in `begroting` versus `begroeting`.

4. The Decision Component

As mentioned above, while the activation component is independent of the listener's task, the decision component is not. In word identification tasks, it assigns the winning word candidate, while in lexical decision tasks, it determines whether the acoustic input forms a real word or a pseudo-word. In these processes, DIANA only takes into account a selection of the activated words and pseudo-words, as described in Section 4.1. Sections 4.2 and 4.3 describe the selection procedures in a simple identification task and in a lexical decision task, respectively. In Section 4.4, we discuss situations in which the activation component does not provide sufficient evidence to take a decision.

4.1. Selecting Promising Word Candidates

Theoretically, for the purpose of word recognition, the number of potential word candidates can be very large, up to 100,000 words or more. This would correspond to the situation in which a participant is presented a randomly chosen real word. From a neurophysiological perspective, however, it is unlikely that so many competing hypotheses are entertained. For this reason, DIANA reduces the number of activated word and pseudo-word candidates; at each time point, hypotheses with activations too far away (determined by a threshold) from the hypothesis with the highest activation are discarded for further consideration. These hypotheses are considered too poor to have a chance to win later.

As mentioned above, DIANA not only assigns activations to complete words but also to parts of words. This implies that, simultaneously, words and their parts, or parts and their parts, may be activated. For instance, the partial input /k ə θ i/ (from "cathedral") may activate hypotheses such as /k ə θ/, /k ə θ i/ and /ə θ i/. In DIANA, such 'nested' candidates (one candidate is a part of the other) are not assumed to be competitors of each other, as they lead to the recognition of the same longer candidate. The decision component therefore ignores all candidates that are part of other candidates with higher activations.

It is worthwhile to observe that this issue is not or cannot be accounted for in the computational models that use a purely symbolic description of the input signal, in which the list of competitors is based on character string comparisons. Neither is it addressed by EARSHOT and LDL-AURIS, because these models do not have a level were such nesting could occur.

4.2. The Decision Strategy for Simple Word Identification

The activation data as presented in the Figures 3–6 show that the difference in activation between the 'correct' word and its best competitor tends to increase (in a non-linear fashion) as the acoustic signal unfolds. Since the activation and the decision components operate in parallel (the decision component receives activations at each time step t), the latter component does not have to wait until the end of the word to make a decision.

The decision component selects the word hypothesis with the highest activation, once this activation differs from the activation from the second best word by a certain amount (a threshold θ). DIANA's use of a decision criterion based on the difference between two activations is commonly used in general models of human decision and RT distributions, for instance, the ballistic accumulation model (BAM) [54] and the linear approach to threshold with ergodic rate models (LATER, [55,56]).

For several reasons, among others between-speaker pronunciation variation and the probabilistic relation between MFCCs and STRFs, it is not guaranteed that the activation for the correct word is always higher than the activation for other words. Inevitably, this may result in word identification errors.

We showed in [2] that an older version of DIANA can predict identification times for words presented in isolation well. In normal running speech, the role of contextual evidence will be higher than for words presented in isolation, and a substantial difference in activation between the correct word and the competing candidate words is likely to be reached earlier than for words presented in isolation. Context may also inhibit the activation of a candidate word, for example, if the word is unexpected given the pre-context, such that a decision is likely be delayed. Via the Bayes' formula the pre-context modulates the exact activations of words unfolding over time, and thereby the moment at which the decision component can decide about potential winning hypotheses. Words that receive bottom-up support in line with top-down expectations are responded to more quickly, while words that receive bottom-up information that conflicts with the top-down expectation are decided upon later (to what extent this takes place depends on the effect of this context-modulation on all competing hypotheses).

Many experiments have shown that there is a speed-accuracy trade off; for example, when participants are faster, they tend to be less accurate, and vice versa. In the literature on decision making (see, e.g., [54,57–62]), this interaction between speed and accuracy is underpinned by neurophysiological and modeling accounts. In DIANA, the speed-accuracy trade off is the result from a parameter (θ) that determines the value of the threshold difference needed between the activations of the best and the second best word for the best word candidate to be selected. Higher values of θ decrease the risk of making a wrong decision, because more evidence has to be gathered before a decision can be made, which implies longer reaction times. Lower values of θ, instead, increase the risk of making a wrong decision, because less evidence has to be gathered before a decision can be made, which implies short reaction times. The exact speed-accuracy relation depends on the nature (e.g., difficulty) of the task. In [2], we discussed how the threshold θ can affect the speed-accuracy trade off.

4.3. The Decision Strategy for Lexical Decision

Participants in a lexical decision experiment may make their lexicality decision, comparing the evidence for the pertinent word to be a real word and the evidence for it to be a pseudo-word. Accordingly, DIANA bases lexicality judgments on the difference in activation between the real word and the pseudo-word with the highest activations. Once this difference has reached a threshold, θ_{ld}, the decision can be made. If the real word has the highest activation, the lexical judgment will be 'real word', otherwise it will be 'pseudo-word'. This decision strategy implies that it is not strictly necessary to decide exactly which word was uttered, but just whether the real word candidate has a higher or lower activation than the pseudo-word candidate.

For a real word as input, DIANA's competition may involve all lexical items that are acoustically close to the input, in combination with pseudo-words that differ from the input in terms of one or more segments. For example, for an input such as 'elephant' (SAMPA: Eləfənt; IPA: ɛləfənt), the number of potential competing pseudo-words may easily reach 100 to 200, which is hard to elucidate in a clear picture. Conceptually, it will be clear that the more acoustic information becomes available, the number of viable lexical candidates that are active in the competition will decrease over time. Simultaneously, the number of potential pseudo-words that may play a role in the competition increases over time, due to the increasing length of the hypotheses.

In a lexical decision experiment, the exact nature of the pseudo-words will influence whether participants will make the lexicality decision as soon as the difference in activation exceeds the threshold. If the experiment contains many stimuli that start as real words but turn into pseudo-words only at their final segments, participants may not do so. Instead, they may adopt the strategy to postpone their decisions until they have heard the complete words [63].

Note that a given real word can receive activation as if it is a real word and as if it is a pseudo-word (i.e., via the non-lexically-constrained activations). Importantly, the top-down activation may make the difference; the activation of pseudo-words is only differentiated by the bottom-up activation (since their top-down activation is stimulus independent) while the activations of real words are modulated by top-down information (e.g., the frequency of occurrence of that word). The precise balance between bottom-up and top-down probabilities depends on the listener's task. In the simulation of a lexical decision experiment during which the listener is confronted with a fifty-fifty proportion of real words and pseudo-words, the priors for 'word' and 'pseudo-word' will be 0.5. With this decision strategy, DIANA thus is also able to explain a lexical bias that is often observed in psycholinguistic experiments. In [2,3], we analyzed accuracy scores and reaction times from large Dutch and north-American-English datasets of lexical decisions. We showed that DIANA's decision strategy can distinguish between real words and pseudo-words well and can predict well the lexical decision times (in terms of the Pearson correlation with participants' reaction times).

4.4. Ambiguity during DIANA's Search Process

As explained above, DIANA can make a decision about the identity of a word or about the lexicality of a stimulus once the difference in activation between two candidates exceeds a certain threshold. In some situations, however, this threshold may not be reached at stimulus offset. DIANA's decision component then selects the candidate with the highest activation and expresses the ambiguity within this selection process in terms of additional reaction time.

DIANA defines the reaction time for a stimulus in these situations as the sum of the duration of the word (during which no decision could be made), a so called 'choice reaction time', and an execution time. The computation of the choice reaction time is based on Hick–Hyman law [64–66], which states that the more choices are available (expressed in terms of entropy), the longer it takes for a decision to be made. In [67], following a number of early and more recent behavioral studies [64,65,68–70], it is shown that the Hick–Hyman law has a neural underpinning in the cognitive control network (CCN) and the default mode network (DMN), which deal with the mental representation of uncertainty and the generation of behavioral responses [71,72] and which support adaptive behavioral control across a broad range of cognitive demands [73–76]. It appeared that the entropy of the decision problem increased the activity of the CCN that is involved in uncertainty processing and response generation, and decreased the activity of the DMN, which is only involved in uncertainty representation. In short, these studies provide a neurophysiological link between entropy in a choice to be made on the one hand, and associated response latencies on the other. From this point of view, entropy may well explain delays in reactions.

The entropy which forms DIANA's basis for the computation of the choice RT takes the activation scores of all candidates (words, pseudo-words, parts of words) into account, after removal of nested variants with lower activations from the competitor list. More details about the entropy computation can be found in Appendix A.4.

5. Future Research Directions

DIANA is transparent about all processes and assumptions, both at the conceptual and computational level. Transparency, in combination with a process-oriented account, provides clarity about what exactly DIANA can explain and account for. Neurological arguments play a guiding role in DIANA's design; both for the activation and the decision components, the conceptual choices are based on neurophysiological findings (such as the role of STRFs in the auditory cortex, and the neurological underpinning of the Hick–Hyman law). In this section, we will illustrate a number of future research directions for improving DIANA, in particular the structure and content of its lexicon, the computation of the top-down information, the aspect of learning, and several implementation choices.

5.1. Lexicon

As described in Section 3.2, the pronunciation of words in DIANA's internal lexicon is defined in terms of phone sequences. Words sharing a phone subsequence in the lexicon share the articulatory model pertaining to that subsequence. This structure is adequate insofar as differences between words can be expressed at the phone level. It cannot model subtle acoustic differences between the phone sequences that words share. For instance, the present structure of the lexicon cannot capture effects due to prosodic lengthening which listeners may be sensitive to (e.g., [30]). Similarly, homophones, such as 'time' and 'thyme', have in DIANA's present lexicon identical representations, and the present structure of the lexicon, therefore, cannot deal with durational differences among the members forming a homophone pair [77]. Note that DIANA can *detect* duration differences in the acoustic signal; duration modeling is performed via the transition and self-loop probabilities of the hidden Markov models. The question then arises of whether and to what extent to incorporate these 'fine phonetic cues' in the mental lexicon, or, in other words, how to make DIANA's word recognition sensitive to fine phonetic details insofar as they are perceptually relevant [78]. One of the options is to completely disentangle the representations of the different lexical entries, such that 'discolor' and 'discover', 'time' and 'thyme', 'ham' and 'hamster', and so on, do not share any common spectro-temporal structure. Such an option raises the question of where the detailed pronunciation information to be incorporated in the lexicon has to come from. It requires the analyses of either speech corpora in which the prosodic and spectral differences can be inferred in a statistically and perceptually significant way, or an implementation of a solid theory about the morphological-acoustics interface.

Another shortcoming of DIANA's present structure of the lexicon is that each word is considered a separate, independent entry. This implies that, for instance, morphological information about shared stems is missing and that DIANA cannot model the influence of family size on the speed with which a word is recognized (e.g., [79]). These types of effects could be accommodated in a model of the lexicon where words are interconnected on the basis of all kinds of similarities (morphological, phonological, pragmatic, syntactical), as proposed by Bybee [80]. One of the future research directions is therefore the enrichment of DIANA's lexicon by designing a network in which words are linked in a weighted fashion on the basis of all these different similarities. This network will modulate both the set of word candidates considered during the search and their activations. Connecting words on the basis of formal similarities (e.g. phonological or morphological) is a relative easy step compared to connecting words on the basis of their semantics. The latter may require that, in DIANA's lexicon, words are coupled with semantic (distributed) representations (see also [81]).

5.2. Generalizing to Other Languages

So far, DIANA has been tested with Dutch and English [2,3]. This raises the question of to what extent it can also perform well with typologically different languages. One challenge is presented by languages that are morphologically more complex than Dutch and English, such as Finnish. They form a challenge because the fact that the same stem may be incorporated in a very high number of words increases the necessity of a flexible way of incorporating morphological structure, moving away from the current 'localist' approach in DIANA in which each word form is represented as a single entry in the lexicon. Testing DIANA on such languages is on our agenda.

Another challenge is formed by tone languages. In the current version, DIANA is insensitive to pitch and to tone. Tone languages will ask for an extension of the acoustic feature extraction with pitch-related vector components (e.g., pitch itself, its first time-derivative). This is feasible since this extension has been incorporated in several speech decoding systems, for example, Mandarin [82]. To what extent the decoding approach in DIANA is compatible with the lexical structure of tone languages is another topic to be investigated in more detail.

In its current implementation, DIANA is monolingual. In principle, DIANA can also simulate multilingual listeners. In a multilingual setting (see, e.g., [83]), the potential number of competitors is much larger than in a monolingual listener, as multiple lexicons are activated simultaneously. As a consequence, the competition will be more involving, especially if stimuli are presented without any pre-context indicating the language. How this could be accomplished in DIANA is a challenging topic of further research.

5.3. *Top-Down Information*

Word activations result in DIANA from a combination of bottom-up information and top-down information. In the present version of DIANA, the top-down information is provided via a conventional statistical language model (SLM, in the form of an N-gram [49]) that estimates the (scaled) log probability of each word given the directly few preceding words (or given its frequency of occurrence when the word is presented out of context). The value of N depends on the available type of text materials; in the case of a list of isolated words, $N = 1$ (unigram). Previous work has shown that these types of models predict reasonably well the following word [48]. However, these models may be argued to be cognitively too simplistic, as these models only consider a few preceding words, ignore the meanings of the words, ignore the syntactic structure of the sentence, and so on.

We aim to enrich the present top-down information in several ways. First, we will expand the number of preceding words that are taken into account by replacing the simple statistical language model by, for example, LSTM-based neural network-based language models (e.g., [49,84]) which can capture longer span word prediction. Second, we aim to produce expectations about the likelihoods of the different parts of speech, extracted from tagged corpora (for example by a modern dependency grammar approach, e.g., [85]). Third, we aim to enrich the top-down information with the meanings of the preceding words, expressed, for instance, in word2vec [33]. In further steps, the likelihoods of words could even be modulated by visual information presented to DIANA, as is done in image–caption retrieval models, such as [86].

5.4. *Is* DIANA *A Learning Model?*

One may require from a model that it not only simulates adult listener's processing, but also how this adult acquired the knowledge to do so (language acquisition) and how this adult can learn new words and pronunciations. Language acquisition is a process mediated by social interaction in a multi-modal context that enables infants and toddlers to infer associations between acoustic forms and meanings with as a side-effect a capability to break up stretches of speech into words, syllables and sounds. It has been shown that all representations currently used in DIANA could be acquired incrementally [87–94], see also [95]. This paves the way to advance DIANA in the direction of an ecologically defensible model of speech comprehension. The present implementation of DIANA, however, lacks the capability of automatically learning new words, or new, deviant pronunciations of words that are already in the lexicon. We consider the aspect of dynamic word learning as a very relevant way to proceed. Conceptually, this word acquisition process could be associated with the detection of a pseudo-word in the sense of an out-of-vocabulary word, in combination with the inclusion and consolidation of the new form into DIANA's lexicon. How this could be achieved is a topic for further research.

The present version of DIANA needs specifications of the probabilistic relations between MFCC feature vectors and STRFs. The question may be raised of how these are 'learned' by DIANA. We derived these low-level parameters by some kind of iterative optimization procedure using a large transcribed speech corpus. Obviously, this iterative, corpus-based approach is not a realistic proxy for language acquisition. DIANA could learn the low-level parameters incrementally, but doing so would be time consuming, and it would most probably contribute little to the insights that can be gathered with the current implementation in adult word recognition.

5.5. Implementation

The activation component in DIANA used in previous experiments (e.g., [2,6]) relied on speech analysis and decoding algorithms in the HTK software package, which can also be applied for automatic speech recognition [96]. The present version of DIANA [7,52] uses algorithms from a different software package, the KALDI toolkit [97]. The most important advantage of KALDI over HTK is the availability of more flexible tools for handling lattices that contain the dynamically changing activations of word and pseudo-word candidates as the acoustic stimulus unfolds. Another advantage of KALDI is that it allows lexicons with a practically unlimited number of entries, whereas the lexicon size in the HTK-based implementation was limited to about 25,000 entries.

Although parts of DIANA are built upon speech decoding algorithms that can also be used for automatic speech recognition, DIANA cannot be regarded as a variant of automatic speech recognition. The way in which activations are computed as functions over time is different, the way in which short input signals may activate longer words is entirely different, and DIANA's activation computations are more neurologically inspired by the use of STRFs. Fully neural-network inspired approaches (such as EARSHOT) may stimulate the development of a variant of DIANA in which not only the representations (e.g., STRFs) but also processes are neurally informed. Considerations, as put forward by [36–38] about restrictions on relations between the implementation on lower and higher Marr levels, will be guiding in this direction.

6. Conclusions

This article presented DIANA, a process-oriented computational model of human word recognition. It differs from many models in that its input is the same acoustic signal as enters the human ear and in that its output are the outputs that can be produced by human participants in psycholinguistic experiments, so that DIANA's plausibility can be directly tested. More importantly, DIANA's design accounts better for the recent findings in neurophysiological and psycholinguistic research than previous models. Most importantly, DIANA does not assume a pre-lexical layer in which hard decisions are made about abstract pre-lexical units before lexical access, but converts the acoustic signal into representations (i.e., spectro-temporal receptive fields) that are neurophysiologically attested. In addition, DIANA resolves ambiguity following the Hick–Hyman law.

These features also imply that DIANA is fundamentally different from ASR models, including those based on deep neural networks. As a consequence, with DIANA we have a cognitively more plausible model of word recognition, which makes it easier to test new hypotheses about the human word recognition process in a cognitively valid way.

DIANA is work in progress. We have published several short papers on older versions of the model, mostly focusing on aspects of its implementation. In the present article, we have focused on the conceptual choices we made for DIANA, which resulted in those implementations (which are described in more detail in Appendix A). In the near future, we hope to further extend DIANA such that it reflects even better everything that is known about the human word-recognition process. We trust that, also in its present version, among the recently proposed computational models, DIANA can play a seminal role for the advancement of process-based accounts of human word recognition.

Author Contributions: Conceptualization, L.t.B., L.B. and M.E.; computation, L.t.B. and L.B.; writing: L.t.B., L.B. and M.E.; funding acquisition: M.E. All authors have read and agreed to the published version of the manuscript.

Funding: This work was funded in part by an ERC starting grant (284108) and an NWO VICI grant awarded to Mirjam Ernestus.

Institutional Review Board Statement: Not applicable.

Informed Consent Statement: Not applicable.

Data Availability Statement: The version of DIANA described in this paper uses several tools from the Kaldi toolkit. The STRFs used in the simulations and their relations to phonetic symbols were obtained using the Spoken Dutch Corpus. Contact the first author for the code.

Conflicts of Interest: The authors declare no conflict of interest.

Abbreviations

The following abbreviations are used in this manuscript:

ANN	artificial neural network
CCN	cognitive control network
DMN	default mode network
DNN	deep neural network
ECoG	electrocortocography
STRF	spectro-temporal receptive field
MFCC	mel-frequency cepstral coefficient
RT	reaction time
LM	language model
SLM	statistical language model
NDL	naive discriminative learner
LDL	linear discriminative learner

Appendix A. Appendices

The four appendices below provide more detailed information about the computational approach underlying DIANA. In Appendix A.1, we discuss the statistical relation between MFCC feature vectors and spectro-temporal receptive fields. In Appendix A.2, we discuss the activation of words, pseudo-words and word cohorts. In Appendix A.3, we discuss the procedure for modeling lexical decisions. Finally, in Appendix A.4, we discuss various aspects of DIANA's entropy.

Appendix A.1. MFCC Feature Vectors and Spectro-Temporal Receptive Fields

In DIANA, the competition between words, parts of words and pseudo-word candidates is determined by the degree of match between the audio input and the internal representations of these candidates. The audio input is represented as a sequence of mel-frequency cepstral coefficient (MFCC) feature vectors [48]. Any candidate is, via its phone sequence representation and the corresponding hidden Markov model, represented as a sequence of so-called spectro-temporal receptive fields (STRFs) (see, e.g., [34]). Each Markov state is associated to an STRF. The match between the signal and the candidate is then defined by the statistical match between the MFCC sequence and the STRF-based Markov model.

Each individual MFCC feature vector is based on a speech segment of 25 ms wide, and is updated with a 10 ms time shift sliding through the speech signal. In this feature vector, both static and dynamic parts in the speech signal are accounted for [48]. For an utterance, this vector sequence forms a trajectory in the MFCC space, sampled 100 times a second. Obviously, new realisations of the same utterance (even when uttered by the same speaker) may lead to slightly deviant trajectories which may be acoustically different but count the same on a phonemic level. This implies that neighboring MFCC feature vectors along a trajectory probably belong to the same speech sound.

This suggests a statistical relation $P(\text{STRF}|\text{MFCC})$ between MFCCs and STRFs. This relation is established outside of DIANA via a conventional forced-alignment procedure between audio recordings and a parallel phone-level annotation. This alignment defines each STRF as a statistical distribution (cluster) in the MFCC space. STRFs pertaining to states belonging to one phone are similar and may have a substantial overlap in MFCC space, while STRFs related to different phones may be very dissimilar. Because STRFs partition the MFCC space in a statistical way, each trajectory in the MFCC space passes though a number of MFCC regions, thereby activating the hidden Markov states via $P(\text{STRF}|\text{MFCC})$ (see Appendix A.2).

The use of phones in the alignment is not essential for creating a useful set of STRFs, neither is the forced alignment procedure (Viterbi algorithm). Other clustering methods, e.g., an unsupervised clustering of MFCC vectors, may yield equally useful STRFs, but the current choice for phones makes it possible to interpret DIANA's output (e.g., the outputs in Section 3.4) and the candidates during competition in phonetic terms.

Appendix A.2. Activation of Words, Pseudo-Words and Word Cohorts

The activation of word candidates, parts of words and pseudo-words is based on a combination of bottom-up evidence from the speech signal and top-down prediction. For the computation of this activation, Bayes is the starting point (similar to [25,98]):

$$P(W|S) = P(S|W)P(W)/P(S) \tag{A1}$$

in which the left-hand side is the probability sought. The speech signal and the word candidate(s) are denoted by S and W, respectively. S is represented as a sequence of MFCC feature vectors, and W is represented by a Markov model, each Markov node modeled by an STRF.

In the computations, DIANA follows the actual practice in which the denominator $P(S)$ is ignored since it is independent of the word candidate W (see [48]). This step, turning probabilities into likelihoods, makes sense for psycholinguistic experiments; what counts is the activation of some word (sequence) W_0 relative to the activation of competing word (sequences) $W_m, i = 1, \cdots, M$. This allows us to ignore the prior probability $P(S)$ of the speech signal from the equations, since it is the same for all candidates. Moreover, all computations are performed in the logarithmic domain, that is, in terms of log-likelihoods.

During the unfolding of the signal S, DIANA must not only be able to deal with gated input signals, such as 'cathedr', but also with partial word candidates and pseudo-words. To compute on-line activations for word cohorts (parts of words starting at the beginning) and pseudo-words W while the acoustic signal S unfolds, DIANA extends Equation (A1) to:

$$\log P(W|S[0:t]) \sim \log P(S[0:t]|W) + \log P(W) \tag{A2}$$

in which $S[0:t]$ denotes the gated part of the signal S up to time t, and W may be a complete word or word part or pseudo-word. The $\sim$-sign is used because this equation is actually in terms of log likelihoods. Due to the assumption of independence of subsequent hidden Markov states, $\log P(S[0:t]|W)$ in the right-hand side of (A2) can be written as a sum over sequences of vectors (MFCC feature vectors) and corresponding states (STRFs):

$$\log P(S[0:t]|W) = \sum_{\text{vector, state}} \log P(\text{vector}|\text{state}) \tag{A3}$$

in which $P(\text{vector}|\text{state})$ is provided by the forced alignment that was used to construct the STRFs (Appendix A.1). As a result, $\log P(W|S[0:t])$ is an accumulation of feature vector-based contributions from 0 to t.

In order to enable DIANA to compare activation scores of hypotheses with different durations, we perform a duration normalization. This accounts for the fact that longer utterances yield more acoustic evidence, but what essentially counts is the amount of evidence per unit time. This has a parallel with human speech processing in which the time window within which a listener may revise an hypothesis cannot be not arbitrarily long. This normalisation is performed by dividing the log likelihood by the duration t of the speech signal $S[0:t]$. By doing so, we obtain a duration-normalized log likelihood $\log(P(W|S[0:t]))/t$. This normalized value is referred to as the activation of a word at time t.

Appendix A.3. Lexical Activation Score and Lexically Unrestricted Activation Score

To simulate a participant's lexicality judgments in a lexical decision experiment, DIANA adopts a strategy for making a word or pseudo-word decision. In a word identification task, words go into competition with other words. This situation is different from lexical decision, in which the competition is between words on the one hand and pseudo-words on the other. Since the criteria for word decisions may be different from the criteria for pseudo-word decisions [99], DIANA uses a strategy based on the comparison of two activation scores that are computed in parallel, a lexical score (based on all words in the lexicon) and a lexically unrestricted score (based on all phonotactically licensed phone sequences), for each gated signal $S[0:t]$ for all t:

$$argmax_{\text{any lexical form}} \log P(\text{form}|S[0:t])/t \tag{A4}$$

the lexical score, and

$$argmax_{\text{any phonotactically licensed form}} \log P(\text{form}|S[0:t])/t \tag{A5}$$

the lexically unrestricted score.

The results of the computation of the first, 'lexical', activation are presented in e.g., Figures 4 and 5. For the computation of the second activation, DIANA's search space is dynamically enlarged to allow all phonotactically licensed word forms, such as 'cath', 'cathedral', 'cathedruke', 'thedruke', etc. DIANA can create such pseudo-words as word hypotheses on the fly, on the basis of a phone network in which phones are represented as nodes, such that only those phone combinations that are phonotactically licensed appear as possible paths through the network. With respect to the computation of activation scores, pseudo-words or word parts do not behave in a principally different way from real words.

Conceptually, both the lexical and lexically unconstrained activation scores make sense, albeit in different ways; the lexical activation is the key ingredient in the word-to-word competition in speech comprehension of known words, while the second activation is at stake when listeners are confronted with unknown words (e.g., new names) or pseudo-words. Recent studies using EEG analyses [17] show how listeners can take recourse to a phonological grammar to process unknown words. This is related to DIANA's network-based strategy (described above) underlying the lexically unrestricted activation score.

If all top-down predictions are equal, the second score is always at least as good as the first, since the lexically constrained phone sequences form a subset of the phonotactically licensed phone sequences. This implies that the difference between these activations is an indication for the lexicality of the stimulus. DIANA's use of a decision criterion based on the difference between two activations is commonly used in general models of human decision and RT distributions, e.g., the ballistic accumulation model (BAM) [54] and the linear approach to threshold with ergodic rate models (LATER, [55,56]). For several reasons, among others the effect of between-speaker pronunciation variation and the probabilistic relation between MFCCs and STRFs, it is not guaranteed that the activation score difference distinguishes lexical from non-lexical inputs in a fully reliable way. Inevitably, this will give rise to lexicality judgment errors.

Figure A1 (see also [1,2]) shows the distributions of the lexical activations for existing words (in blue) and the lexically unconstrained activations for pseudo-words (in red) of the 2780 existing and 2761 pseudo-words in BALDEY [63]. For each stimulus, the difference between the highest lexical and the highest non-lexical activation is displayed on the horizontal axis. The dashed vertical line represents the position of a criterion value θ (a model parameter) that can be used as a threshold to decide the lexical status of a stimulus. If the difference between lexical and non-lexical activation exceeds θ, the stimulus is classified as a real word, otherwise it is assumed to be a pseudo-word. The expected classification error will depend on the structure of the pseudo-word stimuli in a lexical decision experiment and the exact way they violate lexicality. For example, in stimulus sets in which pseudo-words only deviate from real words in one phone, the blue and red distributions might overlap to a large extent (due to the small acoustic difference between the pseudo-word and its closest real word). In the case of clear acoustic differences between pseudo-words and real words, the blue and red distribution would hardly overlap.

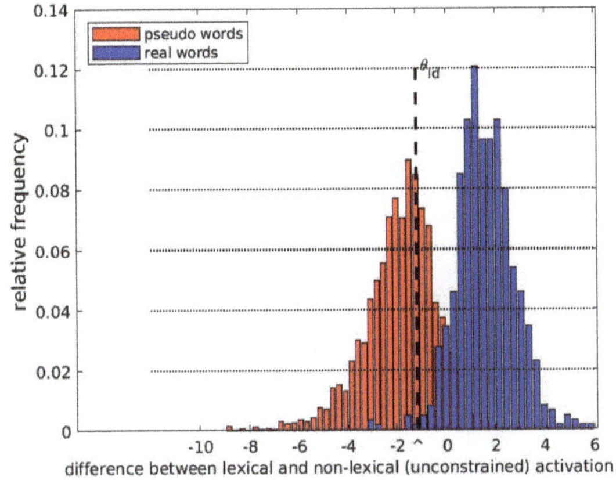

Figure A1. Histogram of the difference of lexical activation and non-lexical activation of 2780 existing words (blue) and 2761 pseudo-words (red) in BALDEY. The dashed vertical line indicates the lexical decision threshold (denoted θ with subscript ld, lexical decision); its optimal value depends on the task. In the figure, it is chosen such that the probability of words erroneously receiving a pseudo-word label would be minimal.

Appendix A.4. Computation of Entropy

In DIANA, entropy, a measure of the 'degree of disorder' in a physical system or of the 'complexity of a decision process', is assumed to be a contributing factor to the long latencies manifest in many psycho-linguistic tasks. During the word search, the entropy is computed from the word probabilities, which are computed from the scaled log likelihoods $\log P(W|S[0:t])$ in Equation (A2). This is performed for each t, such that entropy values are available as a function of time. To compute the probabilities from the scaled log likelihoods, DIANA used a procedure that is similar to the probability normalization step often applied in speech decoding research (A6). In this procedure, the scaled log likelihoods are first transformed into unscaled log likelihoods, by applying a multiplication with a constant c (which is to be estimated from data). Next, the 'softmax' Luce rule is applied to convert the log likelihoods to probabilities, via normalisation to make the sum equal to 1. In total:

$$p_{W,t} = \frac{\exp(c \cdot \log P(W|S[0:t]))}{\sum_W \exp(c \cdot \log P(W|S[0:t]))}, \tag{A6}$$

for each time point t. The sum in the denominator runs over all hypotheses viable at time t. The value of c is estimated to be approximately -0.05, by using the word-confidence estimation approach described in [100]. This value is an approximation; a more precise value can be obtained if the list of candidates is made more precise. The value of c is not critical for the evaluation of the entropy. Similar methods are also applied to compute word confidence measures [101]. In DIANA the entropy was computed after removal of all 'nested' hypotheses from the list of hypotheses (see Section 4). Although the conceptual aspects underlying this procedure are transparent, the required computations are often quite technical in nature.

Appendix A.4.1. Entropy as A Contributing Factor of Reaction Time

Reaction times in a behavioral experiment are a good example of outcomes of a complex cognitive process in which many factors play a role at various time scales (e.g., [3,7,12,102–104]. One of the factors that is likely to play a role in the explanation of reaction time is the complexity of the problem that a participant must solve while making a decision. In lexical decision tasks, the RT distributions are typically very skewed with a long tail towards long RTs [63,105]. For example, in the large-scale Dutch lexical decision database BALDEY [63], which contains over 110,000 lexical decisions and reaction times, only a very small subset of the stimuli receive a valid reaction before the end of a stimulus, about 50% of all RTs (when measured from stimulus offset) exceed 510 ms, and 10% exceed 1560 ms. Moreover, Ref. [105] reports similarly substantial RTs. These long tails indicate that participants may need a substantial amount of time to make a decision, since the time that elapses between stimulus offset and an overt response (e.g., a button press) largely exceeds the neural traveling time required to effectuate an overt response (which is 200 ms at most).

The long tail in RT distributions form a challenge for modeling. In regression models of RT, the dependent variable is often transformed using $\log(RT)$ or the inverse $(1/RT)$, which makes the distribution of the transformed RTs more Gaussian-like. In BAM and similar models [54,56], the skewness is dealt with by putting constraints on the distribution of the drift rate. DIANA, as a process-oriented model, must be able to explain reaction times far beyond the stimulus offset on the basis of activations that are computed before stimulus offset. This is performed by relating the additional reaction time with the complexity of the choice. DIANA takes the Hick–Hyman law as a starting point [64–66]. In the case of N options with equal probability $p = 1/N$, the time necessary to choose one option is linear in the log-transformed number of items:

$$\Delta T = A + B \cdot \log(N) = A + B \cdot \sum_{\text{all } N \text{ items}} -p\log(p) \tag{A7}$$

with A and B constants that depend on the details of the experiment. In DIANA, this situation is translated into the choice a listener has to make among N hypotheses, each with different probabilities (p, from Equation (A6)). To that end, the right-hand term in Equation (A7) is generalized to the entropy $\sum_{i=1}^{N} -p_i \log(p_i) = H(p_1, ..., p_N)$. DIANA's *choice RT*, the contribution to the overall RT due to entropy, is then modeled as

$$\text{choice RT} = \beta \cdot H(p_1, \ldots, p_N) \tag{A8}$$

in which the factor $\beta > 0$ is one of the meta-parameters in DIANA, translating entropy into additional reaction time, and p_i are the probabilities of the hypotheses as derived from Equation (A6). A larger degree of ambiguity (e.g., more close competitors, very similar words in the competition, pseudo-words close to real words) leads to larger entropy and so will increase the reaction time. In the case that there is no ambiguity, H equals zero and so the choice RT vanishes.

When entropy is taken into account, one option to express DIANA's RT predictions beyond stimulus offset reads as follows:

$$\begin{aligned} \text{DIANA RT}_{\text{onset}} = \; & \text{stimulus duration} + \beta \cdot H() + \\ & f(\text{morpho-syntactic factors}) + \\ & \text{execution time} \end{aligned} \tag{A9}$$

in which DIANA's meta-parameter β translates entropy H() to additional choice RT (via choice RT $= \beta \cdot H(p_1, \ldots, p_N)$). Here, f denotes a zero-mean (as yet unspecified) function that modulates the reaction time on the basis of morpho-syntactic factors of the stimulus. In the following subsection we show that this option is supported by regression modeling, by simulating data on a morphologically homogeneous subset of BALDEY, such that the variation in f is limited.

An even more challenging option for integrating entropy would be in line with what has been discussed in the Discussion section concerning the extension of DIANA's lexicon with morphologically and semantically relevant information:

$$\begin{aligned} \text{DIANA RT}_{\text{onset}} = \; & \text{stimulus duration}+ \\ & \beta \cdot H(\text{morpho-syntactic-semantic factors})+ \\ & \text{execution time} \end{aligned} \tag{A10}$$

in which first morpho-syntactic properties, such as stem sharing, family size effects, and semantic relations, modulate the hypotheses' probabilities, on the basis of which a new entropy H() is computed.

Figure A2 illustrates the distribution of the entropy (a density plot) as computed in DIANA's decision component for all 5541 auditory stimuli used in BALDEY [63]. The distribution shows a tail towards 0. This tail is due to relatively long stimuli, some of which have no—or at best very few—competitors left at their offset. The mismatch between the highly asymmetric distribution of entropy values (long tail towards the lower values) and the fairly symmetric distribution of log-transformed RTs shows that entropy on its own may not be a very powerful predictor of RTs, but nevertheless may serve as a significant predictor of RTs in regression models. An example is shown in the next subsection.

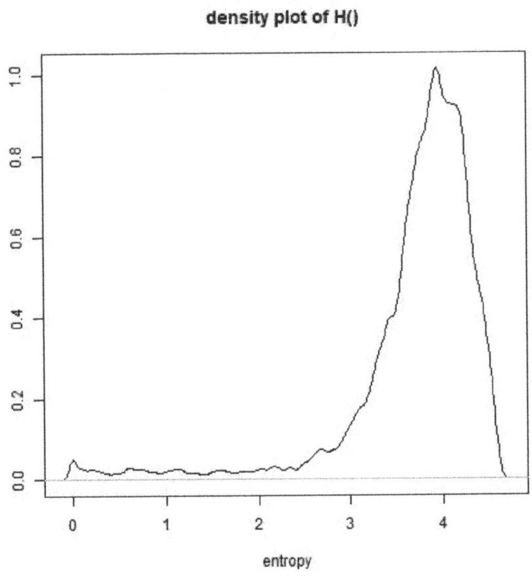

Figure A2. Density of the entropy $H()$ over all 5541 BALDEY stimuli.

Appendix A.4.2. Entropy as a predictor of the RTs in BALDEY

To illustrate the impact of DIANA's entropy computed at offset of the acoustic stimuli in predicting RT values, we present one linear mixed-effects [106] -based regression analysis using the data in BALDEY [63]. In DIANA, it is assumed, via Hick's law, that a larger entropy leads to longer reaction times when all other factors are assumed equal, while taking into account the challenges connected to this interpretation as voiced in [66]. If true, this should correspond to the entropy having a positive regression coefficient in models of RT.

Since RT distributions typically comprise some extremely fast, and a long tail of very slow, reactions that arguably are not representative of the 'normal' cognitive processes, we ignored all trials with an RT measured from stimulus onset (RTonset) in the lowest percentile (pertaining to an RT threshold of 550 ms) and all RTs beyond $\mu + 2\sigma$. This step removed approximately six percent of the BALDEY trials, respectively. In the analysis, we further limited the trials to those BALDEY stimuli that were correctly responded to. The reaction times were log-transformed and served as the dependent variable.

We also added the predictor 'wordiness', defined as the ratio of the activation of the best lexical candidate and the activation of the winning candidate (word or pseudo-word). Furthermore, we added conventional predictors (see also [63]), such as `logwdur` (log-transformed word duration), `logFreq`, `session`, `trial`, `wordclass`, and `compoundtype`. The predictor `wordclass` is a factor with three levels (adjective, noun, and verb, with, in BALDEY 22,154, 48,861 and 32,917 tokens, respectively). 'Adjectives' are on the intercept. `Compoundtype` distinguishes four types of compounds (simple, adj+noun, noun+adj, and noun+noun). Here, 'simple' is on the intercept. Since the correlation between `logwdur` and `entropy` was $r = -0.51$, `entropy` was residualized over log word duration, with `entrlogwdur` as a new predictor. Finally, we added two predictors `prevBVis` and `maRT` [104,107] as control predictors. These predictors model the local trends that exist in human RT sequences that are independent of the stimulus itself but have a mid-term range of about 10–20 stimuli due to, e.g., learning effects, fatigue and fluctuating attention (for details about `prevBVis` and `maRT`, see, e.g., [104]).

Model search was by backward elimination starting from a regression model with all predictors and plausible interactions in the fixed structure and control predictors as random slopes without interactions. The final lmer model reported here has been derived according to the guidelines in [102,108,109]. Insignificant interactions and main predictors are left out of the final model. Random slopes were only included insofar as the resulting models converged and the AIC value was improved. Table A1 presents the results of the final lmer model for RTonset. The AIC of this model (`m`) equals −753.2811.

```
m = lmer(logRT ~ logwdur*logFreq+entrlowdur+wordiness+
    session+trial+wordclass+maRT+prevBVis+compound_type+(1|word)+
    (1 |subject)+(0+maRT |subject),
    data=data4[data4$response == "correct",])
```

Table A1. Output of lmer model modeling RTonset, including the DIANA-based predictors `entrlogwdur`, which is the entropy residualized over log word duration, and `wordiness`. For details see the text.

$\mathbf{RT_{onset}}$			
	Estimate	**Std. Error**	***t* Value**
(Intercept)	6.742×10^{-1}	3.265×10^{-1}	2.065
logwdur	3.306×10^{-1}	6.513×10^{-3}	50.760
logFreq	7.011×10^{-2}	1.135×10^{-2}	6.178
entrlogwdur	2.354×10^{-2}	2.039×10^{-3}	11.548
wordiness	5.559×10^{-2}	9.084×10^{-3}	6.119
session	2.564×10^{-3}	2.919×10^{-4}	8.783
trial	2.672×10^{-5}	4.883×10^{-6}	5.472
wordclass(nom)	6.933×10^{-3}	2.965×10^{-3}	2.338
wordclass(verb)	3.106×10^{-2}	2.951×10^{-3}	10.526
maRT	6.000×10^{-1}	4.299×10^{-2}	13.958
prevBVis	2.651×10^{-3}	$2.883e \times 10^{-4}$	9.196
compoundtype(A+N)	-3.023×10^{-2}	8.855×10^{-3}	−3.414
compoundtype(N+A)	-7.375×10^{-3}	1.018×10^{-2}	−0.724
compoundtype(N+N)	4.197×10^{-3}	4.236×10^{-3}	0.991
logwdur:logFreq	-1.277×10^{-2}	1.785×10^{-3}	−7.153

The final model for RTonset was significantly better (in terms of AIC) than alternative models, including models in which `entrlogwdur` or `wordiness` were left out. Table A1 provides regression coefficients and t-values of this model; under the assumption of the validity of the t-distribution, all values of $|t| > 1.96$ are considered to indicate significance at the level $p < 0.05$. In this model the two-way interaction between `logwdur` and log frequency is kept, and `maRT` serves as a random slope under subject without correlation with the intercept. As expected, (log) word duration (`logwdur`) is one of the most significant predictors of reaction time measured from stimulus onset.

Interestingly, the Table shows that both DIANA-related predictors (`entrlogwdur` and `wordiness`) are significant, with positive βs. A larger entropy (measured at stimulus offset) thus leads to a larger reaction time, all other factors being equal. The positive β for wordiness shows that the more probability mass is attributed to word hypotheses, the slower the decision. This seems counter-intuitive, since one would expect that the larger the probability mass attributed to one hypothesis, the faster the decision should be. However, wordiness is the score of the top-ranking hypothesis, irrespective of whether that is an existing word or a pseudo-word. It appears that the number of cohorts formed by sequences of transcription symbols resulting from trying the 'transcribe' pseudo-word stimuli is much smaller than the number of cohorts that make up the words in a large lexicon.

The other coefficients can be explained on the basis of earlier findings (e.g., [63]). Higher word frequency leads to smaller RTs, but modulated by an interaction with word duration. In general, later sessions and trials lead to slower RTs. Compared to adjectives, nouns and verbs yield longer RTs. Noun-noun compounds produce the slowest RTs among compounds. The 'local trend'-related control predictors maRT and prevBVis are highly significant with $\beta > 0$, again showing the usual substantial local speed effect [104].

Finally, Figure A3 shows the predictions of DIANA when simulating a word identification task. The dataset was chosen to be a morphologically homogeneous subset of BALDEY (real words, morphologically simple, bi-syllabic), such that morphological factors are factored out as much as possible (see also [1–3]).

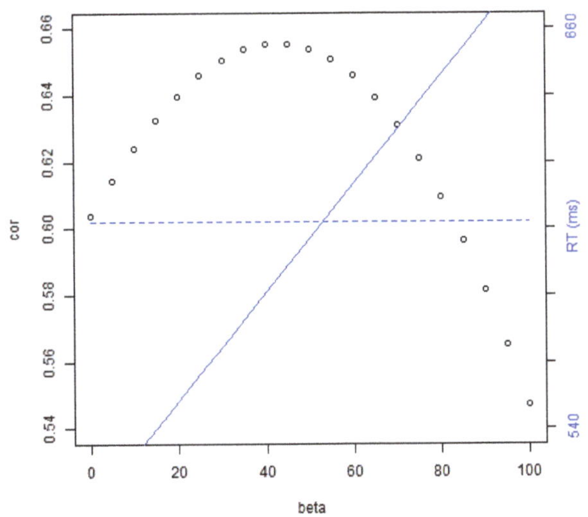

Figure A3. DIANA's simulations for a word identification task. The dataset is a morphologically homogeneous subset of BALDEY. The black dotted curve shows the correlation between DIANA's RTs and the average RT of participants as a function of beta (β), the coefficient of the entropy in the equation for DIANA's RT. The solid blue line shows the average of DIANA's RT as a function of β, while the dashed blue horizontal line indicates the participant's mean RT. The right-hand side vertical axis pertains to the blue curves.

The figure shows the performance of DIANA via Equation (A9) in terms of the correlation between DIANA's RT sequences and the average RTs from participants in BALDEY (the dotted black line) and the average RT (solid blue line), both as a function of the coefficient β in Equation (A9). The black dotted line shows that the highest correlation between DIANA and the participants is obtained for a value of $\beta > 0$ ($\rho_{max} = 0.651$, 95% confidence interval [0.641–0.663]), i.e., significantly higher than the prediction without entropy ($\beta = 0$).

The dashed blue horizontal line indicates the participant's mean RT. The right-hand side vertical axis pertains to the blue curves. The figure suggests that there is no single value for β for which both the correlation is optimal (highest point in black curve) and the average RT prediction is correct (crossing of solid and dashed blue lines). This strongly indicates that the operational definition of entropy, as currently used, can be refined, by, e.g., taking into account more complex (morphologically oriented) word–word relation during DIANA's word competition stage, in line with the discussion of Equations (A9) and (A10).

References

1. ten Bosch, L.; Boves, L.; Ernestus, M. Towards an end-to-end computational model of speech comprehension: Simulating a lexical decision task. In Proceedings of the Interspeech, Lyon, France, 25–29 August 2013.
2. ten Bosch, L.; Ernestus, M.; Boves, L. Comparing reaction times from human participants and computational models. In Proceedings of the Interspeech, Singapore, 14–18 September 2014.
3. ten Bosch, L.; Boves, L.; Tucker, B.; Ernestus, M. DIANA: Towards computational modeling reaction times in lexical decision in North American English. In Proceedings of the Interspeech, Dresden, Germany, 6–10 September 2015.
4. ten Bosch, L.; Boves, L.; Ernestus, M. Combining data-oriented and process-oriented approaches to modeling reaction time data. In Proceedings of the Interspeech, San Francisco, CA, USA, 8–12 September 2016.
5. ten Bosch, L.; Boves, L.; Ernestus, M. The recognition of compounds: A computational account. In Proceedings of the Interspeech, Stockholm, Sweden, 20–24 September 2017.
6. Nenadić, F.; ten Bosch, L.; Tucker, B.V. Implementing DIANA to Model Isolated Auditory Word Recognition in English. *Proc. Interspeech* **2018**, *2018*, 3772–3776. [CrossRef]
7. ten Bosch, L.; Boves, L. Word Competition: An Entropy-Based Approach in the DIANA Model of Human Word Comprehension. *Proc. Interspeech* **2021**, *2021*, 531–535. [CrossRef]
8. Scharenborg, O.; Boves, L. Computational modelling of spoken-word recognition processes: Design choices and evaluation. *Pragmat. Cogn.* **2010**, *18*, 136–164. [CrossRef]
9. Marslen-Wilson, W.; Welsh, A. Processing interactions and lexical access during word recognition in continuous speech. *Cogn. Psychol.* **1978**, *10*, 29–63. [CrossRef]
10. Marslen-Wilson, W. Functional parallellism in spoken word recognition. *Cognition* **1987**, *25*, 71–102. [CrossRef]
11. Marslen-Wilson, W.; Tyler, L. The temporal structure of spoken language understanding. *Cognition* **1980**, *8*, 1–71. [CrossRef]
12. Cutler, A. *Native Listening: Language Experience and the Recognition of Spoken Words*; MIT Press: Cambridge, MA, USA, 2012.
13. Marslen-Wilson, W. Activation, competition and frequency in lexical access. In *Cognitive Models of Speech Processing: Psycholinguistic and Computational Perspectives*; Altman, G.T.M., Ed.; MIT Press: Cambridge, MA, USA, 1990; pp. 148–172.
14. Marslen-Wilson, W.; Brown, C.; Tyler, L. Lexical representations in spoken language comprehension. *Lang. Cogn. Process.* **1988**, *3*, 1–16. [CrossRef]
15. Bard, E.; Shillcock, R.; Altmann, G. The recognition of words after their acoustic offsets in spontaneous speech: Effects of subsequent context. *Percept. Psychophys.* **1988**, *44*, 395–408. [CrossRef]
16. Marr, D. *Vision: A Computational Approach*; Freeman & Co.: San Francisco, CA, USA, 1982.
17. Silva, S.; Vigário, M.; Fernandez, B.L.; Jerónimo, R.; Alter, K.; Frota, S. The Sense of Sounds: Brain Responses to Phonotactic Frequency, Phonological Grammar and Lexical Meaning. *Front. Psychol.* **2019**, *10*, 1–11. [CrossRef]
18. Gow, D.; Olson, B. Lexical mediation of phonotactic frequency effects on spoken word recognition: A Granger causality analysis of MRI-constrained MEG/EEG data. *J. Mem. Lang.* **2015**, *82*, 41–55. [CrossRef]
19. Gwilliams, L.; King, J.R.; Marantz, A.; Poeppel, D. Neural dynamics of phoneme sequencing in real speech jointly encode order and invariant content. **2020**, *preprint*. [CrossRef]
20. Port, R.F. Rich memory and distributed phonology. *Lang. Sci.* **2009**, *32*, 43–55. [CrossRef]
21. McClelland, J.L.; Elman, J.L. The TRACE model of speech perception. *Cogn. Psychol.* **1986**, *18*, 1–86. [CrossRef]
22. Usher, M.; McClelland, J.L. On the time course of perceptual choice: The leaky competing accumulator model. *Psychol. Rev.* **2001**, *108*, 550–592. [CrossRef]
23. Norris, D. Shortlist: A connectionist model of continuous speech recognition. *Cognition* **1994**, *52*, 189–234. [CrossRef]
24. Magnuson, J.S.; You, H.; Luthra, S.; Li, M.; Nam, H.; Escabí, M.; Brown, K.; Allopenna, P.D.; Theodore, R.M.; Monto, N.; et al. EARSHOT: A Minimal Neural Network Model of Incremental Human Speech Recognition. *Cogn. Sci.* **2020**, *44*, e12823. [CrossRef]

25. Norris, D.; McQueen, J. Shortlist B: A Bayesian Model of Continuous Speech Recognition. *Psychol. Rev.* **2008**, *115*, 357–395. [CrossRef]
26. Smits, R.; Warner, N.; McQueen, J.; Cutler, A. Unfolding of phonetic information over time: A database of Dutch diphone perception. *J. Acoust. Soc. Am.* **2003**, *113*, 563–574. [CrossRef]
27. Warner, N.; Smits, R.; McQueen, J.; Cutler, A. Phonological and frequency effects on timing of speech perception: A database of Dutch diphone perception. *Speech Commun.* **2005**, *46*, 53–72. [CrossRef]
28. Scharenborg, O. Modelling fine-phonetic detail in a computational model of word recognition. In *Proceedings of Interspeech*; Causal Productions Pty Ltd.: Brisbane, Australia, 2008.
29. Scharenborg, O. Modeling the use of durational information in human spoken-word recognition. *J. Acoust. Soc. Am.* **2010**, *127*, 3758–3770. [CrossRef]
30. Salverda, A.P.; Dahan, D.; McQueen, J.M. The role of prosodic boundaries in the resolution of lexical embedding in speech comprehension. *Cognition* **2003**, *90*, 51–89. [CrossRef]
31. Shafaei-Bajestan, E.; Moradipour-Tari, M.; Uhrig, P.; Baayen, R.H. LDL-AURIS: A computational model, grounded in error-driven learning, for the comprehension of single spoken words. *Lang. Cogn. Neurosci.* **2021**, 1–28. [CrossRef]
32. Hochreiter, S.; Schmidhuber, J. Long Short-Term Memory. *Neural Comput.* **1997**, *9*, 1735–1780. [CrossRef]
33. Mikolov, T.; Sutskever, I.; Chen, K.; Corrado, G.S.; Dean, J. Distributed representations of words and phrases and their compositionality. In Proceedings of the 26th International Conference on Neural Information Processing Systems, Lake Tahoe, NV, USA, 5–10 December 2013; Volume 2, pp. 3111–3119.
34. Mesgarani, N.; Cheung, C.; Johnson, K.; Chang, E.F. Phonetic Feature Encoding in Human Superior Temporal Gyrus. *Science* **2014**, *343*, 1006–1010. [CrossRef]
35. Bhaya-Grossman, I.; Chang, E.F. Speech Computations of the Human Superior Temporal Gyrus. *Annu. Rev. Psychol.* **2021**, *73*, 1–24. [CrossRef]
36. Love, B.C. The Algorithmic Level Is the Bridge Between Computation and Brain. *Top. Cogn. Sci.* **2015**, *7*, 230–242. [CrossRef]
37. Griffiths, T.L.; Lieder, F.; Goodman, N.D. Rational use of cognitive resources: Levels of analysis between the computational and the algorithmic. *Top. Cogn. Sci.* **2015**, *7*, 217–229. [CrossRef]
38. Cooper, R.P.; Peebles, D. On the Relation Between Marr's Levels: A Response to Blokpoel. *Top. Cogn. Sci.* **2017**, *10*, 649–653. [CrossRef]
39. Aertsen, A.; Johannesma, P.I. The spectro-temporal receptive field. A functional characteristic of auditory neurons. *Biol. Cybern.* **1981**, *42*, 133–143. [CrossRef]
40. Hullett, P.W.; Hamilton, L.S.; Mesgarani, N.; Schreiner, C.E.; Chang, E.F. Human Superior Temporal Gyrus organization of spectrotemporal modulation tuning derived from speech stimuli. *J. Neurosci. Off. J. Soc. Neurosci.* **2016**, *36*, 2014–2026. [CrossRef]
41. Chang, K.; Mitchell, T.; Just, M. Quantitative modeling of the neural representation of objects: How semantic feature norms can account for fMRI activation. *Neuroimage* **2011**, *56*, 716–727. [CrossRef]
42. Joos, M. *Acoustic Phonetics. Language Monograph 23*; Linguistic Society of America: Baltimore, ML, USA, 1948.
43. Talavage, T.; Sereno, M.; Melcher, J.; Ledden, P.; Rosen, B.; Dale, A.M. Tonotopic organization in human auditory cortex revealed by progressions of frequency sensitivity. *J. Neurophysiol.* **2004**, *91*, 1282–1296. [CrossRef]
44. Fant, G. *Speech Sounds and Features*; MIT Press: Cambridge, MA, USA, 1973.
45. Liberman, A.; Delattre, P.; Cooper, F.; Gerstman, L. The Role of Consonant-Vowel Transitions in the Perception of the Stop and Nasal Consonants. *Psychol. Monogr. Gen. Appl.* **1954**, *68*, 1–13. [CrossRef]
46. Gordon-Salant, S. Recognition of Natural and Time/Intensity altered CVs by Young and Elderly Subjects with Normal Hearing. *JASA* **1986**, *80*, 1599–1607. [CrossRef]
47. Davis, S.; Mermelstein, P. Comparison of Parametric Representations for Monosyllabic Word Recognition in Continuously Spoken Sentences. *IEEE Trans. Acoust. Speech, Signal Process.* **1980**, *28*, 357–366. [CrossRef]
48. Holmes, J.; Holmes, W. *Speech Synthesis and Recognition*, 2nd ed.; Taylor and Francis: London, UK; New York, NY, USA, 2002.
49. Jurafsky, D.; Martin, J. *Speech and Language Processing (Online)*, 3rd ed.; Pearson: London, UK, 2021.
50. Riad, R.; Karadayi, J.; Bachoud-Lévi, A.; Dupoux, E. Learning spectro-temporal representations of complex sounds with parameterized neural networks. *J. Acoust. Soc. Am.* **2021**, *150*, 353–366. [CrossRef]
51. Connolly, J.F.; Phillips, N.A. Event-related potential components reflect phonological and semantic processing of the terminal word of spoken sentences. *J. Cogn. Neurosci.* **1994**, *6*, 256–266. [CrossRef]
52. Bentum, M.; ten Bosch, L.; van den Bosch, A.; Ernestus, M. Listening with Great Expectations: An Investigation of Word Form Anticipations in Naturalistic Speech. *Proc. Interspeech* **2019**, *2019*, 2265–2269. [CrossRef]
53. Wells, J. SAMPA computer readable phonetic alphabet. In *Handbook of Standards and Resources for Spoken Language Systems*; Part IV, Section B; Gibbon, D.; Moore, R.; Winski, R., Eds.; Mouton de Gruyter: Berlin, Germany; New York, NY, USA, 1997.
54. Brown, S.; Heathcote, A. The simplest complete model of choice response time: Linear Ballistic Accumulation. *Cogn. Psychol.* **2008**, *57* , 153–178. [CrossRef]
55. Noorani, I.; Carpenter, R.H. The LATER model of reaction time and decision. *Neurosci. Biobehav. Rev.* **2016**, *64*, 229–251. [CrossRef]
56. Nakahara, H.; Nakamura, K.; Hikosaka, O. Extended LATER model can account for trial-by-trial variability of both pre- and post-processes. *Neural Netw.* **2006**, *19*, 1027–1046. [CrossRef]

57. Salinas, E.; Scerra, V.E.; Hauser, C.K.; Costello, M.G.; Stanford, T.R. Decoupling speed and accuracy in an urgent decision-making task reveals multiple contributions to their trade-off. *Front. Neurosci.* **2014**, *8*, 85. [CrossRef] [PubMed]
58. Bogacz, R.; Brown, E.; Moehlis, E.; Holmes, P.; Cohen, J.D. The physics of optimal decision making: A formal analysis of models of performance in two-alternative forced choice tasks. *Psychol. Rev.* **2006**, *113*, 700–765. [CrossRef] [PubMed]
59. Wang, X.J. Decision making in recurrent neuronal circuits. *Neuron* **2008**, *60*, 215–234. [CrossRef] [PubMed]
60. Summerfield, C.; Blangero, A. Chapter 12 - Perceptual Decision-Making: What Do We Know, and What Do We Not Know? In *Decision Neuroscience*; Dreher, J.C.; Tremblay, L., Eds.; Academic Press: San Diego, CA, USA, 2017; pp. 149–162. [CrossRef]
61. Suri, G.; Gross, J.J.; McClelland, J.L. Value-based decision making: An interactive activation perspective. *Psychol. Rev.* **2020**, *127*, 153–185. [CrossRef]
62. Lepora, N.; Pezzulo, G. Embodied Choice: How Action Influences Perceptual Decision Making. *PLoS Comput. Biol.* **2015**, *11*, e1004110 . [CrossRef]
63. Ernestus, M.; Cutler, A. BALDEY: A database of auditory lexical decisions. *Q. J. Exp. Psychol.* **2015**, *68*, 1469–1488. [CrossRef]
64. Hick, W.E. On the Rate of Gain of Information. *Q. J. Exp. Psychol.* **1952**, *4*, 11–26. [CrossRef]
65. Hyman, R. Stimulus information as a determinant of reaction time. *J. Exp. Psychol.* **1953**, *45*, 188–196. [CrossRef]
66. Proctor, R.W.; Schneider, D.W. Hick's law for choice reaction time: A review. *Q. J. Exp. Psychol.* **2018**, *71*, 1281–1299. [CrossRef]
67. Wu, T.; Dufford, A.J.; Egan, L.J.; Mackie, M.A.; Chen, C.; Yuan, C.; Chen, C.; Li, X.; Liu, X.; Hof, P.R.; et al. Hick–Hyman Law is Mediated by the Cognitive Control Network in the Brain. *Cereb. Cortex* **2017**, *28*, 2267–2282. [CrossRef]
68. Usher, M.; Olami, Z.; McClelland, J.L. Hick's law in a stochastic race model with speed-accuracy trade-off. *J. Math. Psychol.* **2002**, *46*, 704–715. [CrossRef]
69. Fan, J.; Guise, K.G.; Liu, X.; Wang, H. Searching for the Majority: Algorithms of Voluntary Control. *PLoS ONE* **2008**, *3*, e3522. [CrossRef]
70. Hawkins, G.; Brown, S.D.; Steyvers, M.; Wagenmakers, E.J. Context Effects in Multi-Alternative Decision Making: Empirical Data and a Bayesian Model. *Cogn. Sci.* **2012**, *36*, 498–516. [CrossRef] [PubMed]
71. Miller, E.K.; Cohen, J.D. An Integrative Theory of Prefrontal Cortex Function. *Annu. Rev. Neurosci.* **2001**, *24*, 167–202. [CrossRef]
72. Fan, J. An information theory account of cognitive control. *Front. Hum. Neurosci.* **2014**, *8*, 680 . [CrossRef]
73. Harding, I.H.; Yücel, M.; Harrison, B.J.; Pantelis, C.; Breakspear, M. Effective connectivity within the frontoparietal control network differentiates cognitive control and working memory. *NeuroImage* **2015**, *106*, 144–153. [CrossRef] [PubMed]
74. Fedorenko, E.; Duncan, J.; Kanwisher, N. Broad domain generality in focal regions of frontal and parietal cortex. *Proc. Natl. Acad. Sci. USA* **2013**, *110*, 16616–16621. [CrossRef] [PubMed]
75. Niendam, T.; Laird, A.; Ray, K.; Dean, Y.; Glahn, D.; Carter, C. Meta-analytic evidence for a superordinate cognitive control network subserving diverse executive functions. *Cogn. Affect. Behav. Neurosci.* **2012**, *12*, 241–268. [CrossRef]
76. Cocchi, L.; Zalesky, A.; Fornito, A.; Mattingley, J. Dynamic cooperation and competition between brain systems during cognitive control. *Trends Cogn. Sci.* **2013**, *17*, 493–501. [CrossRef]
77. Gahl, S. "Thyme" and "time" are not homophones. The effect of lemma frequency on word durations in spontaneous speech. *Languge* **2008**, *84*, 474–496. [CrossRef]
78. Hawkins, S. Roles and representations of systematic fine phonetic detail in speech understanding. *J. Phon.* **2003**, *31*, 373–405. [CrossRef]
79. Balling, L.W.; Baayen, R.H. Probability and surprisal in auditory comprehension of morphologically complex words. *Cognition* **2012**, *125*, 80–106. [CrossRef] [PubMed]
80. Bybee, J.L. Morphology as lexical organization. *Theor. Morphol.* **1988**, *1988*, 119141.
81. Dilkina, K.; McClelland, J.L.; Plaut, D.C. Are there mental lexicons? The role of semantics in lexical decision. *Brain Res.* **2010**, *1365*, 66–81. [CrossRef]
82. Zhao, Y; Li, J.; Wang, X.; Li, Y. The Speechtransformer for Large-scale Mandarin Chinese Speech Recognition. In Proceedings of the ICASSP 2019 IEEE International Conference on Acoustics, Speech and Signal Processing, Brighton, UK, 12–17 May 2019; pp. 7095–7099. [CrossRef]
83. Dijkstra, T. The multilingual lexicon In *Handbook of Psycholinguistics*; Oxford University Press: Oxford, UK, 2007; pp. 251–265.
84. Sundermeyer, M.; Schlüter, R.; Ney, H. LSTM Neural Networks for Language Modeling. *Proc. Interspeech* **2012**, *2012*, 1–4 . [CrossRef]
85. Chen, D.; Manning, C. A fast and accurate dependency parser using neural networks. In Proceedings of the Conference on Empirical Methods in Natural Language Processing, Doha, Qatar, 25–29 October 2014, pp. 74–750.
86. Merkx, D.; Frank, S.L.; Ernestus, M. Language learning using speech to image retrieval. *Proc. Interspeech* **2019**, *2019*, 1841–1845.
87. Tsuji, S.; Cristia, A.; Dupoux, E. SCALa: A blueprint for computational models of language acquisition in social context. *Cognition* **2021**, *213*, 104779. [CrossRef]
88. Boves, L.; ten Bosch, L.; Moore, R.K. ACORNS-towards computational modeling of communication and recognition skills. In Proceedings of the Sixth IEEE International Conference on Cognitive Informatics, Lake Tahoe, CA, USA, 6–8 August 2007; Zhang, D., Wang, Y., Kinsner, W., Eds.; 2007; pp. 349–356. [CrossRef]
89. Driesen, J.; Van hamme, H. Modelling vocabulary acquisition, adaptation and generalization in infants using adaptive Bayesian PLSA. *Neurocomputing* **2011**, *74*, 1874–1882. [CrossRef]
90. Romberg, A.; Saffran, J. Statistical learning and language acquisition. *Wiley Interdiscip. Rev. Cogn. Sci.* **2010**, *1*, 906–914. [CrossRef]

91. McMurray, B.; Horst, J.; Samuelson, L. Word learning emerges from the interaction of online referent selection and slow associative learning. *Psychol. Rev.* **2012**, *119*, 831–877. [CrossRef]
92. Smith, L.; Yu, C. Infants rapidly learn word-referent mappings via cross-situational statistics. *Cognition* **2008**, *106*, 1558–1568. [CrossRef] [PubMed]
93. Räsänen, O.; Rasilo, H. A joint model of word segmentation and meaning acquisition through cross-situational learning. *Psychol. Rev.* **2015**, *122*, 792. [CrossRef] [PubMed]
94. Räsänen, O.; Doyle, G.; Frank, M.C. Pre-linguistic segmentation of speech into syllable-like units. *Cognition* **2018**, *171*, 130–150. [CrossRef] [PubMed]
95. Dupoux, E. Category Learning in Songbirds: Top-down effects are not unique to humans. *Curr. Biol.* **2015**, *25*, R718–R720. [CrossRef]
96. Young, S.; Evermann, G.; Gales, M.; Hain, T.; Kershaw, D.; Liu, X.A.; Moore, G.; Odell, J.; Ollason, D.; Povey, D.; et al. *The HTK Book (for HTK Version 3.4)*; Technical Report; Cambridge University Engineering Department: Cambridge, UK, 2009.
97. Povey, D.; Ghoshal, A.; Boulianne, G.; Burget, L.; Glembek, O.; Goel, N.; Hannemann, M.; Motlicek, P.; Qian, Y.; Schwarz, P.; et al. The Kaldi Speech Recognition Toolkit. In Proceedings of the IEEE 2011 Workshop on Automatic Speech Recognition and Understanding. IEEE Signal Processing Society, Waikoloa, HI, USA, 11–15 December 2011; IEEE Catalog No.: CFP11SRW-USB.
98. Scharenborg, O.; Norris, D.; ten Bosch, L.; McQueen, J. How should a speech recognizer work? *Cogn. Sci.* **2005**, *29*, 867–918. [CrossRef]
99. Nenadić, F.; Tucker, B.V. Computational modelling of an auditory lexical decision experiment using jTRACE and TISK. *Lang. Cogn. Neurosci.* **2020**, *35*, 1326–1354. [CrossRef]
100. Wessel, F.; Schlüter, R.; Macherey, K.; Ney, H. Confidence Measures for Large Vocabulary Continuous Speech Recognition. *IEEE Trans. Speech Audio Process.* **2001**, *9*, 288–298. [CrossRef]
101. Oneata, D.; Caranica, A.; Stan, A.; Cucu, H. An evaluation of word-level confidence estimation for end-to-end automatic speech recognition. *arXiv* **2021**, arXiv:2101.05525.
102. Baayen, H.R.; Milin, P. Analyzing reaction times. *Int. J. Psychol. Res.* **2010**, *3*, 12–28. [CrossRef]
103. Wagenmakers, E.J.; Lodewyckx, T.; Kuriyal, H.; Grasman, R. Bayesian hypothesis testing for psychologists: A tutorial on the Savage-Dickey method. *Cogn. Psychol.* **2010**, *60*, 158–189. [CrossRef]
104. ten Bosch, L.; Boves, L.; Mulder, K. Analyzing reaction time and error sequences in lexical decision experiments. *Proc. Interspeech* **2019**, *2019*, 2280–2284.
105. Tucker, B.V.; Brenner, D.; Danielson, D.K.; Kelley, M.C.; Nenadić, F.; Sims, M. The Massive Auditory Lexical Decision (MALD) database. *Behav. Res. Methods* **2019**, *51*, 1187–1204. [CrossRef] [PubMed]
106. R Core Team. *R: A Language and Environment for Statistical Computing;* R Foundation for Statistical Computing: Vienna, Austria, 2013.
107. Brand, S.; Mulder, K.; ten Bosch, L.; Boves, L. Models of Reaction Times in Auditory Lexical Decision: RTonset versus RToffset. *Proc. Interspeech* **2021**, *2021*, 541–545. [CrossRef]
108. Matuschek, H.; Kliegl, R.; Vasishth, S.; Baayen, H.; Bates, D. Balancing Type I error and power in linear mixed models. *J. Mem. Lang.* **2017**, *94*, 305–315. [CrossRef]
109. Meteyard, L.; Davies, R.A. Best practice guidance for linear mixed-effects models in psychological science. *J. Mem. Lang.* **2020**, *112*, 104092. [CrossRef]

brain sciences

MDPI

Article

Learning to Perceive Non-Native Tones via Distributional Training: Effects of Task and Acoustic Cue Weighting

Liquan Liu [1,2,3,4,*], Chi Yuan [1,5], Jia Hoong Ong [1,6], Alba Tuninetti [1,7], Mark Antoniou [1], Anne Cutler [1,4] and Paola Escudero [1,4]

1 The MARCS Institute for Brain, Behaviour and Development, Western Sydney University, Westmead, NSW 2145, Australia; c.yuan@hfut.edu.cn (C.Y.); jiahoong.ong@reading.ac.uk (J.H.O.); alba.tuninetti@bilkent.edu.tr (A.T.); m.antoniou@westernsydney.edu.au (M.A.); a.cutler@westernsydney.edu.au (A.C.); paola.escudero@westernsydney.edu.au (P.E.)
2 School of Psychology, Western Sydney University, Westmead, NSW 2145, Australia
3 Center of Multilingualism across the Lifespan, University of Oslo, 0316 Oslo, Norway
4 Australian Research Council Centre of Excellence for the Dynamics of Language, Canberra, ACT 2601, Australia
5 School of Foreign Studies, Hefei University of Technology, Hefei 230009, China
6 School of Psychology and Clinical Language Sciences, University of Reading, Reading RG6 6AH, UK
7 Department of Psychology, Bilkent University, Ankara 06800, Turkey
* Correspondence: l.liu@westernsydney.edu.au

Abstract: As many distributional learning (DL) studies have shown, adult listeners can achieve discrimination of a difficult non-native contrast after a short repetitive exposure to tokens falling at the extremes of that contrast. Such studies have shown using behavioural methods that a short distributional training can induce perceptual learning of vowel and consonant contrasts. However, much less is known about the neurological correlates of DL, and few studies have examined non-native lexical tone contrasts. Here, Australian-English speakers underwent DL training on a Mandarin tone contrast using behavioural (discrimination, identification) and neural (oddball-EEG) tasks, with listeners hearing either a bimodal or a unimodal distribution. Behavioural results show that listeners learned to discriminate tones after both unimodal and bimodal training; while EEG responses revealed more learning for listeners exposed to the bimodal distribution. Thus, perceptual learning through exposure to brief sound distributions (a) extends to non-native tonal contrasts, and (b) is sensitive to task, phonetic distance, and acoustic cue-weighting. Our findings have implications for models of how auditory and phonetic constraints influence speech learning.

Keywords: distributional learning; tone; discrimination; identification; oddball-EEG; phonetic distance; acoustic cue-weighting

Citation: Liu, L.; Yuan, C.; Ong, J.H.; Tuninetti, A.; Antoniou, M.; Cutler, A.; Escudero, P. Learning to Perceive Non-Native Tones via Distributional Training: Effects of Task and Acoustic Cue Weighting. *Brain Sci.* **2022**, *12*, 559. https://doi.org/10.3390/brainsci12050559

Academic Editors: Richard Wright and Benjamin V. Tucker

Received: 5 March 2022
Accepted: 20 April 2022
Published: 27 April 2022

1. Introduction

Listening to speech involves identifying linguistic structures in a fast, continuous utterance stream and processing their relative ordering to retrieve an intended meaning. In the native language, multiple sources of relevant information are drawn upon to accomplish this task, including knowledge of the vocabulary and the relative likelihood of different sequential patterns at lower and higher levels of linguistic structure, as well as the rules governing sequential processes at the phonetic level. Non-native languages can differ from a listener's native languages on all these dimensions. We here address the question of whether listeners faced with the task of discriminating a novel non-native contrast show preferences as to which dimension of linguistic information they attend to (or attend to most). To better understand the source of any such preference, we compare behavioural against neurophysiological measures.

There is a considerable literature concerning the perception and learning of non-native contrasts, and it naturally shares ground with the very extensive literature on native contrast learning, with the differences in large part arising from learner constraints, given that in the native situation, the learners are still in infancy. Interestingly, overlap between the native and non-native literatures has been increased by the growing attention paid to statistical accounts of learning processes, due to the extensive evidence showing that such information is not restricted to mature brains, but is processed even by the youngest of learners..

Moreover, statistical learning is not specific to the language domain. It is observed for visual objects [1], spatial structures [2], tactile sequences [3] and music perception [4]. Crucially, two-month-olds exhibit neural sensitivity to statistical properties of non-native speech sounds during sleep [5], indicating that statistical learning is a formative mechanism that may have considerable explanatory power for our understanding of perceptual learning processes.

The language acquisition process can draw on statistical regularities in the linguistic environment ranging from simple frequency counts to complex conditional probabilities, including transitional probabilities [6], non-adjacent dependencies of word occurrences [7] and distributional patterns of speech sounds [8]. Listeners' most basic task, namely segmenting continuous speech streams into words and speech sounds, can also be assisted by attention to the distributional properties of the input [9]. Even infants in their first year attend to such information. Maye and colleagues [8] showed that infants as young as 6 months of age use distributional properties to learn phonetic categories. They exposed infants to manipulated frequency distributions of speech sounds that differed in their number of peaks, producing typically unimodal (one-peak) versus bimodal (two-peak) distributions, with the former thought to support a single, large category, while the latter (a distribution with two peaks) promotes the assumption of two categories.

A major factor affecting distributional learning is age, with learning appearing to be more effective for infants than for adults [10] and for younger than older infants [11]. Another factor is attention, with better results for learners whose task requires them to attend to the stimuli, as opposed to those who do not need to payattention to the stimuli's properties [12]. A third factor is the amount of exposure provided to the learner, with more extended distributional exposure yielding better discrimination results [13]. Fourth, experimental design also plays a role: A learning effect is more likely to appear with a habituation-dishabituation procedure than with familiarization-alternation paradigms [14]. Fifth and not least, the learning target itself can directly impact learning outcomes [15] due to its acoustic properties (a non-native contrast similar to the native inventory is more rapidly acquired through DL than a dissimilar contrast [16]) or its perceptual salience (consonants are relatively less salient acoustically than vowels and are thus less rapidly learned distributionally [17,18]). The last factor further leads to a series of cue-weighting studies demonstrating that statistical information in the ambient environment is not the only cue listeners adopt. For example, when learning a speech sound contrast, learners may focus more on the phonetic/acoustic properties of stimuli than statistical regularities constraining the contrast [19–22].

Most studies in this literature so far have focused on consonants and vowels. But speech has further dimensions, including suprasegmental structure which is manifested on many levels, including the lexical level. Around 60–70% of the world's languages are tonal, i.e., they use pitch to distinguish word meanings [23]. Tone language speakers show categorical perception for tones, but non-tone language speakers report them as psychoacoustic rather than linguistic [24]. However, acoustic interactions of tonal and segmental information affect simple discrimination task performance equivalently for native listeners of a tone language and non-native listeners without tonal experience [25]. Specifically, both Cantonese and Dutch listeners take longer and are less accurate when judging syllables that differ in tones compared to segments, suggesting a slower processing of tonal than segmental distinctions, at least in speeded-response tasks.

Non-tonal language speakers have substantial difficulty discriminating and learning tone categories [26], and studies have shown mixed results after distributional training [27].

For infants, bimodal distributional training enhances Mandarin tone perception for Dutch 11-month-olds, but not for 5- or 14-month-olds [28]. For adults, Ong and colleagues [12] found that Australian-English speakers exposed to a bimodal distribution of a Thai tone contrast showed no automatic learning effect; but when the training involved active listening (through a request for acknowledging the heard stimuli), their sensitivity to tone improved. Interestingly, exposure to a bimodal distribution of Thai tones resulted in enhanced perception of both linguistic and musical tones for Australian-English speakers, demonstrating cross-domain transfer [4,29]. Note that linguistic and musical tone perception also tend to be correlated in non-tone language adults' performance [26,30] and in psycho-acoustic perception [31].

All the above results stem from behavioural research because statistical learning research so far has made little use of neurophysiological methods, despite evidence that neural responses can provide processing information at both pre-attentive and cortical levels [32,33]. Compared to behavioural measures, however, neural techniques such as electroencephalography (EEG) have the benefit of requiring no overt attention or decision processes. They are typically sensitive to early pre-attentive responses, reflecting the neural basis of acoustic-phonetic processing [34,35] and providing another measure of implicit discrimination.

Importantly, evidence from non-native speech perception studies shows that non-native listeners exhibit mismatch negativity (MMN) responses for contrasts they do not discriminate in behavioural tasks [36,37]. The MMN response has been used extensively in speech perception studies to examine how the perceptual system indexes incoming acoustic stimuli pre-attentively, i.e., in a manner not requiring overt attention. Based on the formation of memory traces, the MMN response is a negative-going waveform that is typically observed in the frontal electrodes. It signals detection of change in a stream of auditory stimuli, and specifically indicates differences between stimuli with different acoustics. It is obtained by computing the difference in the event-related potential (ERP) response to an infrequent (termed deviant) stimulus versus a frequent (termed standard) one. This response typically occurs 150 to 250 ms post onset of the stimulus switch [38,39].

Despite the scarcity of neurally-based research on distributional learning, some evidence is available on listeners' perceptual flexibility for the tonal dimension of speech. In tone perception, pitch height and pitch direction are the two main acoustic cues [40]. Nixon and colleagues [41] explored German listeners' neural discrimination of a Cantonese high-mid pitch height contrast, by exposing them to a bimodal distribution of a 13-step tone continuum. Although a prediction of enhanced MMN responses at the two Gaussian peaks was not supported, the listeners' perception of pitch height improved across all steps along the continuum. A follow-up study of listeners' neural sensitivity to cross-boundary differences again showed enhanced sensitivity to overall pitch differences over the course of the tone exposure [42]. In other words, acuity with respect to acoustic pitch differences was increased during distributional learning not only between but also within categories. In contrast to the findings of previous studies testing segmental features, these results indicate that exposure to a bimodal distribution may not necessarily lead to the enhanced discrimination of specific steps or categories along the pitch continuum, but may rather alter listeners' overall sensitivity to tonal changes. A caveat in this interpretation is that these studies did not include a unimodal distribution. Taken together, when EEG studies are adopted to examine tone perception and learning, results often illustrate robust sensitivity not merely restricted to the patterns predicted by frequency distributions as shown in behavioural studies.

On the same note, a recent study examining 5–6-month-old infants' neural sensitivity to an 8-step contrast of flat versus falling pitch (a Mandarin tone contrast) found a surprising enhancement effect after exposure to a unimodal but not a bimodal distribution [43]. This finding was explained in relation to listeners' acoustic sensitivity to frequently heard tokens at peak locations along the tone continuum. The high-frequency tokens had smaller differences in the unimodal (steps 4–5) than in the bimodal (steps 2–7) distributions. The authors of [43] argued that frequent exposure to greater acoustic distance may lead to reduced neural sensitivity to a smaller acoustic distance (steps 3–6). This interpretation highlights the role of the magnitude of the acoustic distinctions in the stimuli when prior training and exposure is insufficient to establish phonetic categories, which can be explained by models of non-native perception that focus on acoustic cue weighting and salience (see for instance, [44]).

In the present study, we examined tone perception and learning for non-tonal language speakers, collecting and directly comparing behavioural and neural responses. Australian English listeners with no prior knowledge of any tone language heard a Mandarin tone contrast. Tone perception ability has long been known to vary as a function of the acoustic properties of the tonal input [45]. We varied tone features and the nature of the input distribution (comparing uni- versus bimodality). Experiment 1 tested listeners' ability to discern the level (T1)–falling (T4) Mandarin tone contrast before and after distributional training, using discrimination and identification tasks, respectively. Experiment 2 then examined listeners' neural sensitivity to the same contrast, in a standard MMN paradigm, assessing amplitude, latency, anteriority, laterality and topographic differences between the two modalities. Previous work has shown differences in factors such as anteriority and laterality depending on stimulus characteristics and participant language background [46–48] while other work has not [33,49]. We included them here to ensure that we captured the most comprehensive set of results for examining potential effects of distributional learning at the neural level.

2. Experiment 1: Mandarin Tone Discrimination and Identification by Australian Listeners before and after Distributional Learning

2.1. Methods

2.1.1. Participants

Forty-eight native Australian English speakers naïve to tone languages took part (36 females; $M_{age} = 23.06$, $SD_{age} = 5.57$). Nine participants reported having musical training (ranging from 1 to 10 years), though only one continued to practise music at the time of testing. All participants reported normal speech and hearing, provided written informed consent prior to testing, and received course credit or a small monetary compensation.

2.1.2. Stimuli

This study focused on the Mandarin Chinese level [T1] versus falling [T4] tonal contrast. A female Mandarin speaker produced natural tokens of /taT1/ ('take') and /taT4/ ('big') in a soundproof booth. Recording used the computer program Audacity and a Genelec 1029A microphone, with a 16-bit sampling depth and a sampling rate of 44.1 kHz. To create the 8-step continuum, equidistant stimulus steps differing in pitch contour were constructed from /taT1/ to /taT4/ (Figure 1) using the following procedure in Praat [50]: First, four interpolation points along the pitch contours (at 0%, 33%, 67% and 100%) were marked (Supplementary Material A). Next, the distances (in Hz) between the corresponding points were divided into seven equal spaces, generating six new layers. New pitch tokens were then created by connecting the four corresponding intermediate points on the same layer. This formed a continuum of eight steps (including the endpoint contours) from /taT1/ (step 1) to /taT4/ (step 8). Stimulus intensity was set to 65 dB SPL and duration was set to approximately 400 ms. Five native Mandarin speakers listened to the stimuli and confirmed that they were acceptable tokens of these Mandarin syllables. The contrast (with the same stimuli) has been used in previous studies [51–54].

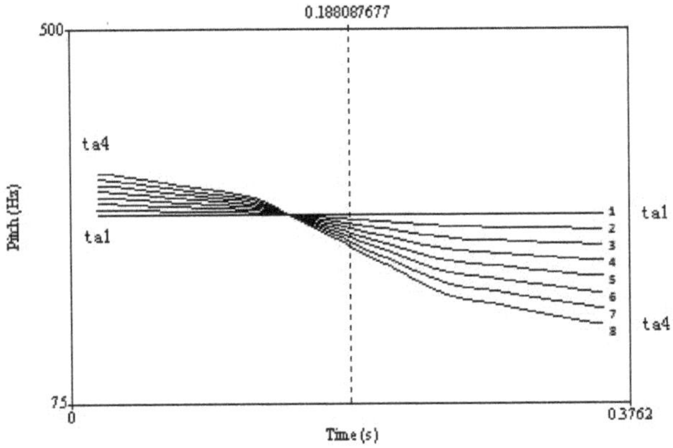

Figure 1. Pitch contours along a /taT1/–/taT4/ continuum (Stimuli and figure from [28]).

2.1.3. Procedure

The experiment consisted of three phases in the following order: pre-test, distributional learning and post-test. Task programming, presentation of stimuli and response recording were conducted using E-Prime (Psychology Software Tools Inc., Sharpsburg, PA, USA) on a Dell Latitude E5550 laptop. Auditory stimuli were presented at 65 dB SPL via Sennheiser HD 280 Pro headphones and were ordered randomly. No corrective feedback was given. The experiment took approximately 40 min to complete.

In the distributional learning phase, participants were randomly assigned to one of the two conditions: unimodal or bimodal distribution (Figure 2). The two conditions had different distributional peaks (a single central category vs. two separate categories) but were equal in terms of the total amount of distributional learning tonal input (256 tokens) and duration (360 s). In the bimodal condition, stimuli from the peripheral positions of the continuum were presented with higher frequency. In other words, participants heard tokens of steps 2 and 7 most frequently. In the unimodal condition, stimuli near the central positions, namely tokens of steps 4 and 5, were presented most frequently. Crucially, stimulus steps 3 and 6 were presented an equal number of times in both conditions.

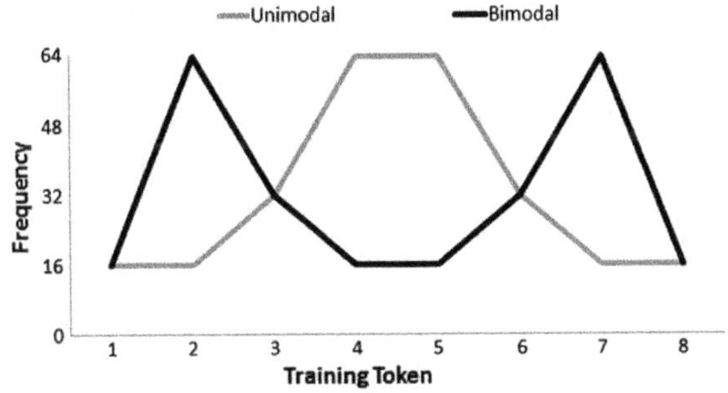

Figure 2. Frequency of occurrence for each training token encountered by listeners in the unimodal (grey line) and bimodal (black line) conditions. Figure from [55].

At pre- and at post-test, participants were first presented with a discrimination task in which they indicated via keypresses whether paired lexical tone stimuli steps were the same or different. Trials included same (e.g., steps 1–1, 2–2) and different pairs (e.g., steps 2–4, 3–6) along the continuum, with each pair presented 10 times. Trials not involving the target contrast functioned as controls. The inter-stimulus interval between tokens was 1000 ms.

Participants then completed an identification task in which they indicated (again via keypresses) whether the tone of each continuum step (e.g., step 3) was a flat tone (indicated by a flat arrow) or a falling tone (indicated by a falling arrow), with each tone presented 6 times. For each task, four auditory examples were played as practice trials prior to testing. Trials were self-paced and presented in random order.

Analyses for the discrimination task targeted the perception of steps 3 to 6, and those for the identification task focused on step 3 and step 6. As an additional control, the Pitch-Contour Perception Test (PCPT; [56,57]) was included in the post-training phase, to examine participants' pitch perceptual abilities (Supplementary Material B). This test required indication of whether isolated tone tokens had a flat, rising or falling contour. The PCPT allows allocation -of participants to high and low aptitude groups, to examine whether ability to perceive pitch affects identification and discrimination responses. No differences either in the identification or in the discrimination tasks were observed between listeners with high versus low aptitude in the present study.

2.2. Results

We first compared listeners' percentage of accurate choices for the target contrast (steps 3–6) in the discrimination task before and after distributional learning to chance (50%) with a one-sample t-test (Table 1). Neither condition showed discrimination above chance before distributional learning, while after training, listeners in both conditions were able to discriminate the contrast. We then conducted a Repeated Measures Analysis of Variance (RM ANOVA) with condition (2-level, unimodal vs. bimodal) as the between-subjects factor and accuracy (2-level, pre- and post-training phases) as the within-subjects variable (Figure 3). The main effect of training was significant ($F(1, 46) = 4.731$, $p = 0.035$, $\eta_g^2 = 0.093$), indicating a difference in accuracy before and after distributional learning. The interaction between condition and test phase was not significant ($F(1, 46) = 0.526$, $p = 0.472$, $\eta_g^2 = 0.011$), suggesting no difference between unimodal and bimodal exposure. In other words, listeners' tone discrimination improved after exposure to either distribution.

Table 1. Mean (SD) accuracy percentage and corresponding t and *p* values in one-sample t-test against the chance level in bimodal and unimodal before and after distributional learning in the discrimination task. (See Supplementary Material C for descriptive statistics of other contrasts.).

		Mean	SD	T	*p*
Bimodal	Pre	60.83%	35.62%	1.490	0.150
	Post	67.50%	33.26%	2.577	0.017
Unimodal	Pre	51.67%	30.59%	0.267	0.792
	Post	65.00%	30.21%	2.432	0.023

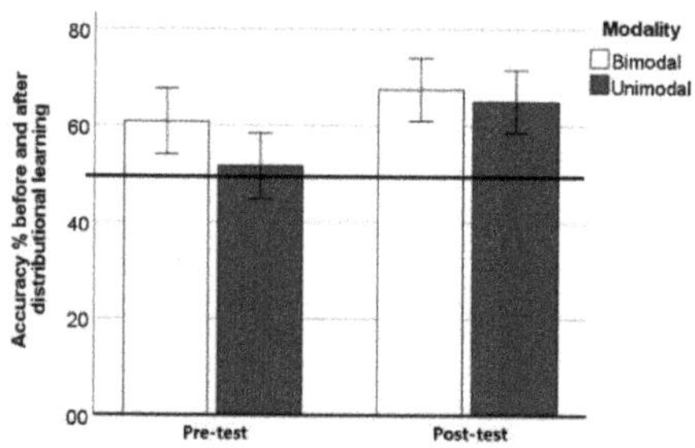

Figure 3. Mean accuracy percentage before and after distributional learning (Error bars = ±1 standard error). The horizontal line indicates chance level (50%) performance.

We then examined listeners' choices in the identification task before and after learning (Table 2). Across participants, Step 6 was more often identified as falling than step 3.

We then conducted a repeated measures ANOVA with condition (2-level, unimodal vs. bimodal) as the between-subjects factor, and with percentage of falling choices across phase (2-level, pre- and post-training) and step (2-level, steps 3 & 6) as the within-subjects variables (Figure 4). The only significant factor was step ($F(1, 46) = 57.343$, $p < 0.001$, $\eta_g^2 = 0.555$). No other factors or interactions were significant ($Fs < 1.936$, $ps > 0.170$, $\eta_g^2 < 0.040$). In contrast to the discrimination outcomes, no trace of improvement was observed in listeners' tone identification after either distributional condition.

Table 2. Mean (SD) percentage of choosing falling over flat tones in bimodal and unimodal before and after distributional learning in the identification task. (See Supplementary Material D for descriptive statistics of other contrasts.).

		Step 3				Step 6			
		Mean	SD	t	*p*	Mean	SD	t	*p*
Bimodal	Pre	34.75%	28.62%	−2.610	0.016	67.29%	31.61%	2.680	0.013
	Post	25.08%	33.30%	−3.665	0.001	65.87%	35.28%	2.204	0.038
Unimodal	Pre	45.83%	28.30%	−0.721	0.478	74.29%	29.79%	3.994	0.001
	Post	42.25%	28.98%	−1.310	0.203	69.38%	29.81%	3.183	0.004

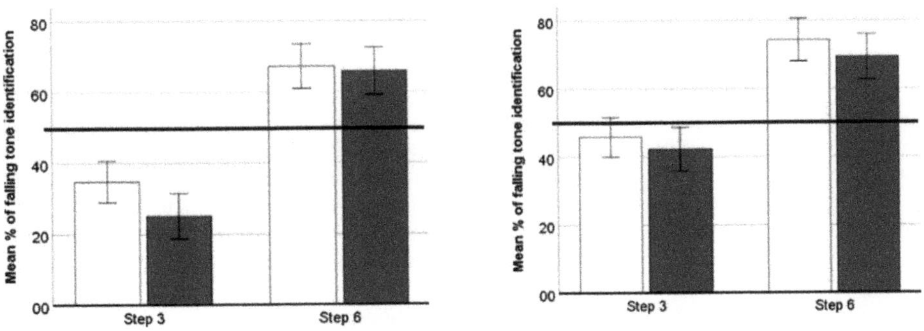

Figure 4. Mean percentage of "falling" classifications for steps 3 and 6 before and after bimodal (**left**) and unimodal (**right**) distributional learning (Error bars = ±1 standard error) White bars indicate performance at pretest, and black bars posttest. The horizontal line indicates chance (50%) performance.

2.3. Discussion

Experiment 1 used behavioural measures to investigate how distributional learning of a Mandarin tone contrast affects listeners' tone discrimination and identification. Listeners showed improved discrimination abilities after exposure to either unimodal or bimodal distributions, and no change was observed in their identification patterns.

In the discrimination task, successful distributional learning would predict enhanced discrimination of steps 3–6 after the bimodal condition, and/or reduced discrimination of this step after the unimodal experience. However, our results showed that listeners' tonal sensitivity was enhanced after distributional learning irrespective of the embedded statistical information. Statistical exposure appears to benefit participants more in acoustic than in statistical cues. This interpretation resembles reported EEG studies with a Cantonese tone contrast [41,42], and agrees with findings showing that listeners attend more to prosodic than statistical cues to segment speech streams [20]. It is worth mentioning that only a single (bimodal) distribution was tested in these cited studies.

In identification, listeners' difficulty in anchoring non-native tones to given categories plausibly reflects their lack of tonal categories in the first place (N.B. performance differences in identification versus discrimination have been attested before.) Non-tone language speakers often find it hard to make recognition responses to tones. Their much better tone discrimination ability, in contrast, reflects sensitivity to general pitch information (note that pitch perception in language and music correlate [26,30]).

To further examine the relationship between tone processing, distributional learning, and the cues listeners pay attention to in these processes, Experiment 2 examined listeners' neural changes induced by distributional learning.

3. Experiment 2: Australian Listeners' Neural Sensitivity to Tones before and after Distributional Learning

3.1. Methods

3.1.1. Participants

A new sample of 32 Australian English speakers naïve to tone languages participated in Experiment 2 (22 females; M_{age} = 22.9, SD_{age} = 7.60). Eight participants reported having musical training (ranging from 1 to 7 years), though only one continued to practise music at time of test. Participants provided written informed consent prior to the experiment and received course credit or a small reimbursement for taking part.

3.1.2. Stimuli

The same stimuli were used as in Experiment 1, except that the duration of all stimulus tokens was reduced to 100 ms to accommodate to the EEG paradigm.

3.1.3. Procedure

As in Experiment 1, there were three phases: pre-test, distributional learning, and post-test. In the distributional learning phase, participants were randomly assigned to either the unimodal or the bimodal condition, with equal numbers of bilingual and monolingual participants in each condition. The two conditions did not differ in the total number of exposure trials (256 tokens) or duration (360 s) but varied in the frequency distribution (one vs. two Gaussian peaks) along the phonetic continuum only. Stimuli near the central positions were presented most frequently in the unimodal condition, whereas those from the peripheral sides of the continuum were presented with the highest frequency in the bimodal condition. Importantly, the frequency of occurrence of tokens from steps 3 and 6 was again identical across both conditions.

An EEG passive oddball paradigm was used for pre- and post-test, in which two separate blocks were presented. Step 3 was standard and step 6 was deviant in one block, and this was reversed in the other. The standard-deviant probability ratio was 80–20%. Since steps 3 and 6 were presented the same number of times in each condition, any potential differences observed in the post-test should be attributed to the condition. The sequence of the blocks was counterbalanced. No fewer than three and no more than eight standard stimuli occurred between deviant stimuli. Each block started with 20 standards and contained 500 trials in total. The inter-stimulus interval was randomly varied between 600 and 700 ms. Following the presentation phase, participants heard as a control (lasting approximately 1 min) 100 instances of the deviant stimuli they had heard in the previous oddball presentation. This design allowed a response comparison of the same number of deviant stimuli in the oddball presentation (N = 100) and the control presentation (N = 100).

Participants were tested in a single session in a sound-attenuated booth at the MARCS Institute for Brain, Behaviour and Development at Western Sydney University. They watched a self-selected silent movie with subtitles during the experiment and were instructed to avoid excessive motor movement and to disregard the auditory stimuli. The stimuli were presented binaurally via Etymotic earphones with the intensity set at 70 dB SPL in Praat [50] and the volume level was set at a comfortable listening level consistent across participants as a result of piloting that showed MMN elicitation. The duration of the EEG experiment was approximately 45 min.

3.1.4. EEG Data Recording & Analysis

EEG data were recorded from a 64-channel active BioSemi system with Ag/AgCl electrodes placed according to the international 10/20 system fitted to the participant's head. Six external electrodes were used: four to record eye movements (above and below the right, on the left and right temple), and two for offline referencing (left and right mastoid). Data were recorded at a 512 Hz sampling rate and we made sure the electrode offset was kept below 50 mV.

Data pre-processing and analysis used EEGLAB [58] and ERPLAB [59]. First, data points were re-referenced to the average of the right and left mastoids. They were then bandpass-filtered with half power cut-offs at 0.1 and 30 Hz at 12 dB/octave. Time windows from 100 to 600 ms post stimulus onset were extracted ("epoched") from the EEG signal and baseline-corrected by subtracting the mean voltage in the 100 ms pre-stimulus interval from each sample in the window. Independent component analyses were conducted to identify and remove noisy EEG channels. Eye-movement components based on the activity power spectrum, scalp topography, and activity over trials were also removed. Noisy EEG channels that were removed were interpolated using spherical spline interpolation. Artefacts above 70 mV were rejected automatically for all channels. Participants with more than 40% of artefact-contaminated epochs were excluded from further analyses (n = 6). The epochs were averaged separately for standards (excluding the first 20 standards and the standards immediately following a deviant stimulus), for each deviant token, and for each control block.

Two difference waves were calculated by subtracting the mean ERP response to each control stimulus from the mean ERP response to its deviant counterpart. These difference waves were then grand-averaged across participants. In the grand-averaged waveform, we sought a negative peak 100 to 250 ms after consonant production (taking the 20 ms consonant portion of the stimulus into consideration) to ensure that we were measuring the neural response to the tone. This resulted in measuring the 120 to 270 ms time window post-stimulus onset. We then centred a 40 ms time window at the identified peak and measured the mean amplitude in that window per individual participant (cf. [47]). These mean individual amplitudes were our measure of MMN amplitude in further statistical analyses. Within the same 40 ms time window, latency was measured by establishing the most negative peak for each participant. These mean individual latencies then became the measure of MMN latency in the further analyses.

3.2. Results

Following previous studies (e.g., [47,60]), MMN amplitudes and latencies were measured at nine channels (Fz, FCz, Cz, F3, FC3, C3, F4, FC4, C4) and were analysed in two separate mixed analyses of variance (ANOVAs) with a between-subject factor of condition (unimodal vs. bimodal) and within-subject factors of phase (pre- vs. post-training), anteriority (Frontal (F) vs. frontocentral (FC) vs. central I), and laterality (left, middle, right). Based on activating different neural populations, the dependent variables including mean amplitude and peak latency may reflect different processing mechanisms [61]: the former may show the robustness of listeners' discrimination as well as the acoustic/phonetic difference between the standard and the deviant stimuli, and the latter may reflect the required time to process the difference between the stimuli. Both variables have been used as auditory perceptual processing measures at early pre-attentive levels for native and non-native speech [36,62]. MMN responses tend to occur at frontal (F) and frontocentral (FC) sites. If distributional training has an effect on tone perception, we predicted an increase in MMN amplitude at those sites between pre- and post-test, as evidence of the auditory change in stimuli initiating an involuntary attentional switch [39,63].

MMN mean amplitude. Figure 5 shows the grand-averaged MMN component in response to the contrast at pre- and post-training. There was a main effect of phase ($F(1, 30) = 4.37$, $p = 0.045$, ${\eta_g}^2 = 0.046$). Specifically, the MMN amplitude at pre-test

($M = -1.67$, $SD = 3.03$) was larger than that at post-test ($M = -0.80$, $SD = 2.56$). No other effects or interactions were significant.

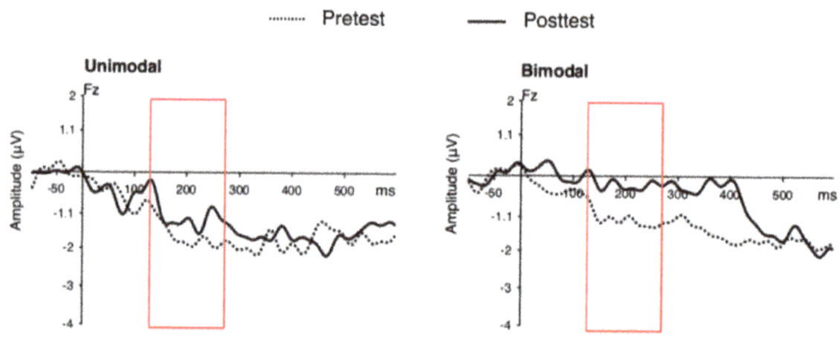

Figure 5. Grand-averaged MMN component by unimodal (**left**) and bimodal (**right**) condition. Dotted lines show the MMN component at pre-training and solid lines represent the MMN component at post-training. The red boxes highlight the time window in which the MMN amplitude peaks were measured (i.e., 120–270 ms post-stimulus onset to account for consonant production).

As condition was our variable of interest, the MMN amplitude response at the Fz electrode site in each phase was compared against zero (Figure 6) following previous literature [49,60,61,64,65]. Participants in the unimodal group exhibited significant MMN amplitudes at pre-test ($t(15) = -4.55$, $p < 0.001$, $d = -1.14$) and at post-test ($t(15) = -2.76$, $p = 0.015$, $d = -0.69$), whereas participants in the bimodal group exhibited a significant MMN amplitude at pre-test ($t(15) = -2.90$, $p = 0.011$, $d = -0.72$) but not at post-test ($t(15) = -0.51$, $p = 0.616$, $d = -0.13$). Paired t-tests comparing the MMN amplitude between pre- and post-test for each condition revealed no difference for the unimodal group ($t(15) = -1.01$, $p = 0.329$, $d = -0.23$) whereas there was a marginal difference for the bimodal group ($t(15) = -2.12$, $p = 0.051$, $d = -0.35$), indicating statistical-learning-induced changes.

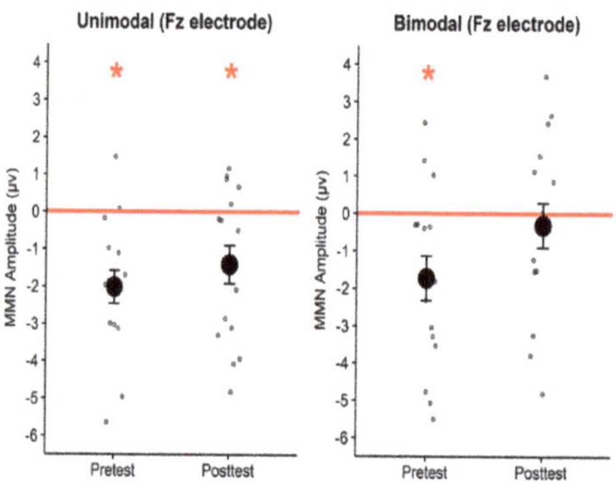

Figure 6. Mean MMN amplitude (large dots) for the two conditions at each test phase. Small dots represent individual data. Error bars represent one standard error. Asterisks represent significant MMN amplitude.

MMN peak latency. A mixed ANOVA with condition (2-level, unimodal vs. bimodal) as the between-subjects factor, and within-subject factors of test (pre- vs. post-test), anteriority (F vs. FC vs. C), and laterality (left vs. middle vs. right) was conducted on the mean MMN peak latency. Main effects of phase ($F(1, 30) = 240.35$, $p < 0.001$, $\eta_g^2 = 0.729$) and condition ($F(1, 30) = 24.57$, $p < 0.001$, $\eta_g^2 = 0.179$) were found, which were qualified by a significant phase × condition interaction ($F(1, 30) = 19.68$, $p < 0.001$, $\eta_g^2 = 0.177$). Post hoc Tukey tests revealed that there were no group differences in latency at pre-test ($M_{diff} = 0.05$, $p = 0.986$)

but at post-test, participants in the bimodal group had significantly earlier MMN peaks than those in the unimodal group (M_{diff} = 20.39, $p < 0.001$). No other effects nor interactions were significant.

3.3. Discussion

Experiment 2 showed learning-induced neurophysiological changes in these listeners' tone perception at pre- and post-test. Reduction in tonal sensitivity was significant in the bimodal condition. In addition, the latency results showed an earlier peak in the post- than in the pre-test, with a larger impact again in the bimodal condition. Although behavioural results indicated limited discrimination prior to training, neural sensitivity was consistent with discrimination. Note that the pre-test sensitivity is not unexpected given listeners' psycho-acoustic sensitivity to non-native tones across ages [51], with sensitivity modulated by tone salience [66].

At post-test, although behavioural outcomes indicated improved discrimination, listeners' neural processing was generally weakened in both conditions. The results from the bimodal distribution may suggest that the increased exposure and familiarity with tones in the neural experiment hindered distributional learning. Alternatively, although the acoustic experience had given listeners more familiarity with tones, the information they received was insufficient to establish tonal categories from the frequency distribution.

Indeed, in the bimodal condition where MMN responses were diminished, frequency peaks in the distribution were near the two ends of the continuum (Figure 2, steps 2–7), whereas the peaks in the frequency distribution were at the midpoint (Figure 2, steps 4–5) in the unimodal condition. To process stimuli efficiently, listeners may focus on the most frequently presented (hence, most salient) stimuli in each condition. In the bimodal condition, steps 2 and 7 were highlighted when played alongside stimuli close to the discrimination boundary (steps 3 and 6), and listeners may expect a similar (or larger) difference to detect deviation compared with post-test. Contrasts with a smaller acoustic difference (i.e., steps 3–6) may then be harder to detect. On the other hand, those who were exposed to the unimodal condition, where steps 4 and 5 were the most frequent, would show neural responses to a contrast of larger acoustic difference (i.e., steps 3–6). In other words, the most frequent and prominent steps, namely the peaks of each distribution, would impact subsequent perception (To explore our proposed hypothesis on the impact of frequency peak, an additional behavioural test was conducted measuring Australian listeners' (N = 12) sensitivity to the designated tonal contrasts with manipulations of the acoustic distance between the tokens. While discrimination of steps 2–7, the acoustically distant contrast mostly presented in the bimodal condition, reached 82% accuracy, discrimination of steps 3–6 was at a chance (p = 0.511 against 0.5) and accuracy in the discrimination of steps 4–5, the acoustically close contrast presented most frequently in the unimodal condition, was 20%. These results reflect contrast salience). The overall pattern suggests that listeners' sensitivity is more auditory or psychophysical than phonetic or phonological at this stage: they can discriminate, but fail to establish categories.

With either an explanation in terms of acoustic salience, or an explanation in terms of perceptual assimilation, certain interactions between acoustic and statistical cues in listeners' neural processing will be assumed. Previous research has not only shown humans' ability to track input frequency distributions from the ambient environment, and their ability to abstract and retain the memory of non-native pitch directional cues, but also clear ability to shift their weighting of acoustic/phonetic cues and to reconfigure their learning strategies [19,20]. In a speech segmentation task when both statistical and prosodic cues are presented, listeners attend more to the latter to acquire speech information [21]. Indeed, when various types of cues (e.g., acoustic feature, frequency distribution) are presented to listeners, the weighting of these cues may be dynamic and change in real-time during experimental training [67,68]. Last but not least, the outcomes also confirm that EEG is more sensitive than behavioral measures in revealing listeners' responses in the course of speech perception [34–37].

4. General Discussion

This study tested the learning of a non-native tone contrast by non-tonal language speakers. Data were collected with both behavioural and neural test methods, statistical distributions were varied (along a single continuum) in the input, and both identification and discrimination data were collected and analysed. Non-native listeners' tone discrimination improved after both unimodal and bimodal exposure, and there was a reduction in sensitivity after training (observed with the more sensitive measures, to wit, MMN amplitude and latency, which were recorded using EEG). In the EEG data, perceptual differences emerged between the two exposure conditions. The bimodal condition appears to have a more negative impact than the unimodal condition after training. The reduced amplitude and early latency in the bimodal condition may reflect inability to establish categories from the frequency distribution, although listeners had become more familiar with tones after training.

Our neurophysiological finding contrasts with results in prior distributional learning studies (see Figure 2), where a bimodal distribution leads listeners to display enhanced distinction of steps midway along the distribution, while theeffect is reversed with the unimodal distribution, where steps within one peak become less distinct perceptually. We proposed that the lack of sensitivity at post-test after bimodal training is due to the exposure to a salient contrast during training that participants then expected to find again at test. In the absence of distributional learning, there was exploitation of salient acoustic cues instead. This interpretation is in line with the phonetic-magnitude hypothesis [44], which holds that the size of acoustic features (or phonetic distinctions, articulatory correlates) plays a more central role than the extent of native language experience, an interpretation consistent with the Second-Language Linguistic Perception model (L2LP; [69–72]), which highlights the interaction between listeners' native phonology and the magnitude of the phonetic difference in auditory dimensions. In a way, the adult listeners in the current study behaved like infants without prior knowledge of lexical tones and like adult L2 learners without prior knowledge of vowel duration contrasts. They concentrated on the most salient phonetic cues, while ignoring or lowering the weighting of other cues.

We further argue that compared to segmental features, pitch is a particularly salient feature. Pitch contrasts, regardless of height [41,42] or direction (as illustrated in the current study), may be more prone to be weighted acoustically compared to statistically, especially among non-tone language speakers who predominantly use pitch in pragmatic but not lexical functions. Note that this finding was only seen neurally and not behaviourally, which speaks to the fact that this might only be detected through the deployment of sensitive measures.

When native listeners process lexical tones, they attend to linguistic features such as pitch, intensity, and duration, based on their existing knowledge of the categorical structure and the phonotactics of their language [73]. When non-native listeners are presented with the same input, they have no relevant categorical or phonotactic knowledge to call upon. But their auditory abilities may be assumed to parallel those of the native listener, so that it is not surprising that simple discrimination elicits a similar pattern of performance for native and non-native listeners [25], even when any task involving recourse to frequency distributions leads to very different results from these two listener groups.

Results conform to a heuristic approach to processing: when hearing a subtle non-native tone contrast, listeners' neural sensitivity may be insufficient to perceive the fine-grained tone steps and turn more towards acoustic information, and especially the most frequently presented (i.e., most salient) stimuli within each type of distribution. The general pattern somewhat conforms to previous findings showing that a bimodal distribution does not necessarily promote distinct discrimination between the two peaks [13,38,39], and a unimodal distribution does not always hinder it [43]. Factors such as acoustic properties or perceptual salience between tokens may play a role.

Furthermore, we argue that listeners who were trained on a bimodal distribution in which the peak contrast (steps 2–7) contains a large, discriminable acoustic distance may exhibit hampered discrimination of contrasts that rest on smaller differences. In consequence, smaller differences along the continuum may be disregarded. On the other hand, training on a unimodal distribution in which the peak contrast (steps 4–5) is extremely difficult to discriminate may ease the processing of contrasts with a larger acoustic difference (i.e., steps 3 and 6). This explanation does not come out of the blue, as similar ideas have been proposed for studies showing that infants are sensitive to acoustically subtle phonetic contrasts [74] and develop their word learning abilities of very small differences in vowels in a speedy fashion [75]. This interpretation assumes that listener sensitivity to ambient acoustic and statistical information can pave the way for perception and learning of new contrasts in a second language. However, our results should not be construed as evidence that native English speakers cannot achieve a level of tone categorization matching that of native tone speakers [76]. Many factors may play a role. For example, past research has shown that although statistical information can prompt the formation of distinct phonemic categories within three minutes for some non-native contrasts, longer duration of exposure may be required to trigger learning in a bimodal condition [13]. In a previous study examining listeners' non-native tone discrimination ability, an overall effect of learning surfaced after around ten minutes of exposure [52]. In the present experiment, embedded statistical information altered perception within six minutes of exposure. The take-home message for this study is that for non-native contrasts presented without context, the acoustic salience of the sounds may initially matter more than the statistical distribution of the sounds. At least in the initial stages of learning, distributional evidence may play only a minor role. Future studies can investigate whether increased input might introduce robust effects of exposure to a bimodal distribution.

Supplementary Materials: The following supporting information can be downloaded at: https://www.mdpi.com/article/10.3390/brainsci12050559/s1, Supplementary Material A: F0 values at 4 points along the /taT1/-/taT4/ continuum; Supplementary Material B: Pitch-Contour Perception Test results; Supplementary Material C: Discrimination task results across all contrasts; Supplementary Material D: Identification task results across all steps.

Author Contributions: Team ACOLYTE proved an irresistible nickname for our collaboration of M.A., A.C., J.H.O., L.L., C.Y., A.T. and P.E.. The specific contributions of each team member however are: Conceptualization, L.L. and P.E.; methodology, J.H.O. and A.T.; data collection, C.Y. and J.H.O.; data curation, L.L.; data analyses, L.L. and J.H.O.; supervision: A.C. and M.A.; writing—original draft preparation, L.L.; writing—review and editing, all authors. All authors have read and agreed to the published version of the manuscript.

Funding: This research was funded by Australian Research Council grants, principally the Centre of Excellence for the Dynamics of Language (CE140100041: A.C., L.L. & P.E.), plus DP130102181 (P.E.), DP140104389 (A.C.) and DP190103067 (M.A. & A.C.). P.E. was further supported by an ARC Future Fellowship (FT160100514), C.Y. by China Scholarship Council Grant 201706695047, and L.L. by a European Union Horizon 2020 research and innovation programme under Marie Skłodowska-Curie grant agreement 798658, hosted by the University of Oslo Center for Multilingualism across the Lifespan, financed by the Research Council of Norway through Center of Excellence funding grant No. 223265.

Institutional Review Board Statement: The study was conducted in accordance with the Declaration of Helsinki and approved by Human Ethics Committee of Western Sydney University (protocol code H11383 in 2015).

Informed Consent Statement: Informed consent was obtained from all subjects involved in the study.

Data Availability Statement: Anonymized data is available upon request to the first author.

Conflicts of Interest: The authors declare no conflict of interest.

References

1. Kirkham, N.Z.; Slemmer, J.A.; Johnson, S.P. Visual statistical learning in infancy: Evidence for a domain general learning mechanism. *Cognition* **2002**, *83*, B35–B42. [CrossRef]

2. Fiser, J.; Aslin, R.N. Unsupervised statistical learning of higher-order spatial structures from visual scenes. *Psychol. Sci.* **2001**, *12*, 499–504. [CrossRef]
3. Conway, C.M.; Christiansen, M.H. Modality-constrained statistical learning of tactile, visual, and auditory sequences. *J. Exp. Psychol. Learn. Mem. Cogn.* **2005**, *31*, 24. [CrossRef] [PubMed]
4. Ong, J.H.; Burnham, D.; Stevens, C.J. Learning novel musical pitch via distributional learning. *J. Exp. Psychol. Learn. Mem. Cogn.* **2017**, *43*, 150. [CrossRef]
5. Wanrooij, K.; Boersma, P.; Van Zuijen, T.L. Fast phonetic learning occurs already in 2-to-3-month old infants: An ERP study. *Front. Psychol.* **2014**, *5*, 77. [CrossRef] [PubMed]
6. Saffran, J.R.; Aslin, R.N.; Newport, E.L. Statistical learning by 8-month-old infants. *Science* **1996**, *274*, 1926–1928. [CrossRef]
7. Newport, E.L.; Aslin, R.N. Learning at a distance I. Statistical learning of non-adjacent dependencies. *Cogn. Psychol.* **2004**, *48*, 127–162. [CrossRef]
8. Maye, J.; Werker, J.F.; Gerken, L. Infant sensitivity to distributional information can affect phonetic discrimination. *Cognition* **2002**, *82*, B101–B111. [CrossRef]
9. Saffran, J.R.; Johnson, E.K.; Aslin, R.N.; Newport, E.L. Statistical learning of tone sequences by human infants and adults. *Cognition* **1999**, *70*, 27–52. [CrossRef]
10. Wanrooij, K.; Boersma, P.; van Zuijen, T.L. Distributional vowel training is less effective for adults than for infants. A study using the mismatch response. *PLoS ONE* **2014**, *9*, e109806. [CrossRef]
11. Reh, R.K.; Hensch, T.K.; Werker, J.F. Distributional learning of speech sound categories is gated by sensitive periods. *Cognition* **2021**, *213*, 104653. [CrossRef] [PubMed]
12. Ong, J.H.; Burnham, D.; Escudero, P. Distributional learning of lexical tones: A comparison of attended vs. unattended listening. *PLoS ONE* **2015**, *10*, e0133446. [CrossRef]
13. Yoshida, K.A.; Pons, F.; Maye, J.; Werker, J.F. Distributional phonetic learning at 10 months of age. *Infancy* **2010**, *15*, 420–433. [CrossRef]
14. Cristia, A. Can infants learn phonology in the lab? A meta-analytic answer. *Cognition* **2018**, *170*, 312–327. [CrossRef] [PubMed]
15. Escudero, P.; Benders, T.; Wanrooij, K. Enhanced bimodal distributions facilitate the learning of second language vowels. *J. Acoust. Soc. Am.* **2011**, *130*, EL206–EL212. [CrossRef] [PubMed]
16. Chládková, K.; Boersma, P.; Escudero, P. Unattended distributional training can shift phoneme boundaries. *Biling. Lang. Cogn.* **2022**, 1–14. [CrossRef]
17. Pons, F.; Sabourin, L.; Cady, J.C.; Werker, J.F. Distributional learning in vowel distinctions by 8-month-old English infants. In Proceedings of the 28th Annual Conference of the Cognitive Science Society, Vancouver, BC, Canada, 26–29 July 2006.
18. Antoniou, M.; Wong, P.C.M. Varying irrelevant phonetic features hinders learning of the feature being trained. *J. Acoust. Soc. Am.* **2016**, *139*, 271–278. [CrossRef] [PubMed]
19. Escudero, P.; Benders, T.; Lipski, S.C. Native, non-native and L2 perceptual cue weighting for Dutch vowels: The case of Dutch, German, and Spanish listeners. *J. Phon.* **2009**, *37*, 452–465. [CrossRef]
20. Lany, J.; Saffran, J.R. Statistical learning mechanisms in infancy. *Compr. Dev. Neurosci. Neural Circuit Dev. Funct. Brain* **2013**, *3*, 231–248.
21. Marimon Tarter, M. Word Segmentation in German-Learning Infants and German-Speaking Adults: Prosodic and Statistical Cues. Ph.D. Thesis, University of Potsdam, Potsdam, Germany, 2019. Available online: https://publishup.uni-potsdam.de/frontdoor/index/docId/43740 (accessed on 1 January 2020).
22. Tuninetti, A.; Warren, T.; Tokowicz, N. Cue strength in second-language processing: An eye-tracking study. *Q. J. Exp. Psychol.* **2015**, *68*, 568–584. [CrossRef]
23. Yip, M. *Tone*; Cambridge University Press: Cambridge, UK, 2002.
24. Kaan, E.; Barkley, C.M.; Bao, M.; Wayland, R. Thai lexical tone perception in native speakers of Thai, English and Mandarin Chinese: An event-related potentials training study. *BMC Neurosci.* **2008**, *9*, 53. [CrossRef] [PubMed]
25. Cutler, A.; Chen, H.C. Lexical tone in Cantonese spoken-word processing. *Percept. Psychophys.* **1997**, *59*, 165–179. [CrossRef]
26. Chen, A.; Liu, L.; Kager, R. Cross-domain correlation in pitch perception: The influence of native language. *Lang. Cogn. Neurosci.* **2016**, *31*, 751–760. [CrossRef]
27. Antoniou, M.; Chin, J.L.L. What can lexical tone training studies in adults tell us about tone processing in children? *Front. Psychol.* **2018**, *9*, 1. [CrossRef] [PubMed]
28. Liu, L.; Kager, R. Statistical learning of speech sounds is most robust during the period of perceptual attunement. *J. Exp. Child Psychol.* **2017**, *164*, 192–208. [CrossRef]
29. Ong, J.H.; Burnham, D.; Stevens, C.J.; Escudero, P. Naïve learners show cross-domain transfer after distributional learning: The case of lexical and musical pitch. *Front. Psychol.* **2016**, *7*, 1189. [CrossRef]
30. Liu, L.; Chen, A.; Kager, R. Perception of tones in Mandarin and Dutch adult listeners. *Lang. Linguist.* **2017**, *18*, 622–646. [CrossRef]
31. Hallé, P.A.; Chang, Y.C.; Best, C.T. Identification and discrimination of Mandarin Chinese tones by Mandarin Chinese vs. French listeners. *J. Phon.* **2004**, *32*, 395–421. [CrossRef]
32. Xu, Y.; Krishnan, A.; Gandour, J.T. Specificity of experience-dependent pitch representation in the brainstem. *Neuroreport* **2006**, *17*, 1601–1605. [CrossRef]

33. Chandrasekaran, B.; Krishnan, A.; Gandour, J.T. Mismatch negativity to pitch contours is influenced by language experience. *Brain Res.* **2007**, *1128*, 148–156. [CrossRef] [PubMed]
34. Näätänen, R.; Winkler, I. The concept of auditory stimulus representation in cognitive neuroscience. *Psychol. Bull.* **1999**, *125*, 826. [CrossRef] [PubMed]
35. Sams, M.; Alho, K.; Näätänen, R. Short-term habituation and dishabituation of the mismatch negativity of the ERP. *Psychophysiology* **1984**, *21*, 434–441. [CrossRef] [PubMed]
36. Kraus, N.; McGee, T.; Carrell, T.D.; King, C.; Tremblay, K.; Nicol, T. Central auditory system plasticity associated with speech discrimination training. *J. Cogn. Neurosci.* **1995**, *7*, 25–32. [CrossRef] [PubMed]
37. Lipski, S.C.; Escudero, P.; Benders, T. Language experience modulates weighting of acoustic cues for vowel perception: An event-related potential study. *Psychophysiology* **2012**, *49*, 638–650. [CrossRef] [PubMed]
38. Näätänen, R. The perception of speech sounds by the human brain as reflected by the mismatch negativity (MMN) and its magnetic equivalent (MMNm). *Psychophysiology* **2001**, *38*, 1–21. [CrossRef]
39. Näätänen, R.; Paavilainen, P.; Rinne, T.; Alho, K. The mismatch negativity (MMN) in basic research of central auditory processing: A review. *Clin. Neurophysiol.* **2007**, *118*, 2544–2590. [CrossRef] [PubMed]
40. Gandour, J. Tone perception in Far Eastern languages. *J. Phon.* **1983**, *11*, 149–175. [CrossRef]
41. Nixon, J.S.; Boll-Avetisyan, N.; Lentz, T.O.; van Ommen, S.; Keij, B.; Cöltekin, C.; Liu, L.; van Rij, J. Short-term exposure enhances perception of both between-and within-category acoustic information. In Proceedings of the 9th International Conference on Speech Prosody, Poznan, Poland, 13–16 June 2018; pp. 114–118.
42. Boll-Avetisyan, N.; Nixon, J.S.; Lentz, T.O.; Liu, L.; van Ommen, S.; Çöltekin, Ç.; van Rij, J. Neural Response Development During Distributional Learning. In Proceedings of the Interspeech, Hyderabad, India, 2–6 September 2018; pp. 1432–1436.
43. Liu, L.; Ong, J.H.; Peter, V.; Escudero, P. Revisiting infant distributional learning using event-related potentials: Does unimodal always inhibit and bimodal always facilitate? In Proceedings of the 10th International Conference on Speech Prosody, Online, 25 May–31 August 2020.
44. Escudero, P.; Best, C.T.; Kitamura, C.; Mulak, K.E. Magnitude of phonetic distinction predicts success at early word learning in native and non-native accents. *Front. Psychol.* **2014**, *5*, 1059. [CrossRef] [PubMed]
45. Burnham, D.; Francis, E.; Webster, D.; Luksaneeyanawin, S.; Attapaiboon, C.; Lacerda, F.; Keller, P. Perception of lexical tone across languages: Evidence for a linguistic mode of processing. In Proceeding of the Fourth International Conference on Spoken Language Processing (ICSLP 1996), Philadelphia, PA, USA, 3–6 October 1996; Volume 4, pp. 2514–2517.
46. Gu, F.; Zhang, C.; Hu, A.; Zhao, G. Left hemisphere lateralization for lexical and acoustic pitch processing in Cantonese speakers as revealed by mismatch negativity. *Neuroimage* **2013**, *83*, 637–645. [CrossRef] [PubMed]
47. Tuninetti, A.; Chládková, K.; Peter, V.; Schiller, N.O.; Escudero, P. When speaker identity is unavoidable: Neural processing of speaker identity cues in natural speech. *Brain Lang.* **2017**, *174*, 42–49. [CrossRef] [PubMed]
48. Tuninetti, A.; Tokowicz, N. The influence of a first language: Training nonnative listeners on voicing contrasts. *Lang. Cogn. Neurosci.* **2018**, *33*, 750–768. [CrossRef]
49. Chen, A.; Peter, V.; Wijnen, F.; Schnack, H.; Burnham, D. Are lexical tones musical? Native language's influence on neural response to pitch in different domains. *Brain Lang.* **2018**, *180*, 31–41. [CrossRef] [PubMed]
50. Boersma, P.; Weenink, D. Praat: Doing Phonetics by Computer (Version 5.1. 05) [Computer Program]. Available online: https://www.fon.hum.uva.nl/praat/ (accessed on 1 May 2009).
51. Huang, T.; Johnson, K. Language specificity in speech perception: Perception of Mandarin tones by native and nonnative listeners. *Phonetica* **2010**, *67*, 243–267. [CrossRef]
52. Liu, L.; Ong, J.H.; Tuninetti, A.; Escudero, P. One way or another: Evidence for perceptual asymmetry in pre-attentive learning of non-native contrasts. *Front. Psychol.* **2018**, *9*, 162. [CrossRef] [PubMed]
53. Liu, L.; Kager, R. Perception of tones by infants learning a non-tone language. *Cognition* **2014**, *133*, 385–394. [CrossRef] [PubMed]
54. Liu, L.; Kager, R. Perception of tones by bilingual infants learning non-tone languages. *Biling. Lang. Cogn.* **2017**, *20*, 561–575. [CrossRef]
55. Ong, J.H.; Burnham, D.; Escudero, P.; Stevens, C.J. Effect of linguistic and musical experience on distributional learning of nonnative lexical tones. *J. Speech Lang. Hear. Res.* **2017**, *60*, 2769–2780. [CrossRef] [PubMed]
56. Wong, P.C.; Perrachione, T.K. Learning pitch patterns in lexical identification by native English-speaking adults. *Appl. Psycholinguist.* **2007**, *28*, 565–585. [CrossRef]
57. Perrachione, T.K.; Lee, J.; Ha, L.Y.; Wong, P.C. Learning a novel phonological contrast depends on interactions between individual differences and training paradigm design. *J. Acoust. Soc. Am.* **2011**, *130*, 461–472. [CrossRef]
58. Delorme, A.; Makeig, S. EEGLAB: An open source toolbox for analysis of single-trial EEG dynamics including independent component analysis. *J. Neurosci. Methods* **2004**, *134*, 9–21. [CrossRef] [PubMed]
59. Lopez-Calderon, J.; Luck, S.J. ERPLAB: An open-source toolbox for the analysis of event-related potentials. *Front. Hum. Neurosci.* **2014**, *8*, 213. [CrossRef] [PubMed]
60. Colin, C.; Hoonhorst, I.; Markessis, E.; Radeau, M.; De Tourtchaninoff, M.; Foucher, A.; Collet, G.; Deltenre, P. Mismatch negativity (MMN) evoked by sound duration contrasts: An unexpected major effect of deviance direction on amplitudes. *Clin. Neurophysiol.* **2009**, *120*, 51–59. [CrossRef] [PubMed]

61. Horváth, J.; Czigler, I.; Jacobsen, T.; Maess, B.; Schröger, E.; Winkler, I. MMN or no MMN: No magnitude of deviance effect on the MMN amplitude. *Psychophysiology* **2008**, *45*, 60–69. [CrossRef] [PubMed]
62. Cheour, M.; Shestakova, A.; Alku, P.; Ceponiene, R.; Näätänen, R. Mismatch negativity shows that 3–6-year-old children can learn to discriminate non-native speech sounds within two months. *Neurosci. Lett.* **2002**, *325*, 187–190. [CrossRef]
63. Escera, C.; Alho, K.; Winkler, I.; Näätänen, R. Neural mechanisms of involuntary attention to acoustic novelty and change. *J. Cogn. Neurosci.* **1998**, *10*, 590–604. [CrossRef] [PubMed]
64. Liu, R.; Holt, L.L. Neural changes associated with nonspeech auditory category learning parallel those of speech category acquisition. *J. Cogn. Neurosci.* **2011**, *23*, 683–698. [CrossRef] [PubMed]
65. Näätänen, R.; Lehtokoski, A.; Lennes, M.; Cheour, M.; Huotilainen, M.; Iivonen, A.; Vainio, M.; Alku, P.; Ilmoniemi, R.J.; Alho, K.; et al. Language-specific phoneme representations revealed by electric and magnetic brain responses. *Nature* **1997**, *385*, 432–434. [CrossRef]
66. Burnham, D.K.; Singh, L. Coupling tonetics and perceptual attunement: The psychophysics of lexical tone contrast salience. *J. Acoust. Soc. Am.* **2018**, *144*, 1716. [CrossRef]
67. Thiessen, E.D.; Saffran, J.R. When cues collide: Use of stress and statistical cues to word boundaries by 7-to 9-month-old infants. *Dev. Psychol.* **2003**, *39*, 706. [CrossRef] [PubMed]
68. Origlia, A.; Cutugno, F.; Galatà, V. Continuous emotion recognition with phonetic syllables. *Speech Commun.* **2014**, *57*, 155–169. [CrossRef]
69. Escudero, P.; Boersma, P. Bridging the gap between L2 speech perception research and phonological theory. *Stud. Second. Lang. Acquis.* **2004**, *26*, 551–585. [CrossRef]
70. Escudero, P. *Linguistic Perception and Second Language Acquisition: Explaining the Attainment of Optimal Phonological Categorization*; Netherlands Graduate School of Linguistics: Amsterdam, The Netherlands, 2005.
71. Escudero, P. The linguistic perception of similar L2 sounds. In *Phonology in Perception*; Boersma, P., Hamann, S., Eds.; Mouton de Gruyter: Berlin, Germany, 2009; pp. 152–190.
72. Yazawa, K.; Whang, J.; Kondo, M.; Escudero, P. Language-dependent cue weighting: An investigation of perception modes in L2 learning. *Second. Lang. Res.* **2020**, *36*, 557–581. [CrossRef]
73. Zeng, Z.; Mattock, K.; Liu, L.; Peter, V.; Tuninetti, A.; Tsao, F.-M. Mandarin and English adults' cue-weighting of lexical stress. In Proceedings of the 21st Annual Conference of the International Speech Communication Association, Shanghai, China, 25–29 October 2020.
74. Sundara, M.; Ngon, C.; Skoruppa, K.; Feldman, N.H.; Onario, G.M.; Morgan, J.L.; Peperkamp, S. Young infants' discrimination of subtle phonetic contrasts. *Cognition* **2018**, *178*, 57–66. [CrossRef] [PubMed]
75. Escudero, P.; Mulak, K.E.; Elvin, J.; Traynor, N.M. "Mummy, keep it steady": Phonetic variation shapes word learning at 15 and 17 months. *Dev. Sci.* **2018**, *21*, e12640. [CrossRef]
76. Xi, J.; Zhang, L.; Shu, H.; Zhang, Y.; Li, P. Categorical perception of lexical tones in Chinese revealed by mismatch negativity. *Neuroscience* **2010**, *170*, 223–231. [CrossRef] [PubMed]

brain sciences

Article

Neural–Behavioral Relation in Phonetic Discrimination Modulated by Language Background

Tian Christina Zhao [1,2]

[1] Institute for Learning & Brain Sciences, University of Washington, Seattle, WA 98195, USA; zhaotc@uw.edu
[2] Department of Speech and Hearing Sciences, University of Washington, Seattle, WA 98195, USA

Abstract: It is a well-demonstrated phenomenon that listeners can discriminate native phonetic contrasts better than nonnative ones. Recent neuroimaging studies have started to reveal the underlying neural mechanisms. By focusing on the mismatch negativity/response (MMN/R), a widely studied index of neural sensitivity to sound change, researchers have observed larger MMNs for native contrasts than for nonnative ones in EEG, but also a more focused and efficient neural activation pattern for native contrasts in MEG. However, direct relations between behavioral discrimination and MMN/R are rarely reported. In the current study, 15 native English speakers and 15 native Spanish speakers completed both a behavioral discrimination task and a separate MEG recording to measure MMR to a VOT-based speech contrast (i.e., pre-voiced vs. voiced stop consonant), which represents a phonetic contrast native to Spanish speakers but is nonnative to English speakers. At the group level, English speakers exhibited significantly lower behavioral sensitivity (d') to the contrast but a more expansive MMR, replicating previous studies. Across individuals, a significant relation between behavioral sensitivity and the MMR was only observed in the Spanish group. Potential differences in the mechanisms underlying behavioral discrimination for the two groups are discussed.

Keywords: linguistic experience; speech perception; mismatch response; magnetoencephalography; individual differences

Citation: Zhao, T.C. Neural–Behavioral Relation in Phonetic Discrimination Modulated by Language Background. *Brain Sci.* **2022**, *12*, 461. https://doi.org/10.3390/brainsci12040461

Academic Editors: Richard Wright and Benjamin V. Tucker

Received: 7 February 2022
Accepted: 24 March 2022
Published: 29 March 2022

1. Introduction

It is a well-demonstrated phenomenon that listeners' sensitivities to speech contrasts with small acoustic differences are affected by their language background. In the original series by Abramson and Lisker, discrimination of pairs of speech sounds along the Voice Onset Time (VOT) continuum was examined. First, they demonstrated that native English speakers showed a single peak in discrimination along the VOT continuum, corresponding to two phonemic categories in English (i.e., voiced vs. voiceless stops). In contrast, with the same VOT continuum, Thai speakers demonstrated two peaks in discrimination, corresponding to three categories in the Thai language (i.e., pre-voiced, voiced and voiceless) [1]. Further, when comparing English speakers to Spanish speakers, they showed that the category boundaries in perception of voice timing are significantly different between the two groups, with much shorter VOT as a boundary in Spanish speakers [2]. Such experiential effects were replicated repeatedly later on with various speech contrasts utilizing different acoustic cues [3–6].

Over the last few decades, researchers have been increasingly interested in the neural mechanisms that may underlie such linguistic effect in speech perception. The mismatch negativity (MMN), or mismatch response (MMR), is one of the most widely studied and used neural measures which has been suggested to index the neural sensitivity to sound change [7]. In its original form, a standard stimulus is repeated for the majority of the sequence (e.g., 80%) while a deviant stimulus is randomly interspersed among the standards (e.g., 20%). A difference wave is then calculated by subtracting the response to standards from the response to deviants. When measured with electroencephalography (EEG), the

MMN can be observed largely in the frontocentral scalp areas, roughly 200 ms after the onset of the change in the difference wave. Both the magnitude and latency of the MMN have been suggested to show the level of sensitivity. For example, a larger change in stimulus (e.g., pure tone changing from 1000 Hz to 1032 Hz) elicited MMN with shorter latency and larger magnitude than a smaller change (e.g., pure tone changing from 1000 Hz to 1016 Hz).

In the realm of speech perception, Näätänen and colleagues were the first to demonstrate a language effect on neural sensitivities to vowel contrasts using MMN [8]. Particularly, for a vowel contrast varying on the second formant that is native to Estonians but nonnative to Finnish speakers, Estonian speakers demonstrated significantly larger MMN than Finnish speakers. This work was later followed to further examine stop consonant processing. In a series of studies focusing on VOT contrasts, researchers also demonstrated that in English speakers, a VOT contrast that crossed the phonemic boundary (i.e., /da/-/ta/) elicited larger MMN than a VOT contrast with the same acoustic distance but within the /ta/ category [9]. Further, by focusing on a VOT contrast (−10 ms vs. −50 ms) that is phonemic in Hindi but not in English, they demonstrated that, indeed, the Hindi speakers had larger MMN than the English speakers along with higher behavioral discrimination [10]. Similar effects have also been documented with many other speech stimuli [11,12].

More recently, the underlying neural generator, or the neural source, of the MMN is of research interest. Particularly, by using magnetoencephalography (MEG), which allows for robust inference of source activity, researchers have suggested that while the main contributor to the MMR is the bilateral temporal region, there is also contribution from the frontal region, likely the inferior frontal region [13,14]. So far, very limited data exist that examine the linguistic effect on MMR with a focus on the underlying sources. Only Zhang and colleagues examined this question by focusing on a pair of speech contrasts (i.e., /ra/-/la/) and measured MMR in Japanese speakers vs. English speakers using MEG [15]. Critically, the /ra/-/la/ contrast is nonnative to Japanese speakers but native to English speakers. Two types of methods were used to model the source-level activities and the results converged and demonstrated that the Japanese speakers demonstrated more widespread and longer activation for the speech contrast than the American English speakers. The authors interpreted the results to reflect a more efficient processing of native contrast in the English speakers, which is consistent with similar research using the fMRI method also using the /ra/-/la/ contrast [16]. However, the authors also cautioned an alternative account that the observed between-group difference could reflect 'rather fundamentally different types of neural processes used by native and nonnative speakers', instead of a difference in neural efficiency.

One way to elucidate this question is to examine the correspondence between the MMR and behavioral discrimination across individuals and compare the correspondence between groups. The rationale is as follows: if two groups of different language backgrounds rely on the same mechanisms but with different efficiency, we can expect the same MMR–behavioral discrimination correlation (i.e., the same slope) for native and nonnative speakers. Alternatively, if the two groups rely on different mechanisms, the slopes for such correlations will be different between the two groups.

However, a neural–behavioral correspondence at the individual level is rarely reported and largely assumed in the MMN/R literature. Much research has only shown parallel results between behavior and MMN/R at the group level. For example, as a group, higher behavioral discrimination was observed for a native speech contrast compared to a nonnative speech contrast. Similarly, as a group, a larger MMN/R was observed for the native speech contrast compared to the nonnative contrast. It is then often assumed that individuals who demonstrate a higher behavioral discrimination score will also have a large MMR. However, this type of analysis and subsequent results were in fact not reported [10,15]. It was therefore unclear whether the lack of neural–behavioral correspondence is due to lack of analysis/results or lack of significant findings. It is then crucial to directly examine and report the MMR–behavioral correspondence. Whether the MMR can explain a significant portion of the variance in behavioral discrimination is key to state whether MMR is indeed part of the neural underpinning of speech discrimination.

It is possible that previous studies have, indeed, conducted such correlational analyses but failed to find significant results due to lack of sensitivity in statistical methods. Indeed, in the most comprehensive review of MMN, the authors suggested that 'in general, the MMN replicability is quite good at the group level but at the individual level, there still is ample space for further improvement before the MMN provides a reliable tool for clinics at the level of individual patients' [7]. To that end, we employed a more exploratory machine-learning-based method to examine the correlation in addition to the traditional parametric regression analysis. The ML-based method takes data across all spatiotemporal points into consideration and, thus, may be more sensitive in detecting correlations between behavior and MMR.

The goal of the current study is thus twofold: (1) to replicate the language effect on MMR at the source level as reported by Zhang and colleagues [15], using a VOT contrast; and crucially, (2) to investigate whether individual differences in MMR at the source level can explain significant portion of variance in behavioral discrimination of speech contrast. In other words, whether there is a significant neural–behavioral correspondence across individuals, and if so, whether such neural–behavioral relation is different for native vs. nonnative speakers.

2. Materials and Methods

2.1. Participants

Monolingual English speakers (n = 15, male = 5, age = 21.4 (s.d. = 1.8)) and Native Spanish speakers (n = 15, male = 5, age = 26.0 (s.d. = 5.0)) were recruited. All participants were healthy adults with no reported speech, hearing or language disorders. All participants were right-handed (Edinburgh Handedness Quotient = 0.99 ± 0.04). All participants completed a short survey on their language and music backgrounds. The Native Spanish speakers all learned Spanish as their first language and still use the language as their predominant language. However, because they have all moved to the U.S., they also speak English to various degrees. Indeed, on the language background survey, the Native Spanish group reported higher efficiency (mean = 3.67, s.d. = 0.95) in foreign languages than the Monolingual English speakers (mean = 1. 70, s.d. = 0.67, t (28) = −6.49, $p < 0.001$). On the other hand, the Monolingual English speakers (mean = 4.67, s.d. = 4.56) reported more musical training experience (i.e., years of private lessons) than Native Spanish speakers (mean = 1.30, s.d. = 1.59, t (17.3) = 2.69, p =0.015, equal variance not assumed). All procedures were approved by the Institute Review Board of the University of Washington and informed consent was obtained from all participants.

2.2. Stimulus

Bilabial stop consonants with varying VOTs were synthesized by the Klatt synthesizer in Praat software [17]. The VOT values were −40 ms and +10 ms. The syllable with 0 ms VOT was first synthesized with a 2 ms noise burst and vowel /a/. The duration of the syllable is 90 ms. The fundamental frequency of the vowel /a/ began at 95 Hz and ended at 90 Hz. The silent gap (10 ms) and the pre-voicing (40 ms) were added after the initial noise burst to create syllables with the positive and negative VOTs. The fundamental frequency for the pre-voicing portion was 100 Hz. The waveforms of the stimulus pair, that is, voiced /ba/ (VOT = +10 ms) and the pre-voiced /ˌba/ (VOT = −40 ms), are shown in Figure 1A.

Critically, this stimulus pair represents a native phonetic contrast in Spanish but not in English. Indeed, the stimulus pair was selected from a VOT continuum between −40 ms and +40 ms that was previously tested and validated [18]. Particularly, Monolingual English speakers demonstrated the category boundary to be above +10 ms while Native Spanish speakers demonstrated the category boundary to be below +10 ms. Therefore, the discrimination of the +10 ms/−40 ms (/ba/ vs. /ˌba/) stimulus pair in this current study should capture the cross-linguistic difference between the two groups with different language backgrounds. That is, it represents a cross-category phonetic contrast only for the Native Spanish speakers and, therefore, should elicit higher sensitivity in them.

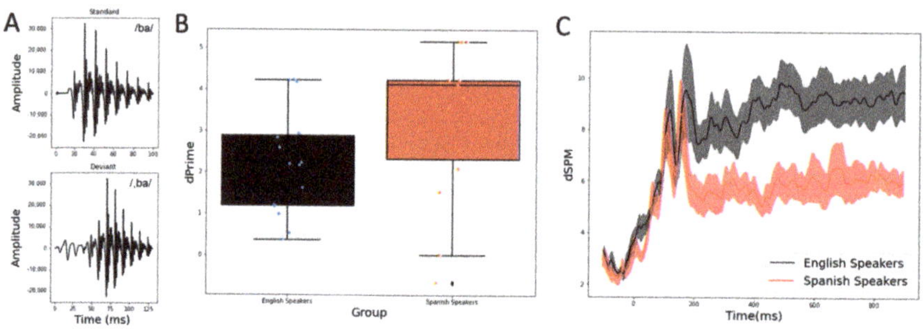

Figure 1. (**A**) Waveforms for the standard stimulus (**top**) and the deviant stimulus (**bottom**). (**B**) Behavioral sensitivity (d′) is different between the two groups, with Spanish speakers exhibiting higher d′ overall as this contrast is native to them. (**C**) Global MMR is different between the two groups with Spanish speakers exhibiting reduced MMR after 200 ms.

2.3. Behavioral Discrimination

An AX behavioral discrimination task was first conducted to assess the sensitivity to the speech contrast. This is the same task as used in previous studies [9,10,15,19], and a cross-linguistic effect needed be established on the behavioral discrimination of this contrast before the neural underpinning could be studied. All participants performed this task on a Dell XPS13 9333 computer running Psychophysical Toolbox [20] in MATLAB version 2016a (MathWorks. Inc., Natick, MA, USA) in a sound-attenuated booth. All sounds were delivered through Sennheiser HDA 280 Headphones at 72 dB SPL.

In an AX discrimination trial, a fixation cross was first presented for 200 ms at the center of the screen to indicate the start of the trial. Then, two speech sounds were played with a 250 ms inter-stimulus interval between them. The two speech sounds can be either the same or different (e.g., /ba/ followed by /ba/ or /ˌba/). The participant was instructed to judge whether the two sounds were the same or different through key presses within 1 s. All 4 possible pairings (i.e., AA (/ba//ba/), AB (/ba//ˌba/), BB (/ˌba//ˌba/), BA (/ˌba//ba/)) were repeated 10 times in a randomized order.

The d′ values for the stimulus pair were calculated for each participant and used as the measure of sensitivity. The d′ measure takes into consideration both hit and false alarm responses, and therefore addresses the issue of response bias [21]. Specifically, the 'hit' is defined as when participants respond 'different' when sounds were different (i.e., for the AB and BA pairs), and the 'false alarm' is defined as when participants respond 'different' when the sounds were the same (i.e., for the AA and BB pairs). Then, d′ is calculated as the normalized hit rateminus the normalized false alarm rate. The hit rate for the Monolingual English speakers was 0.594 (SD = 0.318) and 0.846 (SD = 0.22) for Native Spanish speakers. The false alarm rate for the Monolingual English speakers was 0.076 (SD = 0.117) and 0.107 (SD = 0.169) for Native Spanish speakers.

2.4. MMR Measurement in MEG

MEG recordings were completed inside a magnetically shielded room (MSR) (IMEDCO America Ltd., IN), using a whole-scalp system with 204 planar gradiometers and 102 magnetometers (VectorView^{TM}, Elekta Neuromag Oy, Helsinki, Finland). Five head-position-indicator (HPI) coils were attached to identify head positions under the MEG dewar at the beginning of each block. Three landmarks (LPA, RPA and nasion) and the HPI coils were digitized along with 100 additional points along the head surface (Isotrak data) with an electromagnetic 3D digitizer (Fastrak®, Polhemus, Colchester, VT, USA). In addition, a pair of electrocardiography sensors (ECG) was placed on the front and backside of the participants' left shoulder to record cardiac activity and three pairs of electrooculogram (EOG) sensors were placed horizontal and vertical to the eyes to record saccades and blinks. All data were sampled at 1 kHz.

The sounds were delivered from a TDT RP 2.7 device (Tucker-Davis Technologies, Alachua, FL, USA), controlled by custom Python software on a HP workstation, to insert earphones. The stimulus was processed such that the RMS values were referenced to 0.01 and it was further resampled to 24,414 Hz for the TDT. The sounds were played at the intensity level of 80 dB through tubal insert phones (Model TIP-300, Natus Neurology, Pleasanton, CA, USA). A traditional oddball paradigm was used for stimulus presentation. The syllable with +10 ms VOT was used as the standard (600 trials, 80%), and the syllables with −40 ms VOT were used as deviants (150 trials, 20%) with at least two standards in between deviants. The stimulus onset asynchrony (SOA) values were jittered around 800 ms. The participants listened passively and watched silent videos during recording.

2.5. MEG Data Processing

All MEG data processing was carried out using the MNE-python software [22]. MEG data were first preprocessed using the Oversampled Temporal Projection (OTP) method [23] and the temporally extended Spatial Signal Separation (tSSS) method [24,25] to suppress sensor noise and magnetic interference originating from outside of the MEG dewar. Signal space projection was used to suppress the cardiac and eye movement signals in the MEG data [26]. Then, the data were low pass filtered at 50 Hz. Epochs (−100 to 900 ms) for the standards and deviants were extracted and any epochs with peak-to-peak amplitude exceeding 4 pT/cm for gradiometers or 4.0 pT/cm for magnetometers were rejected. All deviants as well as the subset of standard trials immediately preceding the deviants were averaged to calculate the evoked responses.

To estimate the location of neural generators underlying the evoked responses, each subject's anatomical landmarks and additional scalp points were used with an iterative nearest-point algorithm to rescale the average adult template brain (fsaverage) to match the subject's head shape. FreeSurfer was used to extract the inner skull surface (watershed algorithm) and the cortical and subcortical structures segmented from the surrogate MRI [27]. A one-layer conductor model based on the rescaled inner skull surface was constructed for forward modeling [28]. The surface source space consisted of 20,484 dipoles evenly spatially distributed along the gray/white matter boundary (i.e., 'ico-5'). Because surrogate head models and source spaces were used for each subject, source orientations were unconstrained (free orientation). Baseline noise covariance was estimated using empty room recordings made on the same day of the MEG session. Dipolar currents were estimated from the MEG sensor data using an anatomically constrained minimum-norm linear estimation approach to obtain dSPM values at each source location [29].

The mismatch responses (MMRs) were subsequently calculated at the source level by subtracting the standards from each deviant. That is, the MMR was calculated by subtracting the vectors of standard from the vectors of deviant and then the magnitude of the vectors was calculated.

The Destrieux Atlas (i.e., 'aparc.a2009s' atlas in Freesurfer) was then applied to reduce the data by averaging across vertices within each label [30]. Four regions-of-interest (ROIs) were identified *a prior* based on the existing literature for further statistical analysis: left and right inferior frontal region (i.e., inferior frontal label) and superior temporal region (i.e., superior temporal label). All data reported in this study are publicly available at Open Science Framework.

2.6. Regression Methods

2.6.1. Parametric Multiple Regression

To investigate the correspondence between the behavioral sensitivity (d') and neural sensitivity (MMR), we first took a region-of-interest (ROI) approach to reduce the dimension of the MMR data for the parametric multiple regression analyses (see MEG data processing section for details). Four ROIs were selected *a priori* based on existing research on MMR sources, including the left and right superior temporal gyrus (STG) and inferior frontal gyrus (IFG) [13,14]. In addition, we selected the time window between 200 and 500 ms that captures maximum differences between group. The averaged MMR values were further log-transformed to reduce skewness in preparation for the multiple regression analyses.

In each multiple regression model, the main effects of the MMR value were averaged across the ROI and the selected time window and the language group (i.e., English vs. Spanish speakers) were entered. In addition, the interaction between the MMR and language group was also entered into the model to predict the behavioral sensitivity d′ (IBM SPSS Version 19.0.0).

2.6.2. Machine-Learning-Based Regression

The machine-learning-based regressions were carried out using the open source scikit-learn package [31] in conjunction with the MNE-python software. This method was adopted from a previous study [32]. All spatial and temporal samples were used in this method. Specifically, for each time sample, we employed a support-vector regression (SVR) where the model uses MMR values from all 150 label regions, thus taking the spatial pattern of the MMR into consideration, to predict individual behavioral d′ value [33]. The dataset is first split into a training and a testing set (see below details regarding leave-one-out cross-validation). The MMR spatial–temporal patterns in the training set were first used to fit the model with a linear kernel function (C = 1.0, epsilon = 0.1). Once the model is trained, the MMR spatial–temporal patterns from the testing set were then used to generate predictions of the d′ value. A leave-one-out cross-validation method was used to enhance model prediction. That is, all 15 possible splits of the data (i.e., every one of the 15 participants were assigned as the testing set while the rest of the individuals (14) were the training set) were used to build 15 models and derive an averaged model. The R^2 coefficient of determination between actual measured d′ and model predicted d′ is taken as an index of model performance. The same process was repeated for every time sample of the MMR, which generates a temporal sequence of R^2.

To further evaluate the model performance, within each time sample, we shuffled the correspondence between the d′ and MMR spatial pattern across individuals and then conducted the same SVR analyses. In such cases, the MMR spatial pattern should bear no predictive value to d′ score and the R^2 should reflect a model performing at chance level. We repeated this process 100 times for each time point. Then, we generated an empirical null distribution of R^2 by pooling all the R^2 values from each time point (i.e., 100 permutations for each time point) and we compared our originally obtained R^2 coefficient against this distribution [34]. Specifically, we considered that if our originally obtained R^2 value at a specific time point is larger than the 99th percentile of the empirical null distribution, then the spatial pattern of that time point can significantly predict d′ of an individual. This procedure allows a conservative way to correct for multiple comparison given the explorative nature of the analysis. A final SVR was then fit by using all the time points that were deemed significant by permutation.

3. Results

3.1. Behavioral and Neural Sensitivity to the Speech Pair Is Modulated by Language Background

To examine the effect of language background on behavioral sensitivity to the speech contrast, an independent *t* test was conducted to compare the d′ values between the two groups (IBM SPSS Statistics, version 19.0.0). Supporting our hypothesis and replicating the existing literature, the results revealed (Figure 1B) that the Native Spanish speakers exhibited significantly higher d′ than the Monolingual English speakers ($t(28) = -1.83$, $p = 0.039$, 1-tailed, Cohen's d = 0.668). That is, the contrast being a phonetic contrast in one's native language enhanced the sensitivity to the contrast.

To examine the effect of language background on neural sensitivity, as indexed by the MMR, we first visualized the MMR at the source level averaged across the whole brain for each group (Figure 1C). The first two peaks prior to 200 ms reflect the intrinsic timing difference between the standard and the deviant stimuli (100 ms vs. 130 ms in duration). The divergence between the group largely began after 200 ms and the global MMR amplitude at the source level is much more reduced in the Native Spanish speaker group.

We then further assessed the spatial distribution of MMR within each group as well as the between-group difference. The spatial distribution of the MMR over time for the Monolingual English speakers can be visualized in the left column in Figure 2 and, similarly, the spatial distribution for the Native Spanish speaker can be visualized in the middle column in Figure 2. The spatial distributions for the two groups are largely the same but with much reduced intensity in the Native Spanish group after 200 ms. Complete visualization of MMR for both groups over time can be seen in the Supplementary Materials (Videos S1 and S2).

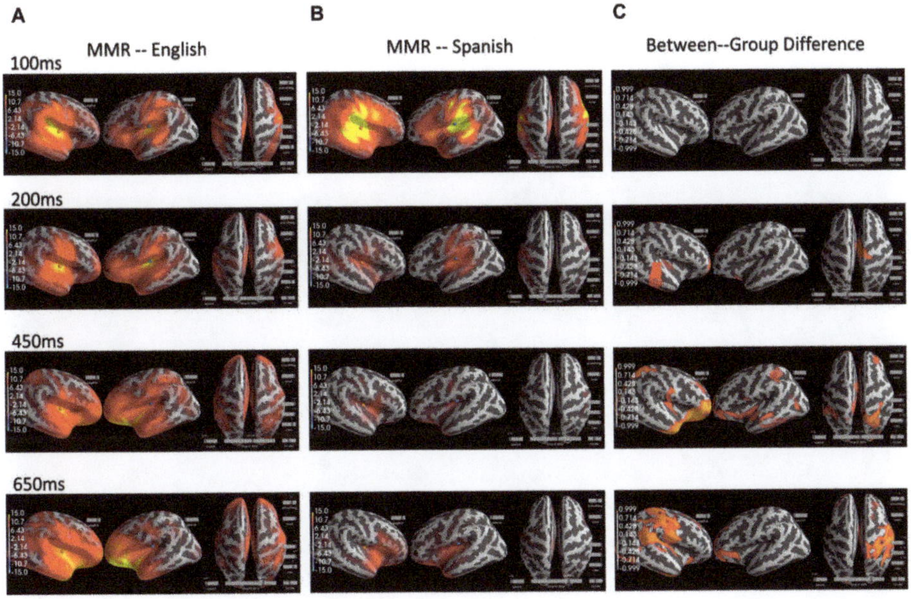

Figure 2. (**A**) MMR across the whole brain over time for the Monolingual English speakers. (**B**) MMR across the whole brain over time for the Native Spanish speakers. (**C**) The spatial regions that are significantly different between the two groups over time. They largely occur after 200 ms and are predominantly in the right hemisphere.

To examine the between-group difference in MMR at the whole-brain level, a spatial–temporal cluster test based on the threshold-free cluster enhancement method (TFCE) was conducted [35]. This test is nonparametric and based on permutation and is designed to allow for improved sensitivity and more interpretable output than the traditional cluster-based method. Specifically, the TFCE values were generated by summing across a series of thresholds, thus avoiding the selection of an arbitrary threshold, and then the *p* values for each spatial temporal sample were calculated through permutation. The og(p) for the between group comparison can be visualized in Figure 2 in the right column. The larger the −log(p), the smaller the p, and the more significant the between-group effect is. As can be seen, the between-group difference is significant largely after 200 ms and becomes more prominent in the right hemisphere than the left hemisphere across a wide range of regions, including the superior temporal regions (200 ms) and the inferior frontal regions (450, 650 ms) that are largely thought to underlie the MMR. Interestingly, the differences were also observed in parietal regions as well as the temporoparietal junction (TPJ) (650 ms). Complete visualization of −log(p) values over time can be seen in the Supplementary Materials (Video S3).

3.2. *Behavioral–Neural Connection Is Affected by Language Background*

3.2.1. Parametric Multiple Regression

Multiple regression analyses were carried out for each ROI (see Section 2.6.1 for details on the models). For the left IFG and left STG, the models show significant fit and marginally significant fit (left IFG: $R^2 = 0.312$, $p = 0.019$, left STG: $R^2 = 0.230$, $p = 0.074$). Crucially, in both models, neither the MMR nor the language background were significant predictors, but the interaction between the two factors was significant and marginally significant (left IFG: B = 4.42, $p = 0.049$, left STG: B = 4.865, $p = 0.095$). As can be visualized in Figure 3A in the top row, in both ROIs, the MMR is significantly predictive of behavioral d′, but only in the Native Spanish group, not in the Monolingual English group. Conversely, similar models with right ROIs yielded nonsignificant fit ($p > 0.1$) (see Figure 3A, bottom row).

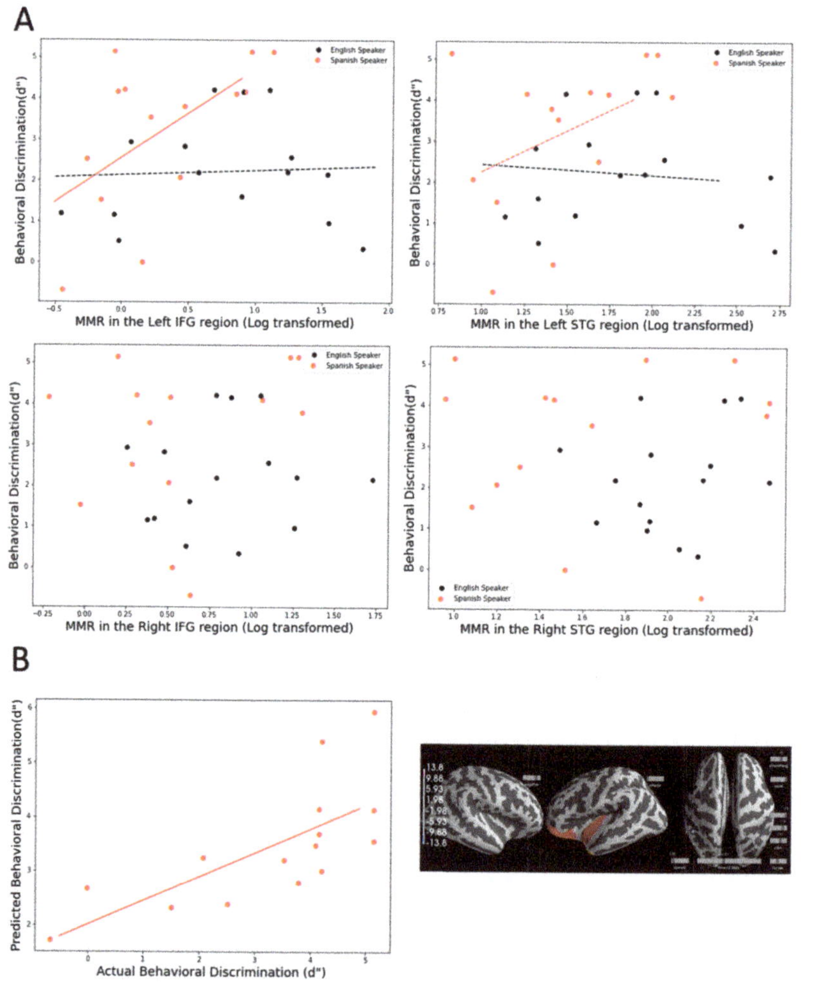

Figure 3. (**A**) Top row: Scatter plot between MMR in the left IFG and behavioral d′ (**left**) and between

MMR in the left (STG) and d′ (**right**). The multiple regression analyses show that MMR in these two regions are significant predictors of behavioral d′ but only in the Native Spanish speakers. Bottom row: Scatter plot between MMR in the right IFG and behavioral d′ (**left**) and between MMR in the right (STG) and d′ (**right**). (**B**) A machine-learning-based regression confirmed the multiple regression based on ROIs. In the Spanish speaker group, a machine-learning-based regression can significantly predict behavioral d′ using the whole brain MMR (left column). The areas that are significant contributors to the model overlap with the a priori selected ROIs. However, no significant prediction was achieved in the Monolingual English speaker group.

3.2.2. Machine-Learning-Based Regression

In order to further explore additional spatiotemporal patterns outside of the a priori selected ones that may also be important in predicting the behavioral sensitivity, we conducted a whole-brain machine-learning-based regression using the whole MMR time series (see Section 2.6.2 for detail on the method). Given the potential differences between the two groups, we ran these analyses separately for the Monolingual English group vs. the Native Spanish group.

Using this method, we examined whether MMR could predict behavioral d′ in Native Spanish speakers and Monolingual English speakers. Consistent with the ROI analyses approach, for the Native Spanish speakers, a time window from 290 to 295 ms was deemed significant and by using that time window, the model can significantly predict individual d′ (Figure 3B, left). That is, the actual measured d′ and the model predicted d′ are significantly correlated ($r = 0.48$, $p = 0.06$). Critically, the areas that significantly contribute to the prediction (Figure 3B, right) show large overlap with the a priori selected ROIs. It is worth noting that the areas involve a larger frontal region including the medial frontal regions.

On the other hand, similar with the ROI analysis, the same ML-based regression did not yield any significant results for the Monolingual English Speaker group, suggesting no spatial–temporal patterns to be a good predictor of the behavioral d′ values.

4. Discussion

The current research extended the rich literature documenting the linguistic effect on speech processing and further examined its neural basis. Particularly, the current study examined the MMR to a speech contrast at the source level and, more importantly, the correlation between the neural MMR measure and the behavioral discrimination of the speech contrast. The speech contrast was a stop consonant contrast based on the Voice Onset Time (i.e., pre-voiced vs. voiced), which is a native phonemic contrast for Spanish speakers, but nonnative to English speakers. Monolingual English speakers and Native Spanish speakers' behavioral discrimination of this contrast was examined along with their MMR, the most widely studied neural signature suggested to index sensitivity to sound change. The MEG-measured MMR allows a focus of examination on the source-level activities. Behaviorally, Native Spanish speakers demonstrated significantly higher sensitivity to this contrast compared to the Monolingual English speakers, demonstrating the expected linguistic effect. For the MMR at the source level, Native Spanish speakers demonstrated significantly widespread reduction compared to the Monolingual English speakers, with the difference predominantly in the right hemisphere. The behavior–MMR relation was further investigated across individuals and the results demonstrated that a significant correlation between MMR and behavioral discrimination was only observed within the Native Spanish group, but not in the Monolingual English group. Additionally, for the Native Spanish group, the cortical regions driving the behavior–MMR correlation are largely in the left frontal region.

The largely reduced MMR across multiple regions at the cortical source level for the native speakers (i.e., Native Spanish speakers), compared to the nonnative speakers (i.e., Monolingual English speakers), replicated a previous study examining linguistic effect on MMR at the source level, using a different speech contrast and populations (i.e., /ra/-/la/, Japanese vs. English speakers) [15]. This is also in line with a subsequent MEG study by

Zhang and colleagues demonstrating that after intensive perceptual training to discriminate the /ra/-/la/ contrast, Japanese speakers' MMR at the source level was also observed to be reduced [19]. These results focusing on the MMR at the source level may seem counter to the EEG-measured MMN results where MMN to native contrasts have repeatedly been shown as larger than MMN to nonnative contrasts [8,10]. However, it is important to keep in mind that while the measurement paradigms are similar for EEG and MEG, the intrinsic differences between the two technologies (i.e., measuring electric potential vs. magnetic field) dictate that they are sensitive to overlapped but different neural populations [36] and are thus picking up different signals. It is important for future research to understand more about the relationship between MEG- vs. EEG-measured MMRs and reconcile the results from these two methods to allow for unified interpretation. For example, simultaneously measured MMR using both M/EEG will allow the investigation of the relationship between MMR in EEG sensors and MMR at the source level.

The whole brain comparison of the group-level MMR between Monolingual English speakers and Native Spanish speakers revealed that the reduction is bilateral in nature, involving regions known to be important for MMR, such as the superior temporal gyrus and inferior frontal regions. Interestingly, the reduction for the Native Spanish group is much more prominent in the right hemisphere than the left hemisphere, particularly in the temporal–parietal junction (TPJ) region (Figure 2). Based on the speech processing model, spectrotemporal analysis of the speech signal at the STG level is bilateral in nature before traveling up the dorsal stream in the left hemisphere where integration of information from other modalities occurs (e.g., sensorimotor, visual) [37]. In this case, the MMR reduction in the left hemisphere and the right STG before 450 ms (Figure 2, right column) may be attributable to a more 'efficient processing' of the acoustic signal in the Native Spanish group. Yet, it remains unclear what the large reduction in the right hemisphere after 450 ms would entail. It was therefore crucial to examine whether any of the MMR was directly relevant for behavioral discrimination of the speech contrasts for the two groups.

As alluded to in the introduction, previous studies have hardly ever reported a correlation between MMR and behavioral discrimination. It was unclear whether it was due to lack of analysis or lack of significant findings. The current study addressed this issue directly and used two methods to evaluate whether MMR is correlated with behavioral discrimination across individuals. The results converged and demonstrated a robust correlation between behaviorally measured discrimination (d') and MMR measured at the source level, but only in the Native Spanish group. The two types of analyses confirmed and complemented each other regarding this behavior–MMR correlation, that is, (1) traditional multiple regression analysis based on MMR extracted from a priori defined ROIs and time windows and (2) a data-driven exploratory machine-learning-based regression that takes the whole brain and the whole MMR time series into consideration. Critically, the behavioral–MMR correlation was only observed in the Native Spanish group, but not in the English speakers. This provides evidence that different mechanisms may be underlying native vs. nonnative speech MMR.

For the Native Spanish group, both types of regression analyses show that regions driving the behavior–MMR correlation are restricted to the left hemisphere and the effect seems larger in the frontal region. Further, the correlation is positive in nature, that is, the larger the MMR, the better the behavior discrimination. This result aligns well with both the theories regarding MMR as well as the speech processing model [7,14,37]. This suggests that native speakers may be processing the speech in a 'phonetic mode' where utilizing information from the motor planning region (e.g., left IFG) is crucial for them to distinguish two acoustically very similar speech sounds. This result is also in line with our recent studies examining the development of MMR at the source level in infants, demonstrating that the most substantial increase in MMR during the sensitive period for phonetic learning is in the left IFG region for native speech contrasts [32].

On the other hand, the lack of any behavior–MMR correlation in the Monolingual English group is surprising and puzzling, especially given the substantially enhanced

MMR across the whole brain at the group level. Nonnative speech processing has generally been considered more to be at the 'acoustic level' of processing, such that one would expect correlation between behavior and MMR in the auditory region (e.g., STG). However, neither type of regression showed indication of such correlation. The multiple regression examined both left and right STG, while the machine-learning-based regression explored the whole brain across all time points over the MMR. One possibility is that the effect for nonnative speech is much smaller and would require a larger sample to detect the correlation. Another possibility is that attention plays a larger role in nonnative speech discrimination as attention is required for the behavioral task while MMR is measured pre-attentively. This may explain the lack of correlation between MMR and behavioral discrimination in the existing literature. Future research will need to replicate with larger samples driven by power analysis and further understand the behavioral relevance of MMR for nonnative speech. On the other hand, future research will also need to better understand the neural mechanisms for nonnative speech processing, compared to native speech processing.

5. Conclusions

The current study extended our current understanding of neural mechanisms underlying the linguistic effect on speech discrimination. It demonstrated that the MMR at the source level is reduced for native speakers, compared to nonnative speakers. Yet, a robust neural–behavior relation was only observed in the native speakers, suggesting potentially different mechanisms to be involved for the nonnative speakers. Future research is warranted to replicate and further elucidate the mechanisms for speech processing, particularly for the nonnative speech.

Supplementary Materials: The following supporting information can be downloaded at: https://www.mdpi.com/article/10.3390/brainsci12040461/s1, Video S1: MMR across the whole brain over time in Monolingual English speakers, Video S2: MMR across the whole brain over time in Native Spanish speakers, Video S3: Difference in MMR between group across the whole brain over time, $-\log(p)$ is plotted.

Funding: The author received partial salary support by the National Institute of Neurological Disorders and Stroke of the National Institutes of Health under Award Number R21NS114343 (PI: Zhao). The content is solely the responsibility of the authors and does not necessarily represent the official views of the National Institutes of Health.

Institutional Review Board Statement: The study was conducted in accordance with the Declaration of Helsinki, and approved by the Institutional Review Board (or Ethics Committee) of University of Washington (STUDY00002368).

Informed Consent Statement: Informed consent was obtained from all subjects involved in the study.

Data Availability Statement: De-identified data and analysis code used to generate the results in this manuscript are available at Open Science Framework (https://osf.io/nyzts/?view_only=7f012b821d2a4764aa2f6a3f0ad4bd93, accessed on 6 February 2022).

Acknowledgments: The author would like to thank Elisabeth DeRichmond and Nikhita Harazi for their help in collecting this dataset.

Conflicts of Interest: The author declares no conflict of interest.

References

1. Abramson, A.S.; Lisker, L. Discriminability along the voicing continuum: Cross language tests. In Proceedings of the Sixth International Congress of Phonetic Sciences, Prague, Czech Republic, 7–13 September 1967; Academia: Prague, Czech Republic, 1970.
2. Abramson, A.S.; Lisker, L. Voice-timing perception in Spanish word-initial stops. *J. Phon.* **1972**, *29*, 15–25. [CrossRef]
3. Zhao, T.C.; Kuhl, P.K. Higher-level linguistic categories dominate lower-level acoustics in lexical tone processing. *J. Acoust. Soc. Am.* **2015**, *138*, EL133–EL137. [CrossRef] [PubMed]
4. Iverson, P.; Kuhl, P.K.; Akahane-Yamada, R.; Diesch, E.; Tohkura, Y.; Kettermann, A.; Siebert, C. A perceptual interference account of acquisition difficulties for non-native phonemes. *Cognition* **2003**, *87*, B47–B57. [CrossRef]

5. Polka, L. Cross-language speech perception in adults: Phonemic, phonetic, and acoustic contributions. *J. Acoust. Soc. Am.* **1991**, *89*, 2961–2977. [CrossRef] [PubMed]
6. Miyawaki, K.; Jenkins, J.J.; Strange, W.; Liberman, A.M.; Verbrugge, R.; Fujimura, O. An effect of linguistic experience: The discrimination of [r] and [l] by native speakers of Japanese and English. *Percept. Psychophys.* **1975**, *18*, 331–340. [CrossRef]
7. Naatanen, R.; Paavilainen, P.; Rinne, T.; Alho, K. The mismatch negativity (MMN) in basic research of central auditory processing: A review. *Clin. Neurophysiol.* **2007**, *118*, 2544–2590. [CrossRef]
8. Näätänen, R.; Lehtokoski, A.; Lennes, M.; Cheour, M.; Huotilainen, M.; Iivonen, A.; Vainio, M.; Alku, P.; Ilmoniemi, R.; Luuk, A.; et al. Language-specific phoneme representations revealed by electric and magnetic brain responses. *Nature* **1997**, *385*, 432–434. [CrossRef]
9. Sharma, A.; Dorman, M.F. Cortical auditory evoked potential correlates of categorical perception of voice-onset time. *J. Acoust. Soc. Am.* **1999**, *106*, 1078–1083. [CrossRef]
10. Sharma, A.; Dorman, M.F. Neurophysiologic correlates of cross-language phonetic perception. *J. Acoust. Soc. Am.* **2000**, *107*, 2697–2703. [CrossRef]
11. Winkler, I.; Lehtokoski, A.; Alku, P.; Vainio, M.; Czigler, I.; Csépe, V.; Aaltonen, O.; Raimo, I.; Alho, K.; Lang, H.; et al. Pre-attentive detection of vowel contrasts utilizes both phonetic and auditory memory representations. *Cogn. Brain Res.* **1999**, *7*, 357–369. [CrossRef]
12. Dehaene-Lambertz, G. Electrophysiological correlates of categorical phoneme perception in adults. *NeuroReport* **1997**, *8*, 919–924. [CrossRef] [PubMed]
13. Alho, K. Ceregral generators of mismatch negativity (MMN) and its magnetic counterpart (MMNM) elicited by sound changes. *Ear Hear.* **1995**, *16*, 38–51. [CrossRef] [PubMed]
14. Garrido, M.I.; Kilner, J.; Stephan, K.E.; Friston, K. The mismatch negativity: A review of underlying mechanisms. *Clin. Neurophysiol.* **2009**, *120*, 453–463. [CrossRef] [PubMed]
15. Zhang, Y.; Kuhl, P.K.; Imada, T.; Kotani, M.; Tohkura, Y. Effects of language experience: Neural commitment to language-specific auditory patterns. *Neuroimage* **2005**, *26*, 703–720. [CrossRef]
16. Callan, D.E.; Jones, J.A.; Callan, A.M.; Akahane-Yamada, R. Phonetic perceptual identification by native- and second-language speakers differentially activates brain regions involved with acoustic phonetic processing and those involved with articulatory–auditory/orosensory internal models. *NeuroImage* **2004**, *22*, 1182–1194. [CrossRef] [PubMed]
17. Boersma, P.; Weenink, D. Praat: Doing Phonetics by Computer (Version 5.1.05). 2009. Available online: https://www.fon.hum.uva.nl/praat/ (accessed on 6 February 2022).
18. Zhao, T.C.; Kuhl, P.K. Linguistic effect on speech perception observed at the brainstem. *Proc. Natl. Acad. Sci. USA* **2018**, *115*, 8716–8721. [CrossRef]
19. Zhang, Y.; Kuhl, P.K.; Imada, T.; Iverson, P.; Pruitt, J.; Stevens, E.B.; Kawakatsu, M.; Tohkura, Y.; Nemoto, I. Neural signatures of phonetic learning in adulthood: A magnetoencephalography study. *NeuroImage* **2009**, *46*, 226–240. [CrossRef]
20. Brainard, D.H. The Psychophysics Toolbox. *Spat. Vis.* **1997**, *10*, 433–436. [CrossRef]
21. Macmillan, N.A.; Creelman, C.D. *Detection Theory: A User's Guide*; Routledge: New York, NY, USA, 2008.
22. Gramfort, A.; Luessi, M.; Larson, E.; Engemann, D.A.; Strohmeier, D.; Brodbeck, C.; Parkkonen, L.; Hämäläinen, M.S. MNE software for processing MEG and EEG data. *NeuroImage* **2013**, *86*, 446–460. [CrossRef]
23. Larson, E.; Taulu, S. Reducing Sensor Noise in MEG and EEG Recordings Using Oversampled Temporal Projection. *IEEE Trans. Biomed. Eng.* **2018**, *65*, 1002–1013. [CrossRef]
24. Taulu, S.; Kajola, M. Presentation of electromagnetic multichannel data: The signal space separation method. *J. Appl. Phys.* **2005**, *97*, 124905. [CrossRef]
25. Taulu, S.; Hari, R. Removal of magnetoencephalographic artifacts with temporal signal-space separation: Demonstration with single-trial auditory-evoked responses. *Hum. Brain Mapp.* **2009**, *30*, 1524–1534. [CrossRef] [PubMed]
26. Uusitalo, M.A.; Ilmoniemi, R.J. Signal-space projection method for separating MEG or EEG into components. *Med. Biol. Eng. Comput.* **1997**, *35*, 135–140. [CrossRef] [PubMed]
27. Dale, A.M.; Fischl, B.; Sereno, M.I. Cortical Surface-Based Analysis: I. Segmentation and Surface Reconstruction. *NeuroImage* **1999**, *9*, 179–194. [CrossRef] [PubMed]
28. Hämäläinen, M.S.; Sarvas, J. Realistic conductivity geometry model of the human head for interpretation of neuromagnetic data. *IEEE Trans. Biomed. Eng.* **1989**, *36*, 165–171. [CrossRef]
29. Dale, A.M.; Liu, A.K.; Fischl, B.R.; Buckner, R.L.; Belliveau, J.W.; Lewine, J.D.; Halgren, E. Dynamic statistical parametric mapping: Combining fMRI and MEG for high-resolution imaging of cortical activity. *Neuron* **2000**, *26*, 55–67. [CrossRef]
30. Destrieux, C.; Fischl, B.; Dale, A.; Halgren, E. Automatic parcellation of human cortical gyri and sulci using standard anatomical nomenclature. *NeuroImage* **2010**, *53*, 1–15. [CrossRef]
31. Pedregosa, F.; Varoquaux, G.; Gramfort, A.; Michel, V.; Thirion, B.; Grisel, O.; Blondel, M.; Prettenhofer, P.; Weiss, R.; Dubourg, V.; et al. Scikit-learn: Machine Learning in Python. *J. Mach. Learn. Res.* **2011**, *12*, 2825–2830.
32. Zhao, T.C.; Kuhl, P. Development of Infants' Neural Speech Processing and Its Relation to Later Language Skills: A MEG Study. 2021. Available online: https://www.biorxiv.org/content/10.1101/2021.09.16.460534v1 (accessed on 6 February 2022).
33. Drucker, H.; Burges, C.J.; Kaufman, L.; Smola, A.; Vapnik, V. Support vector regression machines. In *Advances in Neural Information Processing Systems*; MIT Press: Denver, CO, USA, 1996; pp. 155–161.

34. Xie, Z.; Reetzke, R.; Chandrasekaran, B. Machine Learning Approaches to Analyze Speech-Evoked Neurophysiological Responses. *J. Speech Lang. Hear. Res.* **2019**, *62*, 587–601. [CrossRef]
35. Smith, S.M.; Nichols, T.E. Threshold-free cluster enhancement: Addressing problems of smoothing, threshold dependence and localisation in cluster inference. *NeuroImage* **2009**, *44*, 83–98. [CrossRef]
36. Malmivuo, J. Comparison of the Properties of EEG and MEG in Detecting the Electric Activity of the Brain. *Brain Topogr.* **2012**, *25*, 1–19. [CrossRef] [PubMed]
37. Hickok, G.; Poeppel, D. The cortical organization of speech processing. *Nat. Rev. Neurosci.* **2007**, *8*, 393–402. [CrossRef] [PubMed]

brain sciences

MDPI

Article

Just-Noticeable Differences of Fundamental Frequency Change in Mandarin-Speaking Children with Cochlear Implants

Wanting Huang [1,2], Lena L. N. Wong [2] and Fei Chen [1,*]

1 Department of Electrical and Electronic Engineering, Southern University of Science and Technology, Shenzhen 518055, China; huangwt@sustech.edu.cn
2 Unit of Human Communication, Development, and Information Sciences, Faculty of Education, The University of Hong Kong, Hong Kong 999077, China; llnwong@hku.hk
* Correspondence: fchen@sustech.edu.cn

Abstract: Fundamental frequency (F0) provides the primary acoustic cue for lexical tone perception in tonal languages but remains poorly represented in cochlear implant (CI) systems. Currently, there is still a lack of understanding of sensitivity to F0 change in CI users who speak tonal languages. In the present study, just-noticeable differences (JNDs) of F0 contour and F0 level changes in Mandarin-speaking children with CIs were measured and compared with those in their age-matched normal-hearing (NH) peers. Results showed that children with CIs demonstrated significantly larger JND of F0 contour (JND-C) change and F0 level (JND-L) change compared to NH children. Further within-group comparison revealed that the JND-C change was significantly smaller than the JND-L change among children with CIs, whereas the opposite pattern was observed among NH children. No significant correlations were seen between JND-C change/JND-L change and age at implantation /duration of CI use. The contrast between children with CIs and NH children in sensitivity to F0 contour and F0 level change suggests different mechanisms of F0 processing in these two groups as a result of different hearing experiences.

Keywords: cochlear implants; children; fundamental frequency; demographic factor

Citation: Huang, W.; Wong, L.L.N.; Chen, F. Just-Noticeable Differences of Fundamental Frequency Change in Mandarin-Speaking Children with Cochlear Implants. *Brain Sci.* **2022**, *12*, 443. https://doi.org/10.3390/brainsci12040443

Academic Editor: Manuel Sánchez Malmierca

Received: 16 February 2022
Accepted: 23 March 2022
Published: 26 March 2022

1. Introduction

Fine pitch processing is relatively scarce in non-tonal languages, whereas in tonal languages, which account for 70% of the world's languages [1], rapid pitch variations (i.e., lexical tones) are used to alter the meaning of a syllable. For example, in Mandarin Chinese, each of the four lexical tones has a distinct pattern of pitch inflection: level, mid-rising, dipping, and high-falling, changing the meaning of a syllable such as /ma/ into "mother", "hemp", "horse", or "scold". Prior research findings have consistently suggested that fundamental frequency (F0) provides the primary acoustic information for Mandarin tone recognition [2], while the temporal [3,4] and spectral envelope [5] serve as the secondary acoustic cues.

Cochlear implantation is widely accepted as a life-changing invention for individuals with severe to profound hearing impairment. According to the Sixth National Population Census of the People's Republic of China, the total number of hearing-impaired children aged 0 to 14 was estimated to be more than 4.6 million [6]. It is reported that about 2000 preschool children with congenital severe to profound hearing loss in mainland China (hereafter referred to as Mandarin-speaking children with CIs) underwent cochlear implantation in 2004, increasing at a rate of 30% to 50% per year thereafter [7].

The speech processing strategies currently used in CIs are mainly designed to encode the temporal envelope of sound stimuli. Given the vulnerability of the temporal envelope in noise, it is not surprising that speech perception in noise conditions among Mandarin-speaking children with CIs remains unsatisfactory [8,9] and is even worse than that in English-speaking children with CIs [10,11]. In recent decades, to improve speech perception in CI users speaking tonal languages, speech processing strategies have been developed to provide more F0 information by delivering more spectral information (e.g., HiRes 120) or by enhancing temporal fine structure cues (e.g., Temporal Fine Structure) in CI systems. Unfortunately, these new speech processing strategies have not improved tone perception in CI users who speak tonal languages [12,13]. For instance, using a two-alternative forced-choice paradigm, researchers found that two speech processing strategies (i.e., HiRes or HiRes 120) failed to produce statistically significant differences in Mandarin tone recognition performance among a group of 20 Mandarin-speaking children with CIs, although most of them reported a preference for HiRes 120 [12]. Such negative results have also been found among Cantonese-speaking adult CI users when comparing Cantonese tone recognition using a typical speech processing strategy (i.e., continuous interleaved sampling) to that using a relatively new one [13].

The insignificant improvement in lexical tone recognition using newer speech processing strategies, as reported in previous studies, might have been the result of a lack of understanding of F0 processing in Mandarin-speaking CI users, which has received little attention until recent years. F0s of Mandarin tones comprise two dimensions: F0 level and F0 contour. F0 level, which refers to the height of onset F0 in Mandarin tones, is usually used to identify different talkers (e.g., male vs. female), while F0 contour reflects the trajectory of F0 change over the duration of a single tone and plays a more decisive role in the discrimination of word meaning compared to F0 level. Existing studies on F0 contour processing in Mandarin-speaking children with CIs suggest that these children may use F0 contours differently from their age-matched NH peers as a result of different hearing experiences [14,15]. For example, researchers found that, although Mandarin-speaking children with CIs were able to use F0 contours for tone recognition, they tended to rely more on the temporal envelope than on F0 contours when performing word-level tone recognition tasks [15]. However, the situation was quite different at the sentence level. A recent study investigated the effects of F0 contours on sentence recognition in Mandarin-speaking preschool children with CIs in both quiet and noise conditions. The results showed that when other acoustic cues (e.g., temporal envelope) were neutralized, sentence recognition with flattened F0 contours was significantly worse than that with normal F0 contours in both children with CIs and NH children. While the F0 contour-caused decrease in sentence recognition was only seen in quiet conditions among the NH children, it was seen in both quiet and noise conditions among the children with CIs. Furthermore, the impact of F0 contours on sentence recognition accuracy in children with CIs was significantly more salient than that in NH children [14].

Processing of the F0 level in Mandarin-speaking children with CIs has received much less attention compared to that of the F0 contour. Similar to previous studies on music perception in CI users [16,17], the few studies on F0 level processing in Mandarin-speaking children with CIs have demonstrated deficits when compared to NH controls. For example, researchers measured and compared F0 level discrimination in 24 Mandarin-speaking school-aged children (aged 4.6 to 21.3 years) and found that percent correct performance for F0 level discrimination in the CI group was significantly poorer than that in the NH group, suggesting a compromise in F0 level processing in Mandarin-speaking school-aged children with CIs [18].

The extent to which sensitivity to F0 contour and F0 level change is affected by CI-related demographic factors (e.g., age at implantation, duration of CI use) indicates whether and how F0 processing in Mandarin-speaking children with CIs is shaped by the children's experiences on using CIs. Currently, the relationship between sensitivity to F0 change and CI-related demographic factors is rarely reported in the existing literature. According to the existing literature, while reliance on temporal envelope was significantly correlated with age at implantation, there was no significant relationship between reliance on F0 contours and the duration of CI use [15]. In English-speaking preschool children, researchers found no significant correlations between CI-related demographic factors and sensitivity to F0 contour change [19]. Similarly, no significant correlations between CI-related demographic factors (i.e., age at implantation and duration of CI use) and sensitivity to F0 level change were reported in either English-speaking or Mandarin-speaking preschool children [18].

To date, there is still a lack of understanding of F0 contour and F0 level processing in Mandarin-speaking children with CIs and of the extent to which F0 processing in these children is similar to or different from that in age-matched NH children. In the present study, just-noticeable differences (JNDs) of F0 contour and F0 level change, which were used as indicators of F0 processing, were measured and compared between Mandarin-speaking kindergarten-aged children with normal hearing and children with CIs. Given the well-known deficits in F0 processing of speech sounds in CI systems, it was predicted that children with CIs would be significantly less sensitive to both F0 level and F0 contour change compared to NH children. Based on the previously reported larger JND of pitch contour change in NH adults compared to the JND of pitch level change [20], it was predicted that in NH children, the JND of F0 contour change would also be larger than that of F0 level change. As F0 contour and F0 level perception have only been examined separately in previous studies, the relative sensitivity to F0 contour change and F0 level change has yet to be explored.

To gain a better understanding of the relationship between CI-related demographic factors (e.g., age at implantation, duration of CI use) and F0 change detection in Mandarin-speaking children with CIs, correlation analyses were conducted between CI-related demographic factors and JNDs of F0 contour and F0 level change in the Mandarin-speaking children with CIs. Based on the previously reported results on this issue, it was expected that CI-related demographic factors would not be significantly correlated with the JNDs of F0 contour and F0 level change in the Mandarin-speaking kindergarten-aged children with CIs.

2. Materials and Methods

2.1. Participants

Thirty Mandarin-speaking preschool children with CIs (18 males and 12 females; mean age 4.36 ± 0.70 years old) participated in the current study. The preschool children were chosen for the following reasons: (a) children with normal hearing exhibit protracted development of tonal processing before school age [21–24]; (b) there is a critical period for central auditory system in pediatric CI users before the age of 7 [25,26]; (c) most prelingually deafened children in mainland China receive implantation before the age of six. Therefore, investigations in children before school age provide evidence of tonal development at an early stage, which offers valuable references for the development and improvement of the tonal language-oriented speech processing strategies in CI systems. All participants were recruited from the Beijing Children's Hospital and the China Rehabilitation Research Center for Hearing and Speech Impairment. Children in the CI group met the following inclusion criteria: (a) aged 3–6 years, (b) diagnosed with congenital bilateral severe to profound sensorineural hearing impairment, (c) had received unilateral implantation, and (d) had been using CIs for not less than six months. Children in this group were using CIs from four manufacturers: Advanced Bionics (n = 5), Cochlear (n = 12), MED-EL (n = 12), and Nurotron (n = 1). The speech processing strategies used by these CI manufacturers are HiResolution (HiRes) 120, Advanced Combination Encoder (ACE), Fine Structure (FS) 4, and C-tone, respectively. Among these children, 12 had undergone a unilateral (n = 5) or bilateral (n = 7) hearing aid trial before implantation, and 25 wore a hearing aid on the non-implanted ear after implantation. Details of demographic information and device use are shown in Table 1. Thirty age-matched Mandarin-speaking children (17 male and 13 female, mean age: 4.37 ± 0.48 years old) with audiometric thresholds not worse than 20 dB HL at octave frequencies between 250 and 4000 Hz were included as normal controls in this study. Children in both groups scored within the normal range on the Hiskey-Nebraska Test of Learning Aptitude for children above 3 years of age [27]. The research study was approved by the Human Research Ethics Committee of the University of Hong Kong and Beijing

Children's Hospital. All children participated voluntarily in the study, with informed consent obtained from their parents.

Table 1. Demographics of children with CIs.

No.	Sex	AAT (Years)	HAT (Years)	AAI (Years)	Hearing Device			DCI (Years)	DAVT (Years)
					Left Device	Right Device	CI Speech Processing Strategies		
1	F	5.75	1.75	3.58	Phonak	MED-EL	FS4	2.17	2.17
2	F	5.08	0.42	3.33	Phonak	Cochlear	ACE	1.75	1.75
3	M	4.33	0.33	1.42	Phonak	AB	HiRes120	2.91	2.91
4	F	4.83	0.83	2.92	Cochlear	Phonak	ACE	1.91	1.75
5	M	4.92	3.67	3.67	Cochlear	Widex	ACE	1.25	1.25
6	M	5.33	0	3.17	Phonak	MED-EL	FS4	2.16	2.16
7	M	3.92	0	1.33	Phonak	MED-EL	FS4	2.59	2.5
8	F	4.83	0	2.58	Phonak	Nurotron	C-tone	2.25	1.91
9	F	4.08	1	3.33	MED-EL	Phonak	FS4	0.75	0.75
10	F	4.08	0	2.58	MED-EL	Phonak	FS4	1.5	1.33
11	F	4	0.33	2.75	Phonak	Cochlear	ACE	1.25	1.25
12	M	4.17	0	3	MED-EL	Phonak	FS4	1.17	1.17
13	M	4.33	0.75	3.25	Phonak	Cochlear	ACE	1.08	1.08
14	M	3.75	0	2	AB	Phonak	HiRes120	1.75	1.5
15	M	4.25	0	2.25	Phonak	Cochlear	FS4	2	2
16	M	3.5	0	1.67	Phonak	MED-EL	FS4	1.83	1.25
17	F	4.17	0.67	1.58	Cochlear	Widex	ACE	2.59	1.59
18	M	3.83	0.67	1.92	Cochlear	Phonak	ACE	1.91	1.91
19	M	4	0	2.08	Phonak	Cochlear	ACE	1.92	1.67
20	M	3.58	0	0.67	Phonak	AB	HiRes120	2.91	2.08
21	M	3.5	0	1.58	Phonak	Cochlear	ACE	1.92	1.92
22	F	3.42	0.67	2.08	MED-EL	Phonak	FS4	1.34	1.34
23	M	3.5	0	1.83	Cochlear	Phonak	ACE	1.67	1.5
24	M	4.08	0	2.08	Cochlear	Phonak	ACE	2	1.5
25	F	5.75	0.92	2.5	AB	Null	HiRes120	3.25	1.75
26	M	4.92	0	1.67	MED-EL	Null	FS4	3.25	1.33
27	F	4.33	0	1.17	Null	MED-EL	FS4	3.17	1.5
28	F	5.5	0	2.17	AB	Null	HiRes120	3.33	0.92
29	M	5.33	0	4.17	Phonak	MED-EL	FS4	1.17	0.83
30	M	3.83	0	2.08	Cochlear	Null	ACE	1.53	1.67

AAT: age at test; HAT: hearing aid trial before implantation; AAI: age at implantation; DCI: duration. of CI use; DAVT: duration of auditory-verbal training; AB: Advanced Bionics; FS4: Fine Structure 4; ACE: Advanced Combination Encoder; HiRes 120: HiResolution 120.

2.2. Stimuli

The original stimulus, an isolated Mandarin vowel /a/ with tone 1, was first spoken by an adult female native Mandarin speaker. Using Praat [28], the F0 contour of /a1/ was then replaced with a series of linear F0 contours, with other acoustic features remaining the same. JNDs of F0 contour and F0 level change were measured in two separate blocks. In the block measuring the JND of F0 contour change, the offset F0s were manipulated to vary from 100 to 300 Hz, with the onset of the linear F0 contours being fixed at 100 Hz, resulting in an offset continuum with F0 contours ranging from a level tone to a rising tone. The step size between adjacent offset F0s was set at 1 Hz, and thus, the newly resynthesized stimuli were made up of 201 /a/ carrying different F0 contours (see Figure 1A). The stimuli in the block measuring the JND of F0 level change were the same as those in the F0 contour condition, except that the onset F0 was equal to the offset F0 throughout the resynthesized F0 contours (see Figure 1B). According to the Syllabus of the Chinese Proficiency Test, the duration of a naturally uttered lexical tone is around 200 to 300 ms [29]. To make the stimuli more natural and to guarantee the audibility of stimuli, the duration of each stimulus in both conditions was set at 300 ms.

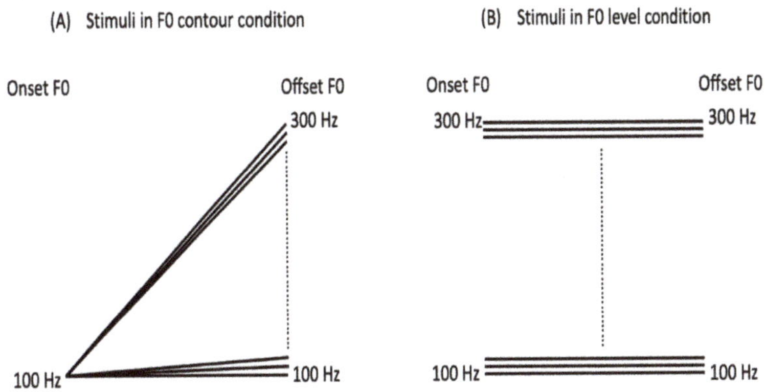

Figure 1. Schematic representation of linear F0 contours in F0 contour (**A**) and F0 level (**B**) conditions.

2.3. Procedure

A three-alternative forced-choice paradigm with a two-down, one-up tracking algorithm was used to measure the JNDs of F0 contour and F0 level change among the children in both groups. Within each trial, two standard stimuli and one deviant stimulus were randomly presented, with the inter-stimulus interval being 400 ms. The probability of the deviant stimulus appearing at each interval was equal to 1/3. In the F0 level condition, 100–100 Hz was selected as the standard stimulus. In contrast, to ensure that pitch contour was the main cue for the detection of F0 change and not pitch height, three flat contours (i.e., 100–100, 200–200, and 300–300 Hz) were used as standard stimuli in the F0 contour condition, two of which were randomly chosen in each trial.

During the test, children in the CI group wore their CIs only. Children in both groups were seated at a rectangular table in a quiet room while performing the task. The sound stimuli were presented at a listening level of 65 dB SPL via a loudspeaker located in front of the children at a 0° azimuth and a distance of one meter from the center of the head of participants. Three identical cartoon dogs were printed on three separate pieces of paper. During the presentation of the three sound stimuli in each trial, the examiner was seated next to the children and pointed at the cartoon dogs one by one with the sound stimuli. Children were asked to indicate which dog's voice sounded different from the other two (see Figure 2). In the F0-contour block, children were first taught with the targeted rising tone (i.e., 100–300 Hz). In each trial of this block, children were presented with the rising tone three times before the onset of sound stimuli.

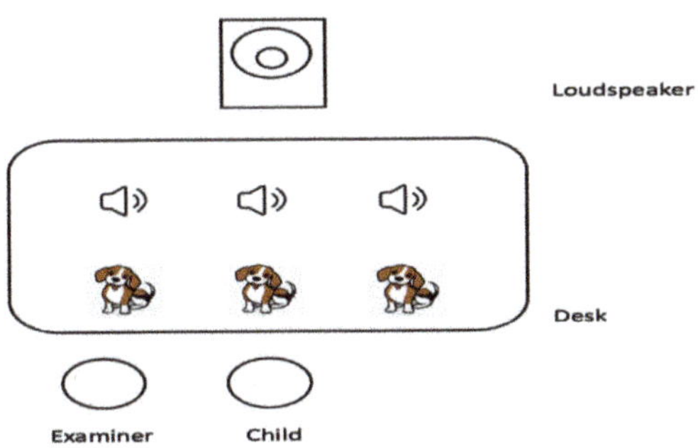

Figure 2. Diagram of the experimental scene. The loudspeaker is one meter away from the child.

Each of the two blocks contained 60 trials. In both blocks, the offset F0s of the deviant stimuli were first started at 300 Hz and went downward and approached the standard stimulus (100 Hz) upon correct responses. Following the successful discrimination of the deviant stimulus in the first trial, the offset F0 change was set at 100 Hz for the second trial. Based on previous studies [20,30], the step size in each trial was adjusted to 5 Hz for the first three reversals and to 1 Hz thereafter. With the exclusion of the first three reversals, the average offset F0 change from the original one (300 Hz) was calculated from the last even number of reversals in the adaptive track. The JND for each condition was defined as the offset F0 difference between the average offset F0 change and the offset F0 of the standard stimulus (i.e., 100 Hz). The children were given practice trials before the test until they were familiarized with the task requirements. The order of the blocks of F0 contour and F0 level measurement was counterbalanced among the children. A break was given every four to six trials.

2.4. Statistical Analysis

To investigate the effects of hearing experience (normal hearing vs. CIs) and F0 dimensions on the sensitivity to F0 change, JNDs of F0 contour change and F0 level change were entered as the dependent variables in a two-way analysis of variance, with group (NH group/CI group) and F0 dimension (F0 level/F0 contour) as the independent variables. A test of simple effects was conducted upon the significant interaction between group and F0 dimension. The Mann–Whitney U test was performed for the comparison between groups.

To investigate the relationship between demographic factors (e.g., age at implantation) and sensitivity to F0 change, Pearson correlation analyses were performed between JNDs of F0 contour/ F0 level change and demographic factors.

The p-value of 0.05 was set as a threshold of statistical significance throughout all tests.

3. Results

The JNDs of F0 contour and F0 level change in children with CIs and NH children are shown in Figure 3. On the whole, the JNDs of F0 change were larger in children with CIs than in their age-matched peers. Within-group comparison between the JND of F0 level change and the JND of F0 contour change revealed that, in the control group, the JNDs of F0 contour change were consistently larger than those of F0 level change, while in children with CIs, contrasting patterns were observed: 22 of the 30 children with CIs exhibited larger JNDs of F0 level change, 6 showed opposite patterns, and 2 demonstrated equal JNDs of F0 contour and F0 level change. The percentage of children exhibiting different patterns of JNDs of F0 level change and F0 contour change for each type of speech processing strategy is shown in Table 2.

Table 2. Percentage of the children with CIs exhibiting different patterns of JNDs of F0 level (JND-L) and F0 contour (JND-C) change regarding each type of speech processing strategy.

CI Speech Processing Strategies	JND-L > JND-C	JND-L = JND-C	JND-L < JND-C
FS4 (n = 12)	66.7%	0%	33.3%
ACE (n = 12)	83.3%	8.3%	8.3%
HiRes 120 (n = 5)	60%	20%	20%
C-tone (n = 1)	100%	0%	0%

CI: cochlear implant; FS4: Fine Structure 4; ACE: Advanced Combination Encoder; HiRes 120: HiResolution 120.

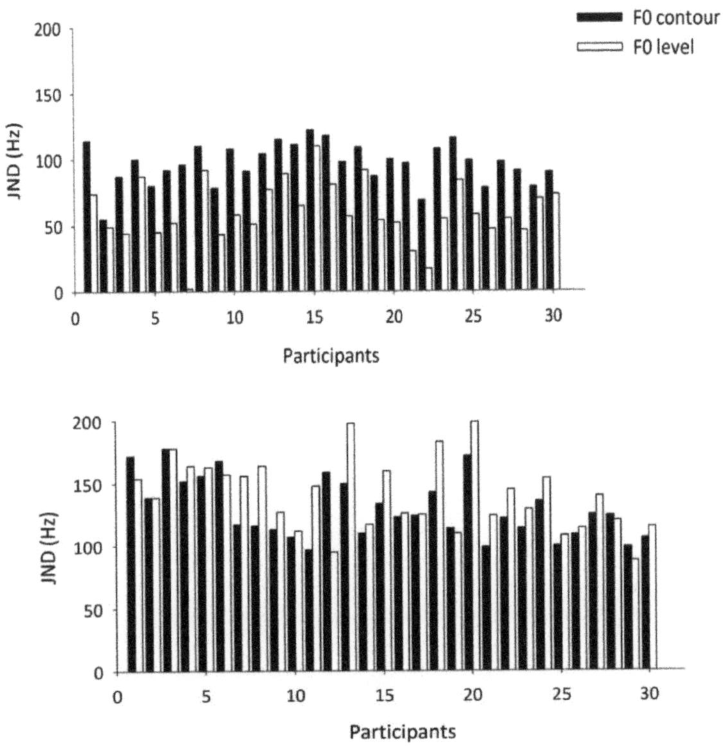

Figure 3. JNDs of F0 contour change and F0 level change in the 30 children with normal hearing (**upper panel**) and in the 30 children with CIs (**lower panel**).

Significant positive correlations between the JND of F0 contour change and the JND of F0 level change were found in the NH group (r = 0.622, $p < 0.001$), the CI group (r = 0.636, $p < 0.001$), and the combination of the two groups (r = 0.793, $p < 0.001$). A two-way analysis of variance (ANOVA), with group (children with CIs/normal hearing) as the between-subject factor and F0 dimension (F0 level/contour) as the within-subject factor (see Figure 4), demonstrated the significant main effects of group (F (1, 58) = 102.12, $p < 0.001$, partial $\eta^2 = 0.64$) and F0 dimension (F (1, 58) = 21.17, $p < 0.001$, partial $\eta^2 = 0.27$) and interaction between group and F0 dimension (F (1, 58) = 4.54, $p < 0.001$, partial $\eta^2 = 0.57$). The ANOVA results are summarized in Table 3. Post hoc analysis revealed significantly larger JNDs of F0 contour change (F (1, 58) = 38.07, $p < 0.001$, partial $\eta^2 = 0.40$) and F0 level change (F (1, 58) = 141.43, $p < 0.001$, partial $\eta^2 = 0.0.71$) in children with CIs than those in NH children. Among NH children, the JND of F0 contour change was significantly larger than that of F0 level change (F (1, 58) = 92.50, $p < 0.001$, partial $\eta^2 = 0.61$). In contrast, among children with CIs, the JND of F0 contour change was found to be significantly smaller than the JND of F0 level change (F (1, 58) = 8.67, $p < 0.01$, partial $\eta^2 = 0.13$). Furthermore, an independent-samples *t*-test, comparing the difference between the JND of F0 contour change and the JND of F0 level change in both groups, suggested that the JND difference between F0 contours and F0 levels was significantly larger in NH children than in children with CIs (t (58) = 4.72, $p < 0.001$).

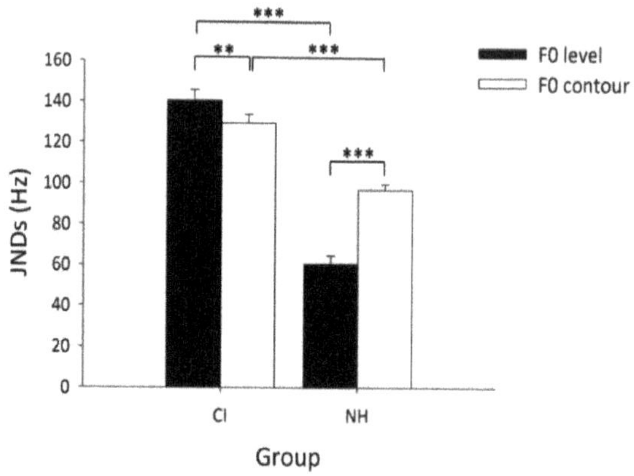

Figure 4. JNDs as a function of F0 dimensions in the CI and NH groups. Error bars indicate standard errors. The level of significance is indicated with asterisks (**: $p < 0.01$; ***: $p < 0.001$).

Table 3. Summary of the ANOVA results.

Source of Variation	df	F Value	Partial Eta^2
group	1	102.12 ***	0.64
F0 dimension	1	21.17 ***	0.27
group × F0 dimention	1	4.54 ***	0.57

Eta^2: Eta squared; ***: $p < 0.001$.

No significant correlation was observed between age at implantation and either the JND of F0 contour change (r = −0.09, p = 0.65) or the JND of F0 level change (r = −0.15, p = 0.42) or between duration of CI use and either the JND of F0 contour change (r = −0.09, p = 0.64) or the JND of F0 level change (r = 0.03, p = 0.86).

4. Discussion

The main purpose of this study is to investigate the processing of F0, including F0 level and F0 contour, in Mandarin-speaking children with CIs as compared to NH peers, which may partially account for the unsatisfactory outcome of Mandarin tone recognition in CI users. To achieve this goal, we investigated the JNDs of F0 contour change and F0 level change in Mandarin-speaking kindergarten-aged children with CIs compared to those of their age-matched NH peers. The results showed that NH children were more sensitive to F0 level changes, as evidenced by the significantly smaller JND in the F0 level condition than that in the F0 contour condition. The higher sensitivity to F0 level change compared to sensitivity to F0 contour change in NH children is consistent with previous findings among NH adults at both behavioral [20] and electrophysiological levels [31]. It should be noted that, although the F0 onset and offset and measurement methods in the F0 level condition in the present study (100 Hz to 300 Hz) were similar to those in NH adults (180 to 250 Hz) [20], the JNDs of F0 level change obtained among NH children in this study was much larger than those that have been found among NH adults [19,20]. The relatively larger JNDs in NH children compared to NH adults are probably the result of the children's less-developed central auditory system, which does not fully mature until about 12 years of age [32,33]. However, since the measurement of JND of F0 level change and JND of F0 contour change was not fully equivalent, it may not be suitable to conclude the relationship between the sensitivity to F0 level change and that to F0 contour change in this group alone.

Although positive correlations between the JNDs of F0 contour and F0 level change were found in both groups of children in the present study, the sensitivity to F0 change in children with CIs was quite different from that in NH children. While NH children consistently showed higher sensitivity to F0 level changes compared to F0 contour changes, large individual variabilities were observed in the CI group. Although the majority of children in the CI group (n = 22) exhibited larger JNDs for F0 level change, four children exhibited the opposite pattern, while two children showed equal sensitivity to F0 contour and F0 level change. Substantial individual variability in speech perception in Mandarin-speaking CI users has been reported in many studies [12,34], which probably results from variations in demographic factors, such as age at implantation [12] and duration of CI use [35], although the results of the present study did not show significant correlations between the JNDs of F0 contour and F0 level change and these demographic factors. This issue is discussed later.

It is known that in CIs, F0 information up to approximately 300 Hz is processed by way of temporal information, which conveys far less fine structure information than can be processed by the normal auditory system [36,37]. Therefore, it is not surprising that children with CIs show a deficit in F0 processing, as reflected by the significantly larger JNDs of both F0 contour and F0 level change compared to NH children. These findings are consistent with those reported in previous studies [15,18,19]. More importantly, as shown in Figure 4, the relationship between sensitivity to F0 contour change and to F0 level change in children with CIs was in contrast to that in the NH children at the group level. While the NH controls demonstrated higher sensitivity to F0 level change, the pediatric CI users were more sensitive to F0 contour change. It is worth noting that the higher sensitivity to F0 contour change compared to F0 level change was widely seen in the CI group, regardless of the type of speech processing strategies used by these children (see Table 2). Therefore, the significant sensitivity to F0 contour change observed in the CI group cannot simply be attributed to speech processing strategies; rather, it is probably a common phenomenon in CI users, resulting from their hearing experience with CIs. Such higher sensitivity to F0 contour change compared to F0 level change could be caused by two factors. First, as reported by the previous study, amplitude modulation depth is usually inconsistently coded by clinical speech processing strategies (even for those produced by the same talker), which results in the inconsistent perception of the F0 level; however, the perception of the F0 contour is not likely to be affected [38]. Under such circumstances, it is not surprising that the inconsistent perception of the F0 level leads to a significantly larger JND of F0 level change than that of F0 contour change. Second, as introduced above, CIs are designed to extract temporal envelope information so that the contour of the frequency fluctuations of speech is maintained. To master a tonal language, in which contour information plays a dominant role in discriminating word meanings, CI users must maximally utilize the contour information conveyed by CIs. Thus, children with CI users in the present study might have developed a unique mechanism of F0 processing that differs from that of their NH peers. In other words, the contour information of speech sounds might have been prioritized compared to F0 level information in speech perception among children with CIs. The difference in sensitivity to F0 level and F0 contour change between children with CIs and NH children suggests that the enhancement of F0 information in the newly developed speech processing strategies (e.g., HiRes 120, Temporal Fine Structure) in Mandarin-speaking children with CIs may not fully satisfy the requirements for F0 processing in speech perception, which partially explains the unsatisfactory improvement in speech perception among this population when switching to new speech processing strategies [12].

No significant correlation was found between JNDs of F0 contour/F0 level and age at implantation/duration of CI use in this study. This finding was consistent with the previous study, in which 23 English-speaking school-aged children with CIs were evaluated [19]. Children in both studies were prelingually deaf, and their hearing during tests depended on the implanted CI on either side of the ear, with the hearing aid on the contralateral side being turned off and removed from the ear. The insignificant relationship between JNDs of F0 contour/F0 level and age at implantation/duration of CI use in this study suggests that age at implantation and duration of CI use may have little impact on the F0 processing in these children. However, it is also possible that given the relatively small sample size, the effect of age at implantation and duration of CI use was overwhelmed by the well-recognized individual variability [39].

There are several limitations in the current study. According to a recent study, different types (i.e., the posterior tympanotomy technique vs. the endomeatal approach) of CI surgery result in different levels of postoperative discomfort [40]. Unfortunately, the type of surgery in the CI group was not well documented in the present study, and it was unclear whether the type of surgery would play a role in the F0 change detection in children with CIs. Technically speaking, the round window approach manages to preserve hearing residues [40] so that F0 change detection in CI recipients will be better. Such an inference is expected to be verified by further studies. In addition, the relatively small sample size ($n = 30$) in this study made it unlikely to demonstrate the relationship between demographic factors and F0 processing in children with CIs. A larger sample size is required to address this question in the future. Additionally, given that the F0 level and F0 contour were processed holistically when perceiving tones [41], an improved paradigm is needed to further confirm the findings in this study.

5. Conclusions

The current study explored F0 processing in Mandarin-speaking children with CIs from a psychophysical perspective. The results revealed significantly compromised sensitivity to F0 change in children with CIs. Specifically, children with CIs demonstrated higher sensitivity to F0 contour change than to F0 level change; this pattern contrasted with the relatively well-developed sensitivity to F0 level change in NH children. However, CI-related demographic factors (i.e., age at implantation and duration of CI use) seemed to have little impact on sensitivity to F0 change in children with CIs. This is the first study to investigate F0 processing in Mandarin-speaking preschool children with CIs by comparing the two dimensions of F0 (i.e., F0 contour and F0 level). The contrast in sensitivity to F0 contour and F0 level change between children with CIs and NH children may suggest different mechanisms of F0 processing in these two groups as a result of their hearing experiences. To the best of our knowledge, this is the first study exploring both dimensions of F0 information in Mandarin-speaking children with CIs. The difference in F0 processing between children with CIs and NH children, revealed in this study, provides new perspectives on the development and improvement of speech processing strategies in CI systems, especially those targeted at tonal language speakers. It is believed that it will significantly improve the life quality of CI users by providing more accurate F0 information in the CI systems.

Author Contributions: Conceptualization, all authors.; methodology, W.H. and F.C.; software, W.H.; validation, F.C.; formal analysis, W.H.; investigation, W.H.; resources, L.L.N.W.; data curation, W.H.; writing—original draft preparation, W.H.; writing—review and editing, all authors; visualization, W.H.; supervision, L.L.N.W. and F.C.; project administration, L.L.N.W. and F.C. All authors have read and agreed to the published version of the manuscript.

Funding: This work of Fei Chen was supported by the National Natural Science Foundation of China (Grant No. 61971212).

Institutional Review Board Statement: The study was conducted in accordance with the Declaration of Helsinki and approved by the Human Research Ethics Committee of The University of Hong Kong (EA1610013; November 2016) and Beijing Children's Hospital (IEC-C-028-A10; March 2017).

Informed Consent Statement: Informed consent was obtained from parents of all subjects involved in the study.

Data Availability Statement: Data are available upon request from the corresponding authors.

Acknowledgments: This work was presented as an abstract in the 177th Meeting of the Acoustical Society of America in Louisville in 2019. This work was submitted as a part of PhD thesis at the University of Hong Kong and was permitted to republish the data. We thank the doctors at the Beijing Children's Hospital and the teachers at the China Rehabilitation Research Center for Hearing and Speech Impairment for helping with participant recruitment. We are also grateful to all participants and their parents.

Conflicts of Interest: The authors declare no conflict of interest.

References

1. Yip, M. *Tone*; Cambridge University Press: Cambridge, UK, 2002; p. 1.
2. Lin, M. The acoustic characteristics and perceptual cues of tones in Standard Chinese. *Chin. Yuwen.* **1988**, *204*, 182–193.
3. Fu, Q.J.; Zeng, F.G. Identification of temporal envelope cues in Chinese tone recognition. *Asia Pac. J. Speech Lang. Hear.* **2000**, *5*, 45–57. [CrossRef]
4. Liu, S.; Samuel, A.G. Perception of Mandarin lexical tones when F0 information is neutralized. *Lang. Speech.* **2004**, *47*, 109–138. [CrossRef] [PubMed]
5. Kong, Y.Y.; Zeng, F.G. Temporal and spectral cues in Mandarin tone recognition. *J. Acoust. Soc. Am.* **2006**, *120*, 2830–2840. [CrossRef] [PubMed]
6. Communiqué on the Main Data of the Sixth National Census in 2010. Available online: http://www.gov.cn/test/2012-04/20/content_2118413.htm (accessed on 20 April 2012).
7. Du, X.; Sun, X.; Huang, Z. Introduction of "Study of Chinese language education in CI postoperative rehabilitation". *China Sci J. Hear. Speech Rehab.* **2005**, *6*, 12.
8. Chen, Y.; Wong, L.L.; Chen, F.; Xi, X. Tone and sentence perception in young Mandarin-speaking children with cochlear implants. *Int. J. Pediatr. Otorhinolaryng.* **2014**, *78*, 1923–1930. [CrossRef]
9. Wei, C.G.; Cao, K.; Jin, X.; Chen, X.; Zeng, F.G. Psychophysical performance and Mandarin tone recognition in noise by cochlear implant users. *Ear Hear.* **2007**, *28* (Suppl. S2), 62S. [CrossRef]
10. Peng, S.C.; Tomblin, J.B.; Turner, C.W. Production and perception of speech intonation in pediatric cochlear implant recipients and individuals with normal hearing. *Ear Hear.* **2008**, *29*, 336–351. [CrossRef]
11. See, R.L.; Driscoll, V.D.; Gfeller, K.; Kliethermes, S.; Oleson, J. Speech intonation and melodic contour recognition in children with cochlear implants and with normal hearing. *Otol. Neurotol.* **2013**, *34*, 490. [CrossRef]
12. Han, D.; Liu, B.; Zhou, N.; Chen, X.; Kong, Y.; Liu, H.; Xu, L. Lexical tone perception with HiResolution and HiResolution 120 sound-processing strategies in pediatric Mandarin-speaking cochlear implant users. *Ear Hear.* **2009**, *30*, 169. [CrossRef]
13. Schatzer, R.; Krenmayr, A.; Au, D.K.; Kals, M.; Zierhofer, C. Temporal fine structure in cochlear implants: Preliminary speech perception results in Cantonese-speaking implant users. *Acta Oto-Laryngol.* **2010**, *130*, 1031–1039. [CrossRef] [PubMed]
14. Huang, W.; Wong, L.L.; Chen, F.; Liu, H.; Liang, W. Effects of Fundamental Frequency Contours on Sentence Recognition in Mandarin-Speaking Children With Cochlear Implants. *J. Speech Lang. Hear. Res.* **2020**, *63*, 3855–3864. [CrossRef] [PubMed]
15. Peng, S.C.; Lu, H.-P.; Lu, N.; Lin, Y.S.; Deroche, M.L.; Chatterjee, M. Processing of acoustic cues in lexical-tone identification by pediatric cochlear-implant recipients. *J. Speech Lang. Hear. Res.* **2017**, *60*, 1223–1235. [CrossRef] [PubMed]
16. Galvin, J.J., III; Fu, Q.J.; Shannon, R.V. Melodic contour identification and music perception by cochlear implant users. *Ann. N. Y. Acad. Sci.* **2009**, *1169*, 518. [CrossRef]
17. Hopyan, T.; Peretz, I.; Chan, L.P.; Papsin, B.C.; Gordon, K.A. Children using cochlear implants capitalize on acoustical hearing for music perception. *Front. Psychol.* **2012**, *3*, 425. [CrossRef]
18. Deroche, M.L.; Lu, H.P.; Limb, C.J.; Lin, Y.S.; Chatterjee, M. Deficits in the pitch sensitivity of cochlear-implanted children speaking English or Mandarin. *Front. Neurosci.* **2014**, *8*, 282. [CrossRef]
19. Deroche, M.L.; Kulkarni, A.M.; Christensen, J.A.; Limb, C.J.; Chatterjee, M. Deficits in the sensitivity to pitch sweeps by school-aged children wearing cochlear implants. *Front. Neurosci.* **2016**, *10*, 73. [CrossRef]
20. Huang, W.T.; Nan, Y.; Dong, Q.; Liu, C. Just-noticeable difference of tone pitch contour change for Mandarin congenital amusics. *J. Acoust. Soc. Am.* **2015**, *138*, EL99–EL104. [CrossRef]
21. Mok PP, K.; Fung HS, H.; Li, V.G. Assessing the link between perception and production in Cantonese tone acquisition. *J. Speech Lang. Hear. Res.* **2019**, *62*, 1243–1257. [CrossRef]
22. Mok, P.P.K.; Li, V.G.; Fung, H.S.H. Development of phonetic contrasts in Cantonese tone acquisition. *J. Speech Lang. Hear. Res.* **2020**, *63*, 95–108. [CrossRef]
23. Xu Rattanasone, N.; Tang, P.; Yuen, I.; Gao, L.; Demuth, K. Five-year-olds' acoustic realization of Mandarin tone sandhi and lexical tones in context are not yet fully adult-like. *Front. Psychol.* **2018**, *9*, 817. [CrossRef] [PubMed]
24. Wong, P. Perceptual evidence for protracted development in monosyllabic Mandarin lexical tone production in preschool children in Taiwan. *J. Acoust. Soc. Am.* **2013**, *133*, 434–443. [CrossRef] [PubMed]
25. Sharma, A.; Dorman, M.F.; Kral, A. The influence of a sensitive period on central auditory development in children with unilateral and bilateral cochlear implants. *Hear. Res.* **2005**, *203*, 134–143. [CrossRef] [PubMed]

26. Sharma, A.; Dorman, M.F.; Spahr, A.J. A sensitive period for the development of the central auditory system in children with cochlear implants: Implications for age of implantation. *Ear Hear.* **2002**, *23*, 532–539. [CrossRef]
27. Hiskey, M.S. A study of the intelligence of deaf and hearing children. *Am. Ann. Deaf.* **1956**, 329–339.
28. Boersma, P. Praat, a system for doing phonetics by computer. *Glot. Int.* **2001**, *5*, 341–345.
29. Liu, Y.L. *Research on the Chinese Proficiency Test*; No. 504; Modern Press Co., Ltd.: Beijing, China, 1989.
30. Liu, C. Just noticeable difference of tone pitch contour change for English-and Chinese-native listeners. *J. Acoust. Soc. Am.* **2013**, *134*, 3011–3020. [CrossRef]
31. Wang, X.D.; Wang, M.; Chen, L. Hemispheric lateralization for early auditory processing of lexical tones: Dependence on pitch level and pitch contour. *Neuropsychologia* **2013**, *51*, 2238–2244. [CrossRef]
32. Moore, J.K.; Linthicum, F.H., Jr. The human auditory system: A timeline of development. *Int. J. Audiol.* **2007**, *46*, 460–478. [CrossRef]
33. Ponton, C.; Eggermont, J.J.; Khosla, D.; Kwong, B.; Don, M. Maturation of human central auditory system activity: Separating auditory evoked potentials by dipole source modeling. *Clin. Neurophysiol.* **2002**, *113*, 407–420. [CrossRef]
34. Wei, C.G.; Cao, K.; Zeng, F.G. Mandarin tone recognition in cochlear-implant subjects. *Hear. Res.* **2004**, *197*, 87–95. [CrossRef] [PubMed]
35. Chen, Y.; Wong, L.L.; Zhu, S.; Xi, X. A structural equation modeling approach to examining factors influencing outcomes with cochlear implant in mandarin-speaking children. *PLoS ONE* **2015**, *10*, e0136576. [CrossRef] [PubMed]
36. Carlyon, R.P.; Van Wieringen, A.; Long, C.J.; Deeks, J.M.; Wouters, J. Temporal pitch mechanisms in acoustic and electric hearing. *J. Acoust. Soc. Am.* **2002**, *112*, 621–633. [CrossRef] [PubMed]
37. Zeng, F.G. Temporal pitch in electric hearing. *Hear. Res.* **2002**, *174*, 101–106. [CrossRef]
38. Vandali, A.; Sly, D.; Cowan, R.; Van Hoesel, R. Pitch and loudness matching of unmodulated and modulated stimuli in cochlear implantees. *Hear. Res.* **2013**, *302*, 32–49. [CrossRef]
39. Xu, L.; Chen, X.; Lu, H.; Zhou, N.; Wang, S.; Liu, Q.; Han, D. Tone perception and production in pediatric cochlear implants users. *Acta Otolaryngol.* **2011**, *131*, 395–398. [CrossRef]
40. Freni, F.; Gazia, F.; Slavutsky, V.; Perello Scherdel, E.; Nicenboim, L.; Posada, R.; Galletti, F. Cochlear implant surgery: Endomeatal approach versus posterior tympanotomy. *Int. J. Environ. Res. Public Health* **2020**, *17*, 4187. [CrossRef]
41. Xu, Y.; Wang, Q.E. Pitch targets and their realization: Evidence from Mandarin Chinese. *Speech Commun.* **2001**, *33*, 319–337. [CrossRef]

brain sciences

MDPI

Article

Phonetic Effects in the Perception of VOT in a Prevoicing Language

Viktor Kharlamov

Department of Languages, Linguistics and Comparative Literature, Florida Atlantic University, Boca Raton, FL 33431, USA; vkharlamov@fau.edu

Abstract: Previous production studies have reported differential amounts of closure voicing in plosives depending on the location of the oral constriction (anterior vs. posterior), vocalic context (high vs. low vowels), and speaker sex. Such differences have been attributed to the aerodynamic factors related to the configuration of the cavity behind the oral constriction, with certain articulations and physiological characteristics of the speaker facilitating vocal fold vibration during closure. The current study used perceptual identification tasks to examine whether similar effects of consonantal posteriority, adjacent vowel height, and speaker sex exist in the perception of voicing. The language of investigation was Russian, a prevoicing language that uses negative VOT to signal the voicing contrast in plosives. The study used both original and resynthesized tokens for speaker sex, which allowed it to focus on the role of differences in VOT specifically. Results indicated that listeners' judgments were significantly affected by consonantal place of articulation, with listeners accepting less voicing in velar plosives. Speaker sex showed only a marginally significant difference in the expected direction, and vowel height had no effect on perceptual responses. These findings suggest that certain phonetic factors can affect both the initial production and subsequent perception of closure voicing.

Keywords: prevoicing; VOT; aerodynamic voicing constraint; perceptual identification; Russian

Citation: Kharlamov, V. Phonetic Effects in the Perception of VOT in a Prevoicing Language. *Brain Sci.* **2022**, *12*, 427. https://doi.org/10.3390/brainsci12040427

Academic Editors: Richard Wright and Benjamin V. Tucker

Received: 20 February 2022
Accepted: 18 March 2022
Published: 23 March 2022

1. Introduction

The current study examined the link between production and perception of voicing in initial voiced plosives in Russian, a prevoicing language. In the domain of production, the amount of closure voicing is known to vary depending on the plosive's place of articulation, its vocalic context, and speaker's biological sex. This research aimed to determine whether similar asymmetries exist in perception, with listeners expecting differential amounts of voicing for anterior versus posterior plosives, low versus non-low vowel contexts, and female versus male voices.

1.1. Aspirating vs. Prevoicing Languages

The contrast between voiced versus voiceless plosives involves a timing difference in the onset of vocal fold vibration relative to the release burst of the plosive. This is known as 'voice onset time (VOT)' [1–3]. In languages such as English and German, both voiced and voiceless plosives are usually articulated with a positive VOT in initial positions, meaning that the onset of voicing occurs after the release burst. The voicing lag is short for voiced segments (approximately 30 ms or less) and long for voiceless segments (usually in excess of 50 ms) [4]. The long-lag VOT is achieved by aspirating the plosive, so such languages are referred to as 'aspirating'.

A different VOT pattern is seen in the so-called 'prevoicing' languages, including Dutch and Russian, where voiced initial plosives are produced with negative VOT [5–7]. This means that voicing starts during closure, and the VOT lead in voiced plosives is usually contrasted with zero or short lag VOT in voiceless plosives that are articulated without aspiration in initial positions. The rates of prevoicing can differ across languages and dialects. Continental French and Russian, for example, have robust prevoicing in initial plosives, with negative VOT found in over 94% of voiced tokens [6–9]. In Spanish, the rate of prevoicing has been reported to be around 86% [9]. In Dutch, negative VOT has been observed in 75% of initial voiced plosives only [5]. There are also languages such as Polish that contrast prevoiced versus (slightly) aspirated plosives [10] as well as languages with optional prevoicing. For example, although English is not a prevoicing language, voicing lead does occur in some speakers and under certain conditions, such as hyperarticulated and laboratory-type speech [4,11,12]. Prevoicing is also common in African American English [13] and in some regional dialects, including Inland California [14].

1.2. Production of Voicing

According to the myloelastic-aerodynamic theory of voice production, voicing is achieved by keeping the vocal folds adducted and allowing air to flow through the glottis, which causes vocal fold vibration [15]. To maintain the airflow, subglottal pressure needs to be higher than the pressure inside the oral cavity. This is difficult to implement for plosives, since they are articulated with a period of complete oral closure that leads to rapid equalization of subglottal and supraglottal pressure. This phenomenon is known as the 'aerodynamic voicing constraint (AVC)' [16–18]. Speakers use a variety of mechanisms to overcome the AVC and facilitate the production of voicing, such as lowering the tongue body to enlarge the vocal tract [19] or allowing nasal venting during oral closure [7,9].

The AVC is especially problematic for utterance-initial plosives when the subglottal pressure has not yet reached its peak level [20] as well as the more posterior places of articulation [20–23]. Anterior plosives are produced with a larger volume of space behind the oral constriction and with access to more compliant surfaces (including most of the tongue surface and the inside walls of the cheeks). This facilitates expansion of the supraglottal cavity in reaction to rising pressure. In contrast, posterior plosives are articulated with a smaller cavity behind the constriction and less compliant surfaces (e.g., the back of the tongue and the pharyngeal wall). As a result, oral pressure increases faster and the difference between oral versus subglottal pressure is neutralized sooner, which inhibits voicing.

Asymmetries in closure voicing duration or its frequency across places of articulation have been noted in several previous production studies. For example, initial velar plosives in English show up to 14ms more prevoicing (when produced with negative VOT) [4]. English velars are also more likely to be devoiced than non-velars [24]. In Polish, velar plosives have up to 22 ms less prevoicing than bilabials and alveolars [10]. In Dutch, bilabial plosives have a higher rate of prevoicing than alveolars (there is no voiced velar plosive in the language), and there is also a non-significant durational difference in the expected direction (possibly due to a small sample size) [5]. In Swedish, there is up to 30 ms less prevoicing for velars than non-velars [25]. In Russian, there is between 12 ms to 18 ms less prevoicing in velar plosives compared to bilabials and alveolars [6,7].

In addition to being affected by the plosive's place of articulation, closure voicing may vary depending on adjacent vowel height, with anticipatory coarticulation affecting the size of the cavity behind the oral constriction and influencing the extent to which air pressure can be lowered [23,24]. However, production findings for the role of vowel height have been less consistent than the effects of consonantal posteriority. Some studies report that vocal fold vibration is easier to produce for plosives followed by high vowels, which may be attributed to a larger pharyngeal cavity during the articulation of high vowels [23,24,26]. Other studies show a general faciliatory effect of adjacency to a vowel but no differences between high versus low vowels [5]. In Russian, the effect of vowel height is in the opposite direction, with up to 12ms more prevoicing seen for plosives followed by non-high vowels [7]. Such a finding may be due to greater exposure of cheek walls for non-high vowels [23] or greater vocal fold tension for high vowels [27].

One more factor known to affect closure voicing is speaker's biological sex (usually referred to as 'gender' in previous studies) [5,6,13,25]. Male speakers tend to have larger vocal tracts [28], which can be expected to make it easier for males to produce and maintain voicing during complete oral occlusion. However, similarly to the effects of vowel height, the findings for speaker sex have been inconsistent. In Dutch, prevoicing is more frequent in male speakers (21% difference) and closure voicing is also longer for males (by 20 ms) but the durational difference is not significant (which may be due to a small sample size of five speakers per sex) [5]. In Norwegian, the effect of speaker sex is reversed, with longer and more frequent prevoicing found in female tokens, which may be related to female speech being more hyperarticuled [29]. In Russian, the speaker sex effect is either not significant (possibly due to averaging across places of articulation) [8] or it is a general effect, with male speakers producing 10 ms longer prevoicing [6], or there is no overall difference but there is an interaction between speaker sex and vocalic context, with Russian males producing up to 22 ms less prevoicing in tokens with a high vowel [7].

Overall, previous production studies have shown that the likelihood of closure voicing and its duration vary across places of articulation, with stronger prevoicing seen for the more anterior plosives. Voicing also appears to be sensitive to the quality of the following vowel but the direction of the effect is not consistent in production, with some studies showing facilitation and other studies reporting inhibition of prevoicing in high vowel contexts. Finally, speaker sex may also play a role, with male speakers producing more frequent or longer closure voicing; however, the effect of speaker sex is also limited and not always consistent in production.

1.3. Perception of Voicing

While the production side of voicing has now been examined for several prevoicing languages, little is currently known about the way listeners perceive negative VOT and whether perceptual judgments are affected by the voicing asymmetries that exist in production. Most previous perception studies have concentrated on aspirating languages, such as English, and the general finding has been a typical categorical perception S-curve with a cross-over in perceptual judgments from 'voiced' to 'voiceless' occurring between +25 ms to +50 ms [10,30,31]. When two or more places of articulation are examined, positive VOT usually shows an effect of posteriority. In English, the cross-over boundary occurs up to 17ms later for velars than non-velars [31]. Speakers who are bilingual in English and a prevoicing language, such as Spanish, also show cross-over boundaries in the positive range as well as the expected posteriority effect of up to 10 ms [32].

For prevoicing languages, most previous perception studies only examined one place of articulation. Cross-over boundaries of −4 ms and −11 ms have been reported for bilabial plosives in Puerto Rican Spanish and Peruvian Spanish, respectively [33]. In Hebrew, which contrasts prevoiced versus (slightly) aspirated stops, perceptual judgments for bilabials show a category shift at around +6 to +7 ms [34]. For alveolars, Polish shows cross-over boundaries ranging from −3.5 ms to +21 ms, with the exact boundary location dependent on the range of values in the VOT continuum [10]. In Russian, the crossover point for alveolars has been observed at around −16 ms [35]. For Dutch, category boundaries are known for both bilabials (−9.5 ms) and alveolars (−3.5 ms) [5], which suggests that less voicing may in fact be needed for the more posterior plosives to be classified as voiced. However, the Dutch values were obtained in a classification tree analysis, so they may not be directly comparable to the boundaries established in perceptual identification studies mentioned above.

1.4. Current Study

The current research focused on determining whether the effects of place of articulation, vowel height, and speaker sex that occur in production can also be found in perception. The study examined perceptual identification responses for plosive-initial CV sequences with VOT values in the ambiguous range around the cross-over boundary. The language of investigation was Russian, a prevoicing language with very limited previous research on VOT perception. Based on the earlier production findings, the study was expected to show that if place of articulation did affect perceptual judgments, less voicing would be needed for the more posterior plosives to be identified as voiced. For vowel height, a difference in responses to tokens with high versus low follow vowels could be expected but the exact direction of the effect could not be predicted due to the inconsistency of previous production findings. For speaker sex, male tokens could be predicted to be judged as voiceless more often than female tokens with the same amount of VOT, although the magnitude of the effect was likely to be limited due to the lack of a consistent pattern in production. If supported by the actual results, such findings would provide evidence for the AVC and related phonetic factors affecting both the initial production of voicing and its subsequent perception.

2. Materials and Methods

2.1. Participants, Stimuli, and Procedures

Participants were monolingual speakers of Russian (n = 60; 32 female, 28 male; age range of 18 to 30, mean age of 20.4). They were recruited and tested at a university campus in Perm, Russia. During the pre-test interview, all participants indicated having beginner-level knowledge of other languages, acquired in a classroom setting, and no regular exposure to non-Russian speech. None self-reported a hearing problem or another issue that may affect speech production or perception.

The stimulus list consisted of 180 CV tokens that differed across consonantal place of articulation, following vowel height, speaker sex, and VOT level. The initial plosive was bilabial ([b/p]), alveolar ([d/t]), or velar ([g/k]). The vowel was low ([a]), mid ([o]), or high ([u]). Other vowels were not used because of the restrictions related to consonantal palatalization in Russian. The CV sequences were produced by two native Russian speakers (male, 22 y.o.; female, 20 y.o.). Depending on the token, the mean pitch ranged from 125 Hz to 130 Hz for the male speaker and between 245 Hz and 255 Hz for the female speaker. For each CV sequence, a VOT continuum was created in Praat [36]. VOT values ranged from −60 ms to +30 ms in 10 ms steps. Tokens that were originally produced with at least 60 ms of prevoicing served as the base. For negative VOT, voicing was removed in 10ms increments, starting at −60 ms. For positive VOT, bursts were extended by splicing in burst noise from CV sequences with initial voiceless plosives with the same place of articulation and the same following vowel. Intensity of the tokens and vowel duration were adjusted to match across places of articulation, vocalic contexts, and speaker sexes. A set of 10 CV sequences with initial fricatives was also prepared for use in a training module ([va, fo, zu], etc.).

In addition to the naturally produced female and male speech, the study used tokens with resynthesized speaker sex. This was done to ensure that differences between female versus male tokens, if observed, were not driven by acoustic properties other than VOT. For this purpose, the original items produced by the male speaker were resynthesized using the 'Change gender' function in Praat [36]. The new median pitch was set to 250 Hz, and formants were shifted by a ratio of 1.1, which adjusts the formants upward to better match female voice characteristics. Pilot testing showed that resynthesis of female tokens into male resulted in an unnatural-sounding voice that participants found distracting, so only the male-to-female resynthesis was used in the study. Sample tokens in original male and resynthesized female voices are provided in Figure 1.

Experimental sessions took place at a psychology lab and lasted approximately 60 min. During the session, participants performed a forced-choice identification task. The ExperimentMFC module in Praat [36] was used to present the stimuli and record listeners' responses. Instructions were given verbally and also printed on screen. Participants were instructed to identify the sequence (e.g., whether they heard 'ba' or 'pa') and to answer as quickly and accurately as possible. The experiment started with a short training stage that utilized words with initial fricatives. This was followed by presenting experimental items in a randomized order. The tokens were presented binaurally through a pair of sound-insulating headphones. The next stimulus was presented 1 second after a response was entered. Each participant responded to a total of 900 tokens (3 places of articulation

× 3 vowels × 2 speaker sexes × 10 VOT levels = 180 tokens × 5 repetitions per token = 900 tokens). The experimental script was set to pause after presenting a set of 50 items, and participants were strongly encouraged to take breaks in-between the sets. Since some CV sequences matched existing Russian words (e.g., [ta] meaning 'that (fem.)' or [da] meaning 'yes'), stimuli were only referred to as 'syllables' throughout the experiment to help mitigate a lexical effect, and the item's lexical status and existence of a lexical competitor were also tested as control variables in statistical modeling.

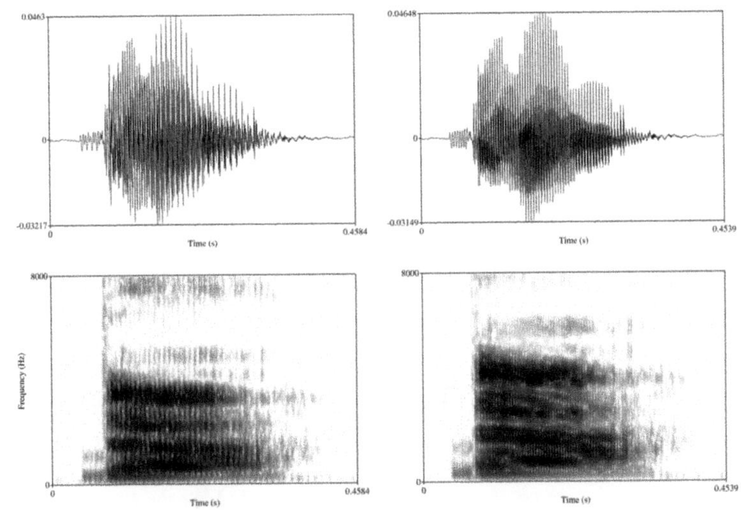

Figure 1. Prevoiced [da] token with −30 ms VOT. Original male voice (**left**) and resynthesized female voice (**right**).

Data were collected over two experiments. In Experiment 1, participants (n = 30; 16 female, 14 male; age range of 18 to 29; mean age of 20.5) listened to the original voices. In Experiment 2, participants (n = 30; 16 female, 14 male; age range of 18 to 30; mean age of 20.4) listened to original male and resynthesized female voice tokens. Since the experiments were performed anonymously and were conducted at the same university, it is possible that some participants took part in both experiments; however, the experimental sessions were two years apart, so no carry-over effects would be expected.

2.2. Data Analysis

Participants' responses were analyzed statistically in R [37]. Given the binary nature of the dependent variable ('voiced' versus 'voiceless'), a logistic mixed-effects model was constructed for each experiment using the *glmer* function of the 'lme4' package [38]. Forest plots illustrating the odds ratios for the models were created with the 'sjPlot' package [39]. Each model examined participants' responses to the tokens with VOT durations in the −20 ms to +20 ms range. This was the time window that encompassed the categorical shift in perception from 'voiced' to 'voiceless' and the adjacent ambiguous regions. The model tested for the fixed effects for VOT duration, consonantal place of articulation, the following vowel, and speaker sex. It also included all 2-way interactions with VOT duration as well as two additional interactions that were significant in production for Russian: (i) consonantal place × vocalic context and (ii) speaker sex × vocalic context [7]. Prior to analysis, VOT duration was rescaled. Place of articulation was coded as 'bilabial' (yes = 1, no = 0) and 'velar' (yes = 1, no = 0). Vowel height was coded as 'high' (yes = 1, no = 0) and 'low' (yes = 1, no = 0). Speaker sex was coded as 'male' (yes = 1, no = 0).

Several control variables were also tested during the model selection process, including listener sex ('male'; yes = 1, no = 0), listener age (in full years), lexical status of the stimulus (i.e., whether the CV sequence matches a real word of Russian; yes = 1, no = 0), and lexical competition (i.e., existence of a voicing-based minimal pair counterpart; yes = 1, no = 0). The lexical factors were coded on the basis of a comprehensive Russian thesaurus [40]. The control variables were not retained in the final model as they did not improve model fit (as indicated by ANOVAs comparing models with and without control variables).

Models with all relevant fixed effects and interactions, participants and items as random effects, and correlated random slopes did not converge, so the random effects structure was gradually simplified until maximal models with non-singular fits were found [41]. For Experiment 1, the maximal model included a random intercept for item. For Experiment 2, the maximal model had a random slope for participant by VOT duration and a random intercept for item.

3. Results

3.1. Experiment 1

Experiment 1 utilized the original voice tokens. Two fixed effects were identified as significant in the *glmer* model: VOT duration ($\beta = -10.96; z = -13.48; p < 0.001$) and consonantal place (velar vs. non-velar; $\beta = 0.41; z = 3.18; p < 0.01$). All other fixed effects and interactions were not significant (all *ps* > 0.1). Full results of statistical modeling are provided in Table A1 in Appendix A.

Figure 2 shows the identification curves for the statistically significant contrast between velars versus non-velars. Tokens with velar plosives are marked with a solid line. Items with non-velars (bilabials, alveolars) are shown with a dashed line. The y-axis represents the mean rates of 'voiced' responses. The x-axis shows the 10 VOT levels (from −60 ms to +30 ms in 10 ms steps). Category boundaries at 50% (cross-over points) are marked with thin dotted lines. The area highlighted in grey was used in statistical modeling. As can be seen in the figure, classic S-shaped identification functions were obtained along the VOT continuum for both velars and non-velars. As prevoicing duration decreased, the rate of 'voiced' responses also decreased. For non-velars, a shift in perceptual judgments from 'voiced' to 'voiceless' occurred at −12 ms. For velars, the category boundary was at −9 ms. Table A2 in Appendix A provides the full results for the mean rates of 'voiced' responses, including the differences across the levels of non-significant factors.

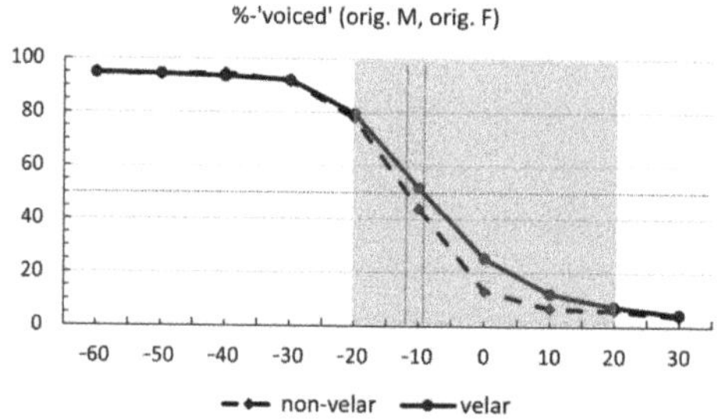

Figure 2. Identification curves for the original voice non-velars (dashed line) and velars (solid line) identified as 'voiced' across 10 VOT levels. Thin dotted lines mark the 50% boundary. Statistical modeling was conducted on the area highlighted in grey.

Figure 3 provides the odds ratios (ORs) for 'voiced' responses in the −20 ms to +20 ms VOT window. OR values represent the odds that an outcome will occur given a particular exposure, compared to the odds of the outcome occurring in the absence of that exposure [42]. The figure shows mean ORs for the fixed effects and interactions, starting with the significant effects of VOT duration and consonantal place (velar vs. non-velar). Whiskers represent the 95% confidence intervals (CIs), with smaller CIs indicating higher precision. OR values above 1.0 (to the right of the vertical gray line) indicate increased occurrence of 'voiced' responses. Values below 1.0 (to the left of the line) signal a decrease in 'voiced' responses. Values of 1.0 or approaching it (on or near the line) show that the rate of 'voiced' responses was unaffected. The further away an OR value is from 1.0, the stronger the causal relationship. As reflected in the figure, VOT duration had the strongest influence on participants' judgments, with higher VOT values corresponding to a prominent decrease in the rate of 'voiced' response. In other words, tokens with little or no prevoicing were significantly more likely to be identified as 'voiceless'. Furthermore, tokens with velars were more likely to be judged as 'voiced' than tokens with non-velars with the same VOT. The magnitude of the place effect was modest. The rest of the factors did not affect participants' responses.

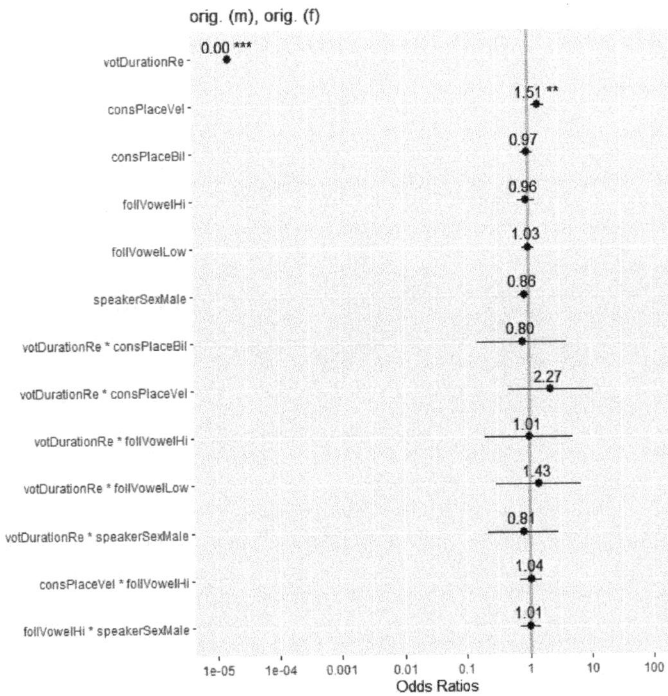

Figure 3. Odds ratios (ORs) of 'voiced' responses across the fixed effects and interactions in the *glmer* model for the original male and female voice tokens in the −20 ms to +20 ms VOT range. Significance is shown with asterisks ('***': $p < 0.001$; '**': $p < 0.01$.). Confidence intervals (CIs: 95%) are marked with whiskers.

While the effects observed in Experiment 1 were in the expected direction, male and female voices may be expected to differ across multiple parameters that are potentially relevant for voicing judgments (e.g., spectral properties of the burst). Such differences may be more salient and may distract participants from focusing on VOT. This may have masked the true extent of the place of articulation effect and may also help explain the absence of significant effects of speaker sex and vocalic environment. To explore this possibility, a second identification experiment was conducted using resynthesized stimuli.

3.2. Experiment 2

In Experiment 2, participants listened to the original male voice and resynthesized female voice tokens. As in the first experiment, VOT duration was significant in the *glmer* model ($\beta = -11.47; z = -13.89; p < 0.001$). Consonantal place (velar vs. non-velar) was also significant ($\beta = 0.36; z = 2.77; p < 0.01$). Unlike Experiment 1, the effect of speaker sex was now marginally significant ($\beta = -0.19; z = -1.67; p = 0.095$). All remaining fixed effects and interactions were not significant (all $ps > 0.1$). Full results of statistical modeling are provided in Table A3 in Appendix A.

Identification curves for the original and resynthesized voice tokens separated by consonantal place (velar vs. non-velar) and speaker sex (female vs. male) are provided in Figure 4. Classic S-shaped identification functions can be seen along the VOT continuum for all four categories. Tokens with less or no prevoicing consistently showed lower rates of 'voiced' responses. The 50% category boundary was at −11 ms for non-velars and −8 ms for velars. For both male and female voice tokens, the boundary was at around −10 ms. Full results for the mean rates of 'voiced' responses are given in Table A4 in Appendix A.

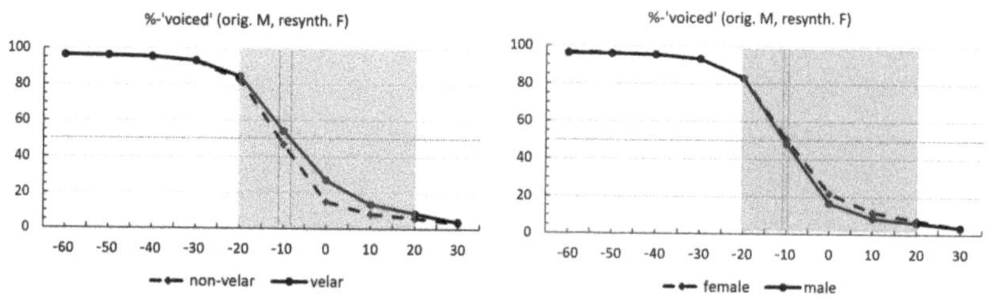

Figure 4. Identification curves for the original voice (male) and resynthesized voice (female) tokens identified as 'voiced' across 10 VOT levels. Left: non-velars (dashed line) vs. velars (solid line). Right: female voice (dashed line) vs. male voice (solid line). Thin dotted lines mark the 50% boundary. Statistical modeling was conducted on the area highlighted in grey.

Figure 5 shows the odds ratios (ORs) for 'voiced' responses in Experiment 2. As can be seen in the figure, VOT duration had the strongest influence on participants' identification of voicing in plosives, with the rate of 'voiced' responses decreasing as VOT increased. Place of articulation had a significant effect of lesser magnitude. Presence of a velar plosive was associated with higher odds of a 'voiced' response compared to tokens with bilabials and alveolars with the same VOT. These findings fully parallel the results for the effects of VOT duration and consonantal place in Experiment 1. Unlike the previous experiment, speaker sex showed a marginally significant effect of small magnitude, which was in the expected direction (i.e., more 'voiceless' responses for the male voice tokens in the −20 ms to +20 ms VOT range). Vowel height did not affect participants' responses and none of the factors interacted. Implications of the current findings are discussed below.

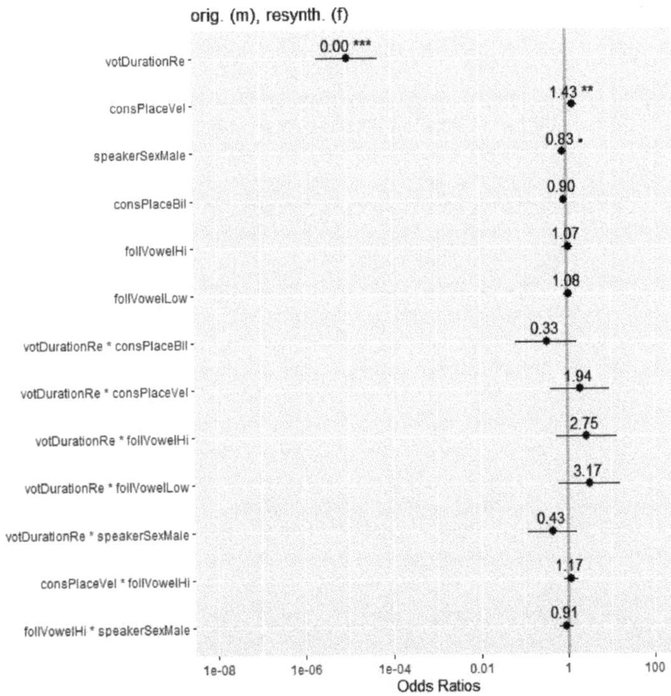

Figure 5. Odds ratios (ORs) of 'voiced' responses across the fixed effects and interactions in the *glmer* model for the original male and resynthesized female tokens in the −20 ms to +20 ms VOT range. Significance is shown with asterisks ('***': <0.001; '**': <0.01; '.': <0.1). Confidence intervals (CIs: 95%) are marked with whiskers.).

4. Discussion

Overall, results from the two experiments indicated that the perceptual boundary for 'voiced' versus 'voiceless' categories is located within the negative VOT space in Russian, with values ranging between −12 ms to −8 ms. This is consistent with the previous findings for other prevoicing languages, such as Dutch and Spanish [5,33], and also matches the −16 ms VOT boundary that has been previously reported for Russian alveolars [35]. The small numeric difference between the current and previous findings for Russian is likely attributable to the differences in VOT continua, which is known to affect the location of the category boundary for voicing judgments [10].

The present study also observed an effect of consonantal place of articulation on participants' responses. The effect was significant in both experiments. As stated in the Introduction, the AVC inhibits the production of voicing in the more posterior plosives [18]. Previous phonetic research has confirmed that the AVC plays a role in the Russian language, with initial velar plosives showing less prevoicing than non-velars [6,7]. The present perceptual findings reveal that the effect of place of articulation is not limited to production. As Russian speakers produce less prevoicing in velars, Russian listeners accept less closure voicing when classifying velars as 'voiced'. This shows that the AVC can affect not only the initial production of VOT in plosives but also its subsequent perception. At the same time, the magnitude of the place effect is quite modest, which suggests that consonantal place of articulation plays only a limited role in perception.

Unlike the place of articulation, adjacent vowel height did not affect participants' judgments in either experiment. In production, voicing duration is known to vary depending on vocalic context [23,24,26]. This has been attributed to coarticulation-related differences in the size of the oral cavity behind the oral constriction, which affects the extent to which air pressure can be lowered [23]; however, the vowel height effect has not been consistent in production. In some studies, vocal fold vibration is facilitated in high vowel contexts [23]. In other investigations, there are no differences across vowel heights [5] or the effect is in the opposite direction, with more prevoicing seen with non-high vowels but only for a subset of plosives and speakers [7]. This lack of consistency in production may in turn explain the absence of the effect in perception. In other words, Russian listeners do not have robust exposure to this type of asymmetry in their own speech and the speech of others, and they do not take the effect of vowel height into account when making voicing judgments.

For speaker sex, the effect was not significant in the first experiment when using original voice tokens that varied across multiple acoustic parameters. Differences between male versus female voice tokens became only marginally significant in the second experiment when acoustic variability was constrained. As noted earlier, production of voicing is thought to be easier for males due to a larger vocal tract [28], and several previous studies have observed an asymmetry in VOT duration or its frequency, with male voice tokens showing longer or more frequent closure voicing [5,6,25]; however, production findings for the role of speaker sex have been inconsistent. Some studies show more voicing for male speakers regardless of the phonetic context [6]. Other investigations report a significant effect for the high vowel context only [7] or even a reversed effect, with more closure voicing seen in female speech [29]. In the current study, listeners showed a possible tendency to perceive male voice tokens as voiceless more often than resynthesized female voice items with the same amount of VOT. However, the effect of speaker sex was not as consistent or prominent as the influence of place of articulation. As in the case of vowel height, this may be due to the lack of consistent exposure to a speaker sex asymmetry in production. The identification task paradigm may also not be sensitive enough to detect the effects of speaker sex or vowel height. Such influences may only exist at the earliest pre-lexical stages of processing, whereas identification responses are largely lexical and they even allow for post-lexical influences from contextual and background knowledge.

The current findings have implications for the role of phonetic influences in speech perception. A link between production and perception has long been advocated for in neurolinguistic and psycholinguistic literature, including the foundational aphasia studies [43] and several prominent theories of speech production and perception, such as the motor theory [44] and the direct-realist theory [45]. The strong version of the motor theory, for example, argues for both production and perception relying on the same phonetic information. In other words, knowing how a speech sound is produced and what kind of acoustic output is generated helps recognize the sound. This view is supported by the apparent activation of motor brain structures during not only articulation but also visual speech processing [46] and auditory perception [47,48]. A strong link between production and perception is also supported by the McGurk effect, which shows that listeners routinely integrate visual cues when identifying speech sounds [49]. It has also been demonstrated that silent articulation affects auditory processing [50] and that temporary disruption of the motor cortex can impair categorical perception [51]. At the same time, we know that individuals with permanent or temporary neurological or physiological impairments that affect speech production do not necessarily show impaired perception [52,53]. This includes previous experience with tracheostomy (intubation) for more than three months that leads to impaired production of vocal fold vibration but does not affect perception of voicing [54]. Thus, the relationship between production and perception cannot be causal [48,53,55]. This view is also shared by theoretical linguists who argue that phonological computation must disregard articulatory and acoustic detail (aka 'substance free' phonology) [56].

In the present study, listeners made voicing judgments for CV sequences with VOT in the ambiguous range. Since most of the syllables did not match existing Russian words, listeners could not simply rely on lexical knowledge, which is well known to affect perceptual responses [57]. They also did not have access to visual or somatosensory feedback. Listeners had to base their responses on acoustic differences in VOT alone or in combination with other cues to voicing that may be present in the token, such as f0 of the following vowel [58]. Perceptual responses showed effects of not only voicing duration but also consonantal place of articulation and, at a marginally significant level, speaker sex. This shows that certain VOT asymmetries exist in both production and perception; however, it cannot be inferred from the current results whether listeners accessed articulatory representations directly (as suggested by the motor theory) or whether they relied on perceptual representations that exist separately and contain fine acoustic detail (e.g., fully specified exemplars of voiced plosives that reflect production differences in voicing) [59,60]. The small magnitude of the place of articulation effect, marginal significance for speaker sex and the absence of differences across vowel heights further suggest that production and perception of voicing are not critically co-dependent, with knowledge of production asymmetries making only a limited contribution to the perceptual decision-making process for VOT. Phonemic identification tasks may also use a different set of mechanisms compared to non-laboratory comprehension [53], so this type of knowledge may only be called upon when other sources of information are limited or not available. Hence, the extent to which the AVC and related phonetic factors play a role in the perception of VOT in everyday communication remains to be determined. Future research will need to address this important question.

5. Conclusions

The present study examined the link between production and perception of the voicing contrast in a prevoicing language, focusing on the effects of consonantal place, adjacent vowel height, and speaker sex. Results of perceptual identification tasks revealed that place of articulation can affect not only the initial production of consonantal voicing but also its subsequent perception. The magnitude of the place effect was small, suggesting that it played only a secondary role in perception. Vowel height did not show a significant effect on perceptual judgments, in contrast to what has been reported in some production studies. Speaker sex showed marginal significance only, with the differences being in the expected direction. These findings are consistent with the theories that advocate for a limited link between production and perception.

Funding: This research received no external funding.

Institutional Review Board Statement: The study was conducted according to the guidelines of the Declaration of Helsinki, and approved by the Institutional Review Board of Florida Atlantic University (protocol code 646702-1; 5/4/2015).

Informed Consent Statement: Informed consent was obtained from all subjects involved in the study.

Data Availability Statement: The original data are available upon request from the author.

Conflicts of Interest: The author declares no conflict of interest.

Abbreviations

The following abbreviations are used in this manuscript:

ANOVA	Analysis of variance
AVC	Aerodynamic voicing constraint
CV	Consonant–vowel sequence
OR	Odds ratio
VOT	Voice onset time

Appendix A

Table A1. Generalized linear mixed model (Experiment 1, original voice tokens).

Fixed Effects	Estimate	Std. Error	z Value	Pr (>[z])
(Intercept)	−1.210611	0.118843	−10.187	<2 × 10^{-16} ***
votDurRe	−10.964394	0.813468	−13.479	<2 × 10^{-16} ***
consPlBil	−0.026452	0.113881	−0.232	0.81632
consPlVel	0.410131	0.128821	3.184	0.00145 **
follVlHi	−0.038413	0.162258	−0.237	0.81286
follVLow	0.029688	0.111306	0.267	0.78968
speakSexM	−0.149305	0.110787	−1.348	0.17776
votDurRe:consPlBil	−0.221077	0.831608	−0.266	0.79036
votDurRe:consPlVel	0.821977	0.800654	1.027	0.30459
votDurRe:follVHi	0.014414	0.822785	0.018	0.98602
votDurRe:follVLow	0.357623	0.802994	0.445	0.65606
votDurRe:speakSexM	−0.216758	0.665062	−0.326	0.74448
consPlVel:follVoHi	0.043204	0.199239	0.217	0.82833
follVHi:speakSexM	0.009037	0.190152	0.048	0.96209

Signif. codes: 0 '***' 0.001 '**' 0.01.

Formula: responseVoi votDurationRe * (consPlaceBil + consPlaceVel + follVowelHi + follVowelLow + speakerSexMale) + consPlaceVel * follVowelHi + speakerSexMale * follVowelHi + (1 | itemName).

Control: glmerControl(optimizer = "bobyqa", optCtrl = list(maxfun = 1 × 10^5)).

Fit by maximum likelihood (Laplace Approximation) [glmerMod].

Family: binomial (logit).

Table A2. Mean rates of 'voiced' responses (%) across 10 VOT levels (Experiment 1, original voice tokens).

VOT	Non-Vel.	Velar	Non-Bilab.	Bilab.	Non-High	High	Non-Low	Low	Female	Male
−60	94.8	94.9	94.8	94.8	94.6	95.2	94.8	94.8	94.8	94.8
−50	94.1	94.4	94.5	93.7	94.3	94.0	94.0	94.7	94.3	94.1
−40	94.6	93.4	94.1	94.4	95.1	92.6	93.5	95.7	94.4	94.0
−30	91.6	92.1	91.9	91.4	91.4	92.4	91.3	92.7	91.8	91.7
−20	77.9	79.6	78.7	78.1	78.3	78.8	78.8	77.9	78.9	78.1
−10	44.0	51.9	48.5	42.9	48.1	43.8	46.5	46.9	47.4	45.9
0	13.2	25.4	18.9	13.9	16.8	18.2	16.5	18.8	19.5	15.0
10	6.7	12.2	9.5	6.6	8.3	9.0	8.9	7.8	9.7	7.3
20	6.0	7.3	6.8	5.7	6.8	5.8	6.2	7.0	6.4	6.5
30	3.8	4.4	4.2	3.7	3.8	4.6	4.2	3.7	3.9	4.1

Table A3. Generalized linear mixed model (Experiment 2, original and resynthesized voice tokens).

Fixed Effects	Estimate	Std. Error	z Value	Pr (>[z])
(Intercept)	−1.06217	0.12179	−8.722	<2 × 10^{-16} ***
votDurRe	−11.47183	0.82623	−13.885	<2 × 10^{-16} ***
consPlBil	−0.10882	0.11508	−0.946	0.34433
consPlVel	0.36090	0.13015	2.773	0.00556 **
follVHi	0.07211	0.16294	0.443	0.65810
follVLow	0.07651	0.11307	0.677	0.49859
speakSexM	−0.18795	0.11257	−1.67	0.09500.
votDurRe:consPlBil	−1.09387	0.84106	−1.301	0.19340
votDurRe:consPlVel	0.66496	0.80869	0.822	0.41092
votDurRe:follVHi	1.01264	0.83036	1.220	0.22265
votDurRe:follVLow	1.15350	0.82802	1.393	0.16360

Table A3. *Cont.*

Fixed Effects	Estimate	Std. Error	z Value	Pr (>[z])
votDurRe:speakSexM	−0.83819	0.67558	−1.241	0.21472
consPlVel:follVHi	0.15926	0.20108	0.792	0.42833
follVHi:speakSexM	−0.09564	0.19227	−0.497	0.61890

Signif. codes: 0 '***' 0.001 '**' 0.01.

Formula: responseVoi votDurationRe * (consPlaceBil + consPlaceVel + follVowelHi + follVowelLow + speakerSexMale) + consPlaceVel * follVowelHi + speakerSexMale * follVowelHi + (1 + votDurationRe | participantNum) + (1 | itemName).

Control: glmerControl(optimizer = "bobyqa", optCtrl = list(maxfun = 1×10^5)).

Fit by maximum likelihood (Laplace Approximation) ['glmerMod'].

Family: binomial (logit).

Table A4. Mean rates of 'voiced' responses (%) across 10 VOT levels (Experiment 2, original and resynthesized voice tokens).

VOT	Non-Vel.	Velar	Non-Bilab.	Bilab.	Non-High	High	Non-Low	Low	Female	Male
−60	96.4	96.3	96.3	96.4	96.9	95.2	96.1	96.9	96.6	96.1
−50	96.1	96.1	96.3	95.7	95.9	96.3	95.8	96.6	96.2	95.9
−40	95.4	95.1	95.4	95.1	95.1	95.9	94.9	96.1	95.3	95.4
−30	93.2	93.2	93.6	92.4	93.3	92.9	92.8	94.0	93.2	93.2
−20	82.5	84.4	83.3	82.8	83.2	83.1	83.5	82.4	83.3	83.0
−10	46.9	54.3	50.9	46.3	49.8	48.6	50.1	48.0	50.6	48.2
0	15.2	27.3	21.2	15.3	19.3	19.1	18.4	21.0	21.9	16.7
10	8.2	13.7	11.1	7.9	9.4	11.2	10.2	9.6	11.6	8.4
20	5.7	8.6	7.7	4.6	6.4	7.0	6.2	7.4	7.3	6.0
30	3.2	3.8	3.6	3.2	3.7	3.0	3.4	3.7	3.4	3.6

References

1. Cho, T.; Ladefoged, P. Variation and universals in VOT: Evidence from 18 languages. *J. Phon.* **1999**, *27*, 207–229. [CrossRef]
2. Abramson, A.S.; Whalen, D.H. Voice Onset Time (VOT) at 50: Theoretical and practical issues in measuring voicing distinctions. *J. Phon.* **2017**, *63*, 75–86. [CrossRef] [PubMed]
3. Cho, T.; Whalen, D.H.; Docherty, G. Voice onset time and beyond: Exploring laryngeal contrast in 19 languages. *J. Phon.* **2019**, *72*, 52–65. [CrossRef] [PubMed]
4. Lisker, L.; Abramson, A.S. A cross-language study of voicing in initial stops: Acoustical measurements. *Word* **1964**, *20*, 384–422. [CrossRef]
5. Van Alphen, P.M.; Smits, R. Acoustical and perceptual analysis of the voicing distinction in Dutch initial plosives: The role of prevoicing. *J. Phon.* **2004**, *32*, 455–491. [CrossRef]
6. Kulikov, V. Voicing and Voice Assimilation in Russian Stops. Ph.D. Thesis, The University of Iowa, Iowa City, IA, USA, 2012.
7. Kharlamov, V. Prevoicing and prenasalization in Russian initial plosives. *J. Phon.* **2018**, *71*, 215–228. [CrossRef]
8. Ringen, C.; Kulikov, V. Voicing in Russian stops: Cross-linguistic implications. *J. Slav. Linguist.* **2012**, *20*, 269–286. [CrossRef]
9. Solé, M.J. Articulatory adjustments in initial voiced stops in Spanish, French and English. *J. Phon.* **2018**, *66*, 217–241. [CrossRef]
10. Keating, P.A.; Mikos, M.J.; Ganong, W.F. A cross-language study of range of voice onset time. *J. Acoust. Soc. Am.* **1981**, *70*, 1261–1271. [CrossRef]
11. Docherty, G.J. *The Timing of Voicing in English Obstruents*; Foris Publications: Berlin, Germany, 1992.
12. Schertz, J. Exaggerating featural contrasts in clarifications of misheard speech in English. *J. Phon.* **2013**, *41*, 249–263. [CrossRef]
13. Ryalls, J.; Zipprer, A.; Baldauff, P. A preliminary investigation of the effects of gender and race on voice onset time. *J. Speech Lang. Hear. Res.* **1997**, *40*, 642–645. [CrossRef] [PubMed]
14. Podesva, R.J.; Eckert, P.; Fine, J.; Hilton, K.; Jeong, S.; King, S.; Pratt, T. Social influences on the degree of stop voicing in Inland California. *Univ. Pa. Work. Pap. Linguist.* **2015**, *21*, 19.
15. Van den Berg, J. Myoelastic-aerodynamic theory of voice production. *J. Speech Hear. Res.* **1958**, *1*, 227–244. [CrossRef] [PubMed]
16. Ohala, J.J. The origin of sound patterns in vocal tract constraints. In *The Production of Speech*; MacNeilage, P.F., Ed.; Springer: Berlin, Germany, 1983; pp. 189–216.
17. Ohala, J.J. Aerodynamics of phonology. In Proceedings of the 4th Seoul International Conference on Linguistics, Seoul, Korea, 11–15 August 1997; pp. 92–97.

18. Ohala, J.J. Accommodation to the aerodynamic voicing constraint and its phonological relevance. In Proceedings of the 17th International Congress of Phonetic Sciences, Seoul, Korea, 11–15 August 1997; pp. 64–67.
19. Westbury, J. Enlargement of the supraglottal cavity and its relation to stop consonant voicing. *J. Acoust. Soc. Am.* **1983**, *73*, 1322–1336. [CrossRef] [PubMed]
20. Westbury, J.; Keating, P. On the naturalness of stop consonant voicing. *J. Acoust. Soc. Am.* **1986**, *22*, 145–166. [CrossRef]
21. Houde, R.A. *A Study of Tongue Body Motion during Selected Speech Sounds*; Speech Communication Research Laboratory: Santa Barbara, CA, USA, 1968.
22. Rothenberg, M. *The Breath-Stream Dynamics of Simple-Released-Plosive Production*; Kager: Syracuse, NY, USA, 1968.
23. Ohala, J.J.; Riordan, C.J. Passive vocal tract enlargement during voiced stops. In *Speech Communication Papers*; Wolf, J.J., Klatt, D.H., Eds.; Acoustical Society of America: New York, NY, USA, 1979; pp. 89–92.
24. Smith, B. Effects of place of articulation and vowel environment on voiced stop consonant production. *Glossa* **1978**, *12*, 163–175.
25. Helgason, P.; Ringen, C. Voicing and aspiration in Swedish stops. *J. Phon.* **2008**, *36*, 607–628. [CrossRef]
26. Pape, D.; Mooshammer, C.; Hoole, P.; Fuchs, S. Devoicing of word-initial stops: A consequence of the following vowel? In Proceedings of the 6th International Seminar on Speech Production, Sydney, Australia, 7–10 December 2006; pp. 207–212.
27. Koenig, L.L.; Fuchs, S.; Lucero, J.C. Effects of consonant manner and vowel height on intraoral pressure and articulatory contact at voicing offset and onset for voiceless obstruents. *J. Acoust. Soc. Am.* **2011**, *129*, 3222–3244. [CrossRef]
28. Docherty, G.J. *Acoustic Phonetics*; The MIT Press: Cambridge, MA, USA, 1998.
29. Ringen, C.; van Dommelen, W.A. Quantity and laryngeal contrasts in Norwegian. *J. Phon.* **2013**, *41*, 479–490. [CrossRef]
30. Abramson, A.S.; Lisker, L. Discriminability along the voicing continuum: Cross-language tests. In Proceedings of the 6th International Congress of Phonetic Sciences, Prague, Czech Republic, 7–13 September 1967; pp. 569–573.
31. Lisker, L.; Abramson, A.S. The voicing dimension: Some experiments in comparative phonetics. In Proceedings of the 6th International Congress of Phonetic Sciences, Prague, Czech Republic, 7–13 September 1967; pp. 563–567.
32. Abramson, A.; Lisker, L. Voice-timing perception in Spanish word-initial stops. *J. Phon.* **1973**, *1*, 1–8. [CrossRef]
33. Williams, L. The voicing contrast in Spanish. *J. Phon.* **1977**, *5*, 169–184. [CrossRef]
34. Horev, N.; Most, T.; Pratt, H. Categorical perception of speech (VOT) and analogous non-speech (FOT) signals: Behavioral and electrophysiological correlates. *Ear Hear.* **2007**, *28*, 111–128. [CrossRef] [PubMed]
35. Kazanina, N.; Phillips, C.; Idsardi, W. The influence of meaning on the perception of speech sounds. *Proc. Natl. Acad. Sci. USA* **2006**, *103*, 11381–11386. [CrossRef] [PubMed]
36. Boersma, P.; Weenink, D. Praat: Doing Phonetics by Computer [Computer Program]. 2022. Available online: http://www.praat.org/ (accessed on 11 March 2022).
37. R Core Team. *R: A Language and Environment for Statistical Computing [Computer Program]*; R Foundation for Statistical Computing: Vienna, Austria, 2021. Available online: https://www.R-project.org/ (accessed on 11 March 2022).
38. Bates, D.; Mächler, M.; Bolker, B.; Walker, S. Fitting linear mixed-effects models using lme4. *J. Stat. Softw.* **2015**, *67*, 1–48. [CrossRef]
39. Lüdecke, D. sjPlot: Data Visualization for Statistics in Social Science [R Package]. 2021. Available online: https://CRAN.R-project.org/package=sjPlot (accessed on 11 March 2022).
40. Kuznetsov, S.A. *Grand Thesaurus of the Russian Language*; Norint: St. Petersburg, Russia, 1998.
41. Barr, D.J.; Levy, R.; Scheepers, C.; Tily, H.J. Random effects structure for confirmatory hypothesis testing: Keep it maximal. *J. Mem. Lang.* **2013**, *68*, 255–278. [CrossRef]
42. Szumilas, M. Explaining odds ratios. *J. Can. Acad. Child Adolesc. Psychiatry* **2010**, *19*, 227–229.
43. Wernicke, C. The symptom complex of aphasia: A psychological study on an anatomical basis. In *Boston Studies in the Philosophy of Science*; Cohen, R.S., Wartofsky, M.W., Eds.; D. Reidel Publishing Company: Dordrecht, The Netherlands, 1969.
44. Liberman, A.M.; Mattingly, I.G. The motor theory of speech perception revised. *Cognition* **1985**, *21*, 1–36. [CrossRef]
45. Fowler, C.A. An event approach to the study of speech perception from a direct–realist perspective. *J. Phon.* **1986**, *14*, 3–28. [CrossRef]
46. Hagoort, P.; Indefrey, P.; Brown, C.; Herzog, H.; Steinmetz, H.; Seits, R. The neural circuitry involved in the reading of German words and pseudowords: A PET study. *J. Cogn. Neurosci.* **1977**, *11*, 383–398. [CrossRef]
47. Wilson, S.M.; Saygin, A.P.; Sereno, M.I.; Iacoboni, M. Listening to speech activates motor areas involved in speech production. *Nat. Neurosci.* **2004**, *7*, 701–702. [CrossRef]
48. Pulvermüller, F.; Huss, M.; Kherif, F.; del Prado Martin, F.; Hauk, O.; Shtyrov, Y. Motor cortex maps articulatory features of speech sounds. *Proc. Natl. Acad. Sci. USA* **2006**, *103*, 7865–7870. [CrossRef] [PubMed]
49. McGurk, H.; MacDonald, J. Hearing lips and seeing voices. *Nature* **1976**, *264*, 746–748. [CrossRef] [PubMed]
50. Sams, M.; Möttönen, R.; Sihvonen, T. Seeing and hearing others and oneself talk. *Brain Res. Cogn. Brain Res.* **2005**, *23*, 429–435. [CrossRef] [PubMed]
51. Möttönen, R.; Watkins, K.E. Motor representations of articulators contribute to categorical perception of speech sounds. *J. Neurosci.* **2009**, *29*, 9819–9825. [CrossRef]
52. Bishop, D.V.; Brown, B.B.; Robson, J. The relationship between phoneme discrimination, speech production, and language comprehension in cerebral-palsied individuals. *J. Speech Hear. Res.* **1990**, *33*, 210–219. [CrossRef]

53. Hickok, G. The cortical organization of speech processing: Feedback control and predictive coding the context of a dual-stream model. *J. Commun. Disord.* **2012**, *45*, 393–402. [CrossRef]
54. Hill, B.P.; Singer, L.T. Speech and language development after infant tracheostomy. *J. Speech Hear. Disord.* **1990**, *55*, 15–20. [CrossRef]
55. Hale, M.; Kissock, M. The perception-production link and linguistic theory. *Loquens* **2019**, *62*, e066. [CrossRef]
56. Hale, M.; Reiss, C. *The Phonological Enterprise*; Oxford University Press: Oxford, UK, 2008.
57. Ganong, W.F. Phonetic categorization in auditory word perception. *J. Exp. Psychol. Hum. Percept. Perform.* **1980**, *6*, 110–125. [CrossRef]
58. Kirby, J.; Ladd, D.R. Effects of obstruent voicing on vowel f0: Evidence from 'true voicing' languages. *J. Acoust. Soc. Am.* **2016**, *140*, 2400–2411. [CrossRef]
59. Johnson, K. Speech perception without speaker normalization: An exemplar model. In *Talker Variability in Speech Processing*; Johnson, K., Mullennix, J.W., Eds.; Academic Press: Cambridge, MA, USA, 1997; pp. 145–165.
60. Pierrehumbert, J.B. Exemplar dynamics: Word frequency, lenition and contrast. In *Frequency and the Emergence of Linguistic Structure*; Bybee, J., Hooper, P., Eds.; John Benjamins: Amsterdam, The Netherlands, 2001.

brain sciences

MDPI

Article

Computational Modelling of Tone Perception Based on Direct Processing of f_0 Contours

Yue Chen [1], Yingming Gao [2] and Yi Xu [1,*]

[1] Department of Speech, Hearing and Phonetic Sciences, University College London, London WC1N 1PF, UK; yue.chen.1@ucl.ac.uk
[2] Institute of Acoustics and Speech Communication, TU Dresden, 01069 Dresden, Germany; yingming.gao@mailbox.tu-dresden.de
* Correspondence: yi.xu@ucl.ac.uk

Abstract: It has been widely assumed that in speech perception it is imperative to first detect a set of distinctive properties or features and then use them to recognize phonetic units like consonants, vowels, and tones. Those features can be auditory cues or articulatory gestures, or a combination of both. There have been no clear demonstrations of how exactly such a two-phase process would work in the perception of continuous speech, however. Here we used computational modelling to explore whether it is possible to recognize phonetic categories from syllable-sized continuous acoustic signals of connected speech without intermediate featural representations. We used Support Vector Machine (SVM) and Self-organizing Map (SOM) to simulate tone perception in Mandarin, by either directly processing f_0 trajectories, or extracting various tonal features. The results show that direct tone recognition not only yields better performance than any of the feature extraction schemes, but also requires less computational power. These results suggest that prior extraction of features is unlikely the operational mechanism of speech perception.

Keywords: speech perception; Mandarin tones; tone recognition; tone features

Citation: Chen, Y.; Gao, Y.; Xu, Y. Computational Modelling of Tone Perception Based on Direct Processing of f_0 Contours. *Brain Sci.* **2022**, *12*, 337. https://doi.org/10.3390/brainsci12030337

Academic Editor: Yang Zhang

Received: 29 January 2022
Accepted: 1 March 2022
Published: 2 March 2022

1. Introduction

How exactly speech perception works is still a mystery. It is widely assumed that multiple acoustic cues are needed for the perception of segments (consonants and vowels) and suprasegmentals (tone, intonation, etc.), and a major goal of research is to find out which cues are relevant for the recognition of these units [1,2]. For example, formants may provide primary cues for signaling different vowel categories, VOT is useful for distinguishing between voiced and voiceless plosives [3], pitch contour and pitch height are useful for differentiating lexical tones [4,5], etc. During speech perception, those cues are detected and then combined to identify specific contrastive phonetic units such as consonants, vowels, or tones [6]. This assumed mechanism, therefore, consists of two phases: feature detection, and phonetic recognition. No research so far, however, has demonstrated how exactly such a two-phase process can achieve the recognition of phonetic units in the perception of continuous speech. At the same time, another possibility about speech perception has rarely been theoretically contemplated, namely, a mechanism in which raw acoustic signals are processed to directly recognize phonetic units, without the extraction of intermediate featural representations. It may be difficult to test this hypothesis with existing accounts, however, because conventional behavioral and neural studies can only allow us to explore what acoustic cues are important for perception, but not how the whole perception process may work. What is needed is a way to explore the operation of speech perception by simulating it as a step-by-step procedural process, starting from acoustic signals as input, and ending with identified phonetic categories as output. This can be done through computational modeling that implements each proposed perception model as a phonetic recognition system. The present study is an attempt to apply this paradigm by making a computational comparison of direct phonetic perception to various two-phase perception models. In order to avoid the pitfall, often seen in computational modeling, of allowing multiple hidden layers that do not correspond to specific aspects of theoretical models of perception, here we try to construct computational models with transparent components that correspond explicitly to specific aspects of the related theoretical models.

1.1. Feature-to-Percept Theories

1.1.1. Distinctive Feature Theories

A major source of the feature-based view of speech perception is the classic theory of distinctive features [7]. The theory was proposed as an attempt to economize the representation of speech sounds beyond segmental phonemes [8,9]. In a pursuit to identify the most rudimentary phonetic entities, an even smaller set of featural contrasts than phonemes, aka distinctive features, was proposed [10]. Jakobson et al. [7] proposed a system with only 12 pairs of features, each making a binary contrast based predominantly on acoustic properties. An alternative system was proposed by Chomsky and Halle [11] with a much larger set of binary features (around 40) that are predominantly based on articulatory properties. Some phonological theories have even claimed that distinctive features are the true minimal constituents of language [12,13]. Most relevant for the current discussion, it is often assumed that the detection of the discrete features [14,15], be it binary or multivalued [1], is the key to speech perception [16,17]. This is seen in both the auditory theories and motor theories of speech perception, two competing lines of theories that have been dominating this area of research, as will be outlined next.

1.1.2. Auditory Theories

Auditory theories as a group assume that perceptual cues of phonetic contrasts are directly present in the acoustic signals of speech [17–20]. These theories assume that it is the distinctive acoustic properties that listeners are primarily sensitive to, and that speech perception is achieved by either capturing these properties [21] or extracting distinctive features [22]. These theories are often presented in opposition to motor theory, to be discussed next, in that they assume no intermediate gestural representations between acoustic cues and perceived categories. They recognize a role of distinctive features, and assume a need for extracting them in perception [17]. This need is elaborated in the Quantal Theory [23–26] based on the observation that auditory properties are not linearly related to continuous changes of articulation, but show some stable plateau-like regions in the spectrum. The plateaus are argued to form the basis of universal distinctive features. In addition, it is further proposed that there are enhancing features to augment the distinctive features [15,18,26–28].

1.1.3. Motor Theories

The motor theory [29–31], in contrast, assumes that the peripheral auditory processing phase of speech perception is followed by an articulatory recognition phase, in which articulatory gestures such as tongue backing, lip rounding, and jaw raising are identified. The motor theory is mainly motivated by the observation of the lack of one-to-one relations between acoustic patterns and speech sounds [32,33]. It is argued that invariance must lie in the articulatory gestures that generate the highly variable acoustic patterns. Therefore, gestures would serve as intermediate features that can match the auditory signals on the one hand, and the perceived phonemic or lexical units on the other hand.

Counter evidence to the motor theory comes from findings that speech perception can be achieved without speech motor ability in infants [34], non-human animals [35], and people suffering from aphasia [36–38]. However, there is also increasing evidence that perceiving speech involves neural activity of the motor system [39,40], and the motor regions are recruited during listening [41–44]. A range of brain studies using methods like TMS also showed evidence for the motor theory [45–47].

However, perceptual involvement of the motor system is not direct evidence for gesture recognition as the necessary prerequisite to phonetic recognition. For one thing, it is not clear whether motor activations occur before or after the recognition of the perceived categories. For another thing, there is increasing evidence that motor area activation mostly occurs only under adverse auditory conditions [45,48,49], which means that motor involvement may not be obligatory for normal perception tasks.

An issue that has rarely been pointed out by critics of motor theory is that gestures are actually not likely to be as invariant as the theory assumes. While a consonantal gesture could be in most cases moving toward a constricted vocal tract configuration, a vowel gesture could be either a closing or opening movement, depending on the openness of the preceding segment. In this respect, a greater degree of articulatory invariance is more likely found in the underlying phonetic targets in terms of vocal tract and/or laryngeal configuration rather than the target approximation movements [50–52].

1.2. Feature-to-Percept vs. Direct Phonetic Perception

As mentioned above, what originally motivated the motor theory, which has also permeated much of the debate between the motor and auditory theories is the apparent and pervasive variability in the speech signal. This is an issue fundamental for any theory of speech perception, namely, how is it possible that speech perception can successfully recover the phonetic categories intended by the speaker despite the variability? Note that, however, the question can be asked in a different way. That is, despite the variability, is there still enough within-category consistency in the acoustic signal that makes speech perception effective? If the answer is yes, the next question would be, what is the best way to capture the within-category consistency?

The answer by all the theories reviewed above would be that a feature-based two-phase process is the best way to capture the within-category consistency. They differ from each other only in terms of whether the extracted features are primarily auditory or articulatory, or a mixture of both as in the case of distinctive features. This commonality is nicely illustrated in Figure 1 from Fant [53]. Here, after being received by the ear, and having gone through the primary auditory analysis, the acoustic signals of speech are first turned into featural patterns that are either auditory (intervals CD) or motor (interval GF). Either way, they are both sub-phonemic and featural, and need to be further processed to identify the categorical phonemes, syllables, words, etc.

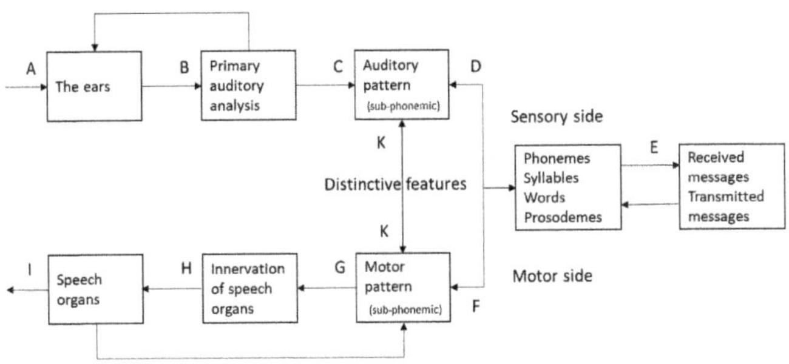

Figure 1. Hypothetical model of brain functions in speech perception and production (adapted from Fant, 1967 [53]).

A key to this two-phase concept is the assumption that for each specific phonetic element only certain aspects of the speech signal are the most relevant, and what needs to be theoretically determined is whether the critical aspects are auditory or motor in nature. This implies that the non-critical properties (i.e., those beyond even the enhancing features) are redundant and are therefore not taken into account in perception. Even though there is also recognition that certain minor cues can be useful [54], it is generally felt that the minor cues are not nearly as important. There is little discussion, however, as to how the perception system can learn what cues to focus on and what cues to ignore.

An alternative possibility, as explored in the present study, is that raw acoustic signals of connected speech, after an initial segmentation into syllable sized chunks, can be processed as a whole to directly recognize the relevant phonetic elements, such as consonants, vowels, and tones. This process does not consist of a phase in which auditory or articulatory features are explicitly identified and then used as input to the recognition of the units at the phonemic level. There are a number of reasons why such a direct perception of phonetic categories may be effective. First, given that speakers differ extensively in terms of static articulatory configurations such as vocal tract length, articulator size, and length and thickness of the vocal folds, greater commonality could be in the dynamics of the articulatory trajectories, which is constrained by physical laws. The dynamic nature of continuous speech [55–58] means that it is intrinsically difficult to find the optimal time points at which discrete features can be extracted from the continuous articulatory or acoustic trajectories. Second, there is evidence that continuous articulatory (and the resulting acoustic) movements are divided into syllable-sized unidirectional target approximation movements [59–61] or gestures [62]. This suggests that processing syllable-sized acoustic signals could be an effective perceptual strategy to capture the full details of all the relevant information about contrastive phonetic units such as consonants, vowels, and tones. Finally, a seemingly trivial but in fact critical reason is that, if detailed acoustic signals are all available to the auditory system, should perception throw away any part of the signal that is potentially helpful? The answer to this question would be no, according to the data processing theorem, also known as data processing inequality [63]. This is an information theoretic concept that states that the information content of a signal cannot be increased via data processing:

$$\text{If } X \rightarrow Y \rightarrow Z \text{ (Markov chain), then } I(X;Y) \geq I(X;Z),\ I(Y;Z) \geq I(X;Z). \tag{1}$$

$$\text{Equality if } I(X;Y \mid Z) = 0.$$

where X, Y and Z form a Markov Chain (a stochastic process consisting of a sequence of events, where the probability of each event depends on the state of the previous event). X is the input, Z is the processed output, and Y is the only path to convert X to Z. What this says is that whenever data is processed, some information is lost. In the best-case scenario, the equality could still largely hold when some information is lost but no processing can increase the amount of original information. In general, the more data processing, the greater the information loss. An extraction of intermediate features before phonetic recognition, therefore, would necessarily involve more processing than direct recognition of phonetic categories from raw acoustic signals.

There have already been some theories that favor relative direct or holistic speech perception. Direct realism [64], for example, argues that speech perception involves direct recognition of articulatory gestures, without the intermediatory of explicit representation of auditory features. However, because gestures are also sub-phonemic (Figure 1), an extra step is still needed to convert them to syllables and words. The exemplar theories [65–67] also postulate that in both production and perception, information about particular instances (episodic information) as a whole is stored. Categorization of an input is accomplished by comparison with all remembered instances of each category. It is suggested that people use already-encountered memories to determine categorization, rather than creating an additional abstract summary of representations. In exemplar models of phonology, phonological structures, including syllables, segments and even sub-segmental features emerge from the phonetic properties of words or larger units [67–69], which implies that units larger than phonemes are processed as a whole. The exemplar theories, however, have not been highly specific on how exactly such recognition is achieved.

In fact, there is a general lack of step-by-step procedural account of any of the theoretical frameworks on speech perception that starts from the processing of continuous acoustic signals. Auditory theories have presented no demonstration of how exactly continuous acoustic signals are converted into auditory cues in the first phase of perception, and how these representations are translated into consonants, vowels, and tones. Motor theory has suggested that gestures can be detected from acoustic signals through analysis by synthesis [31], but this has not yet been tested in perception studies, and has remained only as a theoretical conjecture. What is needed is to go beyond purely theoretical discussion of what is logically plausible, and start to test computationally what may actually work. For this purpose, it is worth noting that computational speech recognition has been going on for decades, in research and development in speech technology. As a matter of fact, automatic speech recognition (ASR) has been one of the most successful areas in speech technologies [70,71]. Zhang et al. [72], for example, reported word-error-rates (WERs) as low as 1.4%/2.6%.

The state of the art in speech recognition, however, does not use intermediate feature extraction as a core technology. Instead, units like diphones or triphones are directly recognized from continuous acoustic signals [73–77]. There have been some attempts to make use of features in automatic speech recognition. For example, the landmark-based approach tries to extract distinctive features from around acoustic landmarks such as the onset or offset of consonant closure, which can then be used to make choices between candidate segments to be recognized [14,78–80]. In most cases, however, systems using landmarks or distinctive features are knowledge-based, and the detected features are used as one kind of feature added on top of other linguistic features and acoustic features to facilitate the recognition of phonemes [81,82]. In this kind of process, there is no test of the effectiveness of distinctive features relative to other features. Some other automatic speech recognition systems use acoustic properties around the landmarks, but without converting them to any featural representations [83–86]. What is more, those recognition systems still use phoneme as the basic unit, which implies that phoneme is the basic functional speech unit, and units under phonemes do not have to be categorical. There have also been systems that make some use of articulatory features. However, there is no system that we know of that performs phonetic recognition entirely based on articulatory gestures extracted from acoustic signals.

Feature-to-percept, therefore, is questionable as a general strategy of speech perception, especially given the possibility of direct phonetic perception as an alternative. There has not yet been any direct comparisons of the two strategies, however. In light of the data processing theorem in Equation (1), both strategies would involve data processing that may lead to information loss. In addition, a strategy that can generate better perceptual accuracy could be computationally too costly. Therefore, in this study, a set of modelling experiments are conducted to compare the two perceptual strategies, measured in terms of recognition accuracy and computational cost. As the very first such effort, the object of recognition is Mandarin tones, because they involve fewer acoustic dimensions than consonants and vowels, as explained next.

1.3. Tone Recognition: A Test Case

In languages like Mandarin, Yoruba, and Thai, words are distinguished from each other not only by consonants and vowels, but also by pitch patterns known as tones. Tone in these languages therefore serves a contrastive function like consonants and vowels. Syllables with the same CV structure can represent different words when the pitch profiles in the syllable vary. Although tonal contrasts are sometimes also accompanied by differences in consonants, vowels and voice quality, pitch patterns provide both sufficient and dominant cues for the identification of tones [59,87].

How to define tonal contrasts is a long-standing issue for tone language studies. In general, efforts have predominantly focused on establishing the best featural representation of tones, starting from Wang's [88] binary tone features in the style of distinctive features of Jakobson et al. [7]. Later development has moved away from simple binary features. For East Asian languages, a broadly accepted practice is to use a five-level system [89] which assumes that five discrete levels are sufficient to distinguish all the tones of many languages. Also different from the classical feature theory, the five-level system represents pitch changes over time by denoting each tone with two temporal points. The four tones of Mandarin, for example, can be represented as 55—Tone 1, 35—Tone 2, 214—Tone 3, and 51—Tone 4, where a greater number indicates a higher pitch. Two points per tone is also widely used for African tone languages [90–92], although for those languages usually only up to three pitch levels, High, Mid, Low, are used. Morén and Zsiga [93] and Zsiga and Nitisaroj [94] even claimed that for Thai, only one target point per tone is needed for connected speech. There has also been a long-standing debate over whether pitch level alone is sufficient to represent all tones, or slope and contour specifications are also needed as part of the representation [4,5]. There are also alternative schemes that try to represent tone contours, such as the T value method, LZ value method [95–97], etc., but they also focus on abstracting the pitch contours into several discrete levels.

Under the feature-to-percept assumption, the two-point + five-level tone representation would mean that, to perceive a tone, listeners need to first determine if the pitch level is any of the five levels at each of the two temporal locations, so as to derive at a representation in the form of, e.g., 55, 35, 21 or 51. Those representations would then lead to the recognition of the tones. In such a recognition process, the key is to first detect discrete pitch levels at specific temporal locations before tone recognition. A conceivable difficulty with such tone feature detection is the well-known extensive amount of contextual variability. For example, due to inertia, much of the pitch contours of a tone varies heavily with the preceding tone, and it is only near the end of the syllable that the underlying tonal targets are best approached [98,99]. This would make tone level detection hard, at least for the first of the two temporal locations.

An alternative to the feature-to-tone scheme, based on the direct phonetic perception hypothesis, is to process continuous f_0 contour of each tone-carrying syllable as a whole without derivation of intermediate featural representations. The plausibility of holistic tone processing can be seen in the success of tone recognition in speech technology. The state-of-the-art automatic tone recognition can be as accurate as 94.5% on continuous speech [100], with no extraction of intermediate tonal features. In fact, the current trend is to process as many sources of raw acoustics as possible, including many non-f_0 dimensions in complex models [100,101]. This suggests that maximization of signal processing rather than isolation of distinctive cues may be the key to tone recognition.

Tone recognition would therefore serve as a test case for comparing direct and two-phase perception. But the speech-technology-oriented approach of using as many acoustic properties as possible makes it hard to isolate the key differences between the two approaches. Given that f_0 alone is sufficient to convey most tonal contrasts in perception as mentioned above, in this study we will use a computational tone recognition task that processes raw f_0 contours in connected Mandarin speech, with the aim to test if the perception of phonetic categories is more likely a two-phase feature-to-percept process or a single-phase direct acoustic decoding process. We will apply two machine learning algorithms to process Mandarin tones from syllable-sized f_0 contours extracted from a connected speech corpus in ways that parallel different tone perception hypotheses, including direct perception, pitch level extraction, pitch profile features, and underlying articulatory targets.

2. Methods and Materials

The overall method is to use computational models to simulate tone perception as a semi-automatic recognition task by training them with syllable-sized f_0 contours in connected speech. The perception strategies under comparison are simulated by different ways of processing the raw f_0 contours, and the efficacy of each strategy is estimated in terms of recognition rate.

2.1. Recognition Models

Two recognition models are applied to recognize Mandarin tones. One is a supervised model, Support Vector Machine (SVM), which can be used to simulate conditions where learners already know the tone inventory of the language. As we only have f_0 values as input, there is no need to use very complex models to train the data. The other model is an unsupervised model, Self-Organizing Map (SOM), which can be used to simulate conditions where learners have no knowledge of the tonal inventory in the language.

As shown in the experimental results to be reported later, the recognition rates achieved by the SOM model were much lower than those achieved by the SVM model, despite the promising results reported previously [102]. In addition, although SOM can simulate clustering of patterned data like f_0 trajectories, it is difficult to simulate the extraction of abstract features. Therefore, SOM is applied only in the tone recognition experiments based on pitch contours (full f_0 contours and pitch level detection).

2.1.1. Support Vector Machine (SVM)

SVM is a supervised machine learning model developed for binary classification tasks. An SVM model is a representation of the examples as points in space, mapped so that the examples of the separate categories are divided by a clear hyperplane or gap that is as wide as possible. New examples are then mapped into that same space and predicted to belong to a category based on which side of the gap they fall. Visually, f_0 contours of each tone may consist of different patterns and so can be mapped in different spaces as a whole in the training phase. During the testing phase, every sampled contour will get a probability of each tone category and be predicted as a certain tone. Then, we could get an average accuracy of each tone. In the application of SVM, all the training samples and testing samples are converted to D-dimensional vectors and labelled with +1 or -1. A simple functional margin can be: $f(x) = sign(W^t x + b)$. If the label is +1, $W^t x + b$ is expected to be larger than 0, otherwise it is smaller than 0. The weight W is a combination of a subset of the training examples and shows how each dimension of the vectors is used in the classification process. The vectors in this subset are called support vectors. In our experiment, one f_0 contour is one sample consisting of 30 sample points, and treated as a 30-dimention vector. This is done with the LibSVM tool [103] with RBF kernel. It generalizes the binary classification to a n-class classifier that splits the task into $n(n-1)/2$ binary tasks and the solutions are combined by a voting strategy [104]. Five-fold cross-validations were applied on the training set automatically and randomly during training to optimize the model, and the classification accuracy of the testing set will be shown in the Results section to compare the performance of each model. The training and testing set will be introduced later in Section 2.2.

2.1.2. Self-Organizing Map (SOM)

In contrast to SVM, SOM is an unsupervised machine learning algorithm that projects high-dimensional input space onto a discrete lower dimensional array of topologically ordered processing units. During this training process, the SOM model compresses the information while keeping the geometric relationships among input data. In the tone recognition task based on full f_0 contours, the networks were designed to contain 100 units/prototypes, and all the f_0 contours were put into the training model. After many iterations, each contour tends to approximate a certain unit and all the f_0 contours are finally mapped onto the 10×10 prototypes. Observing the trained units, we could see that neighboured units are gradually varied and the clusters of units share similar characteristics based on f_0 contours.

After training, every unit will have a tone property calculated by a firing frequency matrix. A unit with the probability of 68% or above for a tone is considered as categorized as that tone. During the testing phase, each f_0 contour in the testing data was mapped onto a unit which means this contour was recognized as that tone. This categorization process is done with the "kohonen-package" in R [105].

For any highly abstract features to work, one of the first critical steps is to extract them from observations through identification and naming. This is not a trivial task, and its effectiveness can be shown only in terms of the ultimate rate of recognition of the phonetic category. For the five-level tone representation system and the two-level distinctive feature system mentioned earlier, pitch levels can be detected or recognized using SVM or SOM from f_0 contours and then transformed into tone by simple mapping. For more abstract features like pitch profile features and underlying articulatory targets, features need to be extracted first in a particular model and then put into the tone recognition system (SVM).

2.2. Material

The data were syllable-sized f_0 contours produced by four female and four male Mandarin speakers [98]. Each token is a 30 equidistant (hence time-normalized) discrete point vector taken from either the first or second syllable of a disyllabic tone sequence in the middle position of a carrier sentence. There was no differentiation of the tokens from the first and second syllables, leaving the information of syllable position in word/phrase unrepresented. Two frequency scales were used to represent f_0, Hertz and semitones. The latter was converted from Hertz with the following equation:

$$\text{semitone} = log_2(f_0) \times 12 \tag{2}$$

where the reference f_0 is assumed to be 1 Hz for all speakers. Note that this kind of raw data (i.e., without applying a normalization scheme such as Z-score transformation, c.f., [106]) leave most of the individual differences in pitch height intact, particularly between the female and male speakers, as can be seen in the plots of f_0 contours in Figures 2 and 3.

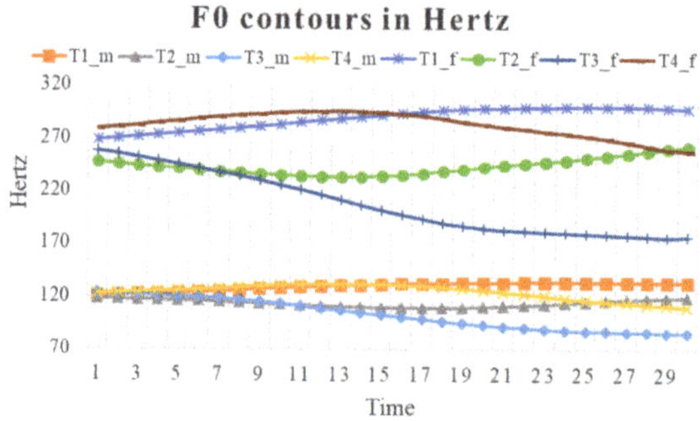

Figure 2. Mean time-normalized syllable-sized f_0 contours of four Mandarin tones, averaged separately for female and male speakers.

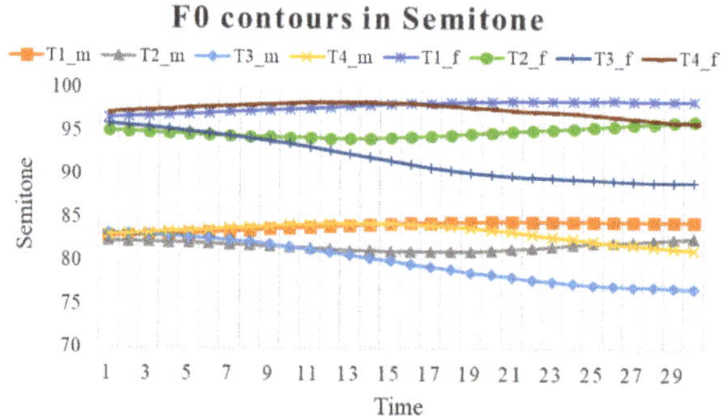

Figure 3. Mean time-normalized syllable-sized semitone contours of four Mandarin tones, averaged separately for female and male speakers.

There were a total of 1408 tokens of Tone 1 (high-level), 1408 tokens of Tone 2 (rising), 1232 tokens of Tone 3 (low), and 1408 tokens of Tone 4 (falling). The fewer tokens of Tone 3 were because those of the first syllable followed by another Tone 3 were excluded to circumvent the problem of the well-known tone sandhi rule, which turns the first Tone 3 to be similar [89,98] though not identical [107] to Tone 2. The whole dataset was then divided into a training subset and a testing subset randomly, with a ratio of 2:1.

2.3. Overall Design

Five modelling experiments were set up to compare the efficacy of different tone recognition schemes, both cross-gender and cross-speaker. The first experiment used raw f_0 contours of the four Mandarin tones to train an SVM model and an SOM model, respectively, which were then used to classify the tone categories. The second experiment, pitch level detection, again used raw f_0 contours to train a SVM model and a SOM model, respectively. But this time, the models were used to detect pitch levels, which were then mapped to tone categories. In the two subsequent experiments, different tonal features were extracted from the raw f_0 contours, and were then used to train the SVM models. The extracted features were then used to recognize the tones. The extracted tonal features were ordered from the most to the least adherent to the feature-to-percept paradigm:

1. Pitch height (2 levels and 5 levels);
2. f_0 profile (slope etc.);
3. Underlying pitch targets (quantitative target approximation (qTA) parameters).

3. Results

3.1. Experiment 1—Full f_0 Contour

In the first experiment, the raw f_0 contours were used to both train the tone recognition model and test the performance. As shown in Table 1, with raw f_0 data, tone recognition rates based on SVM model were very high. In the mixed-gender condition, the recognition rates were 97.4% for contours in semitones and 86.3% for contours in Hertz. In the male-only condition, the recognition rate reached 99.1% for contours in semitones. In the female-only condition, the recognition rate was 96.6%. The much lower recognition rates for contours in Hz is not surprising, as the logarithmic conversion in calculating semitones has effectively normalized the vertical span of the pitch range, with only individual differences in pitch height still retained. The performances of the SOM model are lower than that of SVM, but even the rates in the mixed-gender condition were all above 70%. In later experiments, we will only focus on mixed-gender conditions.

Table 1. Tone recognition rates using raw f_0 contours based on SVM and SOM models.

	SVM		SOM	
	Hertz	**Semitone**	**Hertz**	**Semitone**
Male	96.0%	99.1%	89.7%	90.8%
Female	76.7%	96.6%	76.7%	77.6%
All	86.3%	97.4%	72.8%	72.0%

SVM: Support Vector Machine; SOM: Self-Organizing Map.

Table 2 is the tone confusion matrix of f_0 contours in semitones. The performance of the tone classification is similar to the human tone recognition reported by McLoughlin et al. [108] shown in Table 3. The corpus they used has a context-free carrier sentence structure that is similar to that used in the present study. Similar to the results in Table 2, their recognition rate is the highest for Tone 3 and lowest for Tone 4.

Table 2. Tone confusion matrix using semitone of mixed-gender based on SVM.

	T1	T2	T3	T4
T1	98.2%	0.2%	0.5%	1.1%
T2	0.9%	96.6%	0.9%	1.6%
T3	0.3%	1.0%	98.4%	0.3%
T4	2.7%	0.0%	0.7%	96.6%

T means Tone here.

Table 3. Tone confusion matrix context-free words in AWGN-corrupted spoken sentences [108].

	T1	T2	T3	T4
T1	95.68%	2.03%	2.08%	0.21%
T2	1.24%	97.92%	0.08%	0.76%
T3	0.35%	0.42%	99.02%	0.21%
T4	2.63%	1.72%	1.38%	94.27%

3.2. Experiments 2–3: Pitch Level Representation

In this experiment, we abstracted the f_0 contours into a two-position height representation. We tested both a two-level (distinctive-feature style) and a five-level abstraction that would correspond to two popular featural representations of tones [88,89].

3.2.1. Experiment 2—Distinctive Feature Style (Two Level) Representation

In a distinctive feature system, Mandarin tones can be represented by two levels: high and low. The four lexical tones of Mandarin can be represented as 11—Tone 1, 01—Tone 2, 00—Tone 3, and 10—Tone 4. '1' means high and '0' means low. In our implementation of this featural representation system, each f_0 contour was split into two halves and each was labelled high or low. The first 15 points of the contours were labelled as 1—Tone 1, 0—Tone 2, 0—Tone 3, 1—Tone 4, and the later 15 points are labelled as 1—Tone 1, 1—Tone 2, 0—Tone 3, 0—Tone 4. Two sub-experiments were conducted. One is training and testing the two halves separately, and the other is training and testing the two halves together. The models used were SVM and SOM. In the first sub-experiment, after the classification, the results of the two halves are combined and checked.

Table 4 shows tone recognition rates of this experiment. The rates are 93.67% and 92.90% for the separate and together sub-experiments, respectively, based on the SVM model. Assuming that distinctive features of tones are unknown knowledge until after learning, SOM is more comparable to human tone perception than SVM. The recognition rates are 80.59% and 82.82% for the separate and together sub-experiments, respectively.

Table 4. Tone recognition rates using two-level abstraction based on SVM and SOM models.

	Separate		Together	
	Hertz	Semitone	Hertz	Semitone
SVM	93.5%	93.7%	91.9%	92.9%
SOM	80.8%	80.6%	81.0%	82.8%

SVM: Support Vector Machine; SOM: Self-Organizing Map.

3.2.2. Experiment 3—Five-Level Representation

In a five-level, hence non-binary, pitch level representation, the four lexical tones of Mandarin can be represented as 55—Tone 1, 35—Tone 2, 21—Tone 3, and 53—Tone 4, as shown in Figure 4. In the featural representation of this system, each f_0 contour is again split into two halves, each consisting of 15 points. Again, two sub-experiments were conducted. One is training and testing the two halves separately, and the other is training and testing them together. The models used were SVM and SOM. In the first sub-experiment, after the classification, the results of the two halves are combined and checked.

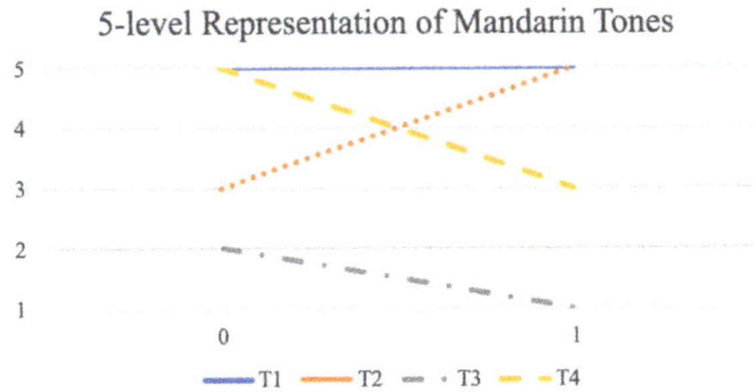

Figure 4. Five-level Representation of Mandarin Tones.

Table 5 shows recognition rates of both SVM and SOM. For SVM, when the two halves of the f_0 contours are trained separately, the recognition rate reached around 90%. When the two halves are trained together, the rate dropped to just above 80%. The results of SOM are even lower than SVM. The recognition rates are 58.9% and 43.7% for the separate and together sub-experiments, respectively.

Table 5. Tone recognition rates using five-level abstraction based on SVM and SOM models.

	Separate		Together	
	Hertz	**Semitone**	**Hertz**	**Semitone**
SVM	88.8%	90.3%	80.5%	84.1%
SOM	56.7%	58.9%	43.5%	43.7%

SVM: Support Vector Machine; SOM: Self-Organizing Map.

3.3. Experiment 4—f_0 Profile Representation

Besides the discrete representations tested so far, there are also schemes that use mathematical functions to represent f_0 profiles with parameters with continuous values. A recent study explored fitting the tone contours with two mathematical functions, parabola and broken-line (BL) and concluded that three of the cues obtained in the parabola fitting were broadly effective: mean f_0, slope, and curve [109]. In this experiment we tested the effectiveness of fitting both functions to the f_0 contours in the test corpus in the least-squared sense. The expression of the parabola is as follows:

$$f(t) \approx c_0 + c_1\left(t - \frac{1}{2}\right) + c_2[\left(t - \frac{1}{2}\right)^2 - 1/12] \tag{3}$$

The expression of BL is as follows:

$$f(t) \approx \begin{cases} a_1 + b_1 t, & t < d \\ a_2 + b_2 t, & t \geq d \end{cases} \quad (d \text{ is the position of breakpoint}) \tag{4}$$

The features we used for testing were the top five pairs of features reported in Tupper et al. [109] for maximizing classification accuracy, as follows:

- Slope: c_1 in the parabola fit;
- Curve: c_2 in the parabola fit, which is one half the second derivative of the fitted f_0 contour;
- Onglide: difference between f_0 at contour onset and breakpoint in the BL fit;
- Offglide: difference between f_0 at breakpoint and contour offset in the BL fit;
- Overall: difference between f_0 at contour onset and offset in BL fit.

The features extracted from f_0 contours are trained by the SVM model. Table 6 shows the recognition rates of the top five pairs of features used in Tupper et al. [109]. The best results for mixed genders are 92.3% in semitones and 89.3% in hertz, both of which are based on slope + curve.

Table 6. Tone recognition rates using f_0 profile features based on SVM model.

	Herz	Semitone
Slope + Curve	89.3%	92.3%
Curve + Overall	85.9%	90.4%
Slope + Onglide	68.3%	66.7%
Onglide + Offglide	71.8%	75.7%
Offglide + Overall	70.4%	75.1%

3.4. *Experiment 5—qTA Articulatory Feature Extraction*

qTA is another mathematic function also capable of representing tonal contours [59]. The model is based on the assumption that speech articulation is a mechanical process of target approximation that can be simulated by a critically damped spring-mass system (similar to the command-response model [110] and the task dynamic model [111]). The model can be fitted to not only tonal contours in continuous speech, but also intonational contours carrying multiple communicative functions [107]. QTA's ability to simulate the articulatory process of generating tonal contours makes it an ideal model to test the motor theory, according to which speech perception is a process of detecting the articulatory gestures that generate the speech signals [31] through analysis-by-synthesis [112]. Analysis-by-synthesis is a process of analysing a signal by reproducing it, and it has been successfully applied in previous modelling works with qTA [59,107,113]. In this experiment, we used analysis-by-synthesis to fit qTA to the tonal contours in the same corpus used in the other experiments.

qTA assumes that the f_0 contour of each syllable is generated with a single pitch target, defined by the linear function,

$$x(t) = mt + b \tag{5}$$

where m (in st/s) and b (in st) denote the slope and offset of the underlying pitch target, respectively. The surface f_0 is modeled as the system response driven by the pitch target,

$$f_0\,(t) = (mt + b) + \left(c_0 + c_1\,t + \cdots + c_{N-1}t^{N-1}\right)e^{-t/\tau} \tag{6}$$

where the time constant τ (in s) represents the strength of the target approximation movement.

The values of m, b, and τ (referred to as qTA parameters) can be determined by fitting the original pitch contour in the least-squares sense. We used Target Optimizer [114] to extract qTA parameters. The Target Optimizer internally converts the f_0 samples from Hz scale to semitone scale and normalizes them by subtracting the mean values of the whole utterance. The three estimated qTA parameters were then used as input to a tone recognizer.

In qTA, the offset f_0 of the preceding syllable is transferred to the current syllable to become its onset f_0 to simulate the effect of inertia. Therefore, the onset f_0 of a syllable is expected to be potentially relevant to tone recognition, as it carries contextual tonal information. In the second training condition, therefore, this onset f_0 was added to the qTA parameters to form a four-dimensional input feature for each syllable. These four-dimensional features are then used as input to the SVM model for tone recognition.

Table 7 shows the performance of qTA features from the two training conditions. With the three-dimensional features, the recognition accuracy was 90.7%. With the four-dimensional features, which included the f_0 onset parameter, the accuracy increased to 97.1%.

Table 7. Tone recognition rates using qTA (quantitative target approximation) features.

Features	Accuracy
3-dim qTA parameters	90.7%
3-dim qTA parameters plus f_0 onset value	97.1%

4. Discussion

In the five experiments, we tested whether direct phonetic perception or feature-to-percept is a more likely mechanism of speech perception. All the tests were done by applying the SVM and/or SOM model with either full f_0 contours or various extracted f_0 features as the training and testing data. Except for qTA (due to model-internal setting), all the models were tested with f_0 in both Hz and semitone scales. The performance of the models was assessed in terms of tone recognition rate. In all the experiments the recognition was consistently better for the SVM model than for the SOM model, and better with the semitone scale than the Hz scale. To make a fair comparison of the all the models, a summary of the best performances in all five experiments based on SVM in semitones is shown Figure 5. As can be seen, the highest recognition rate, 97.4%, was achieved in the full f_0 contour condition in Experiment 1. With pitch level features extracted in Experiments 2–3, recognition rates of 93.7% and 90.3% were achieved for the two-level and five-level conditions, respectively. These are fairly good, but are well below the top recognition rate in Experiment 1. Experiment 4 tested two mathematical (parabola and broken-line) representations of f_0 profiles, which, unlike the discrete pitch level features in Experiments 2–3, have continuous values. The highest recognition rate of 92.3% was achieved for the combination of slope and curve. This is very close to the best recognition rate of 93.7% with the two-level condition in Experiment 2. Another continuous parametric representation tested in Experiment 5, namely, qTA parameters based on [59], achieved a very high recognition rate of 97.1% when initial f_0 was included as a fourth parameter, which is almost as high as the benchmark of 97.4% in Experiment 1. Without the initial f_0, however, the recognition rate was only 90.7%. It is worth noting, however, that the initial f_0 is actually included in the full f_0 contour in Experiment 1, as it is just the first f_0 point in an f_0 contour.

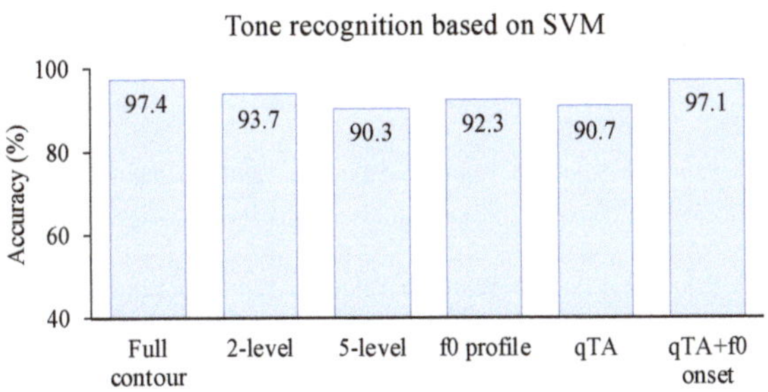

Figure 5. Summary of tone recognition rates based on SVM model for full f_0 contour (Experiment 1), two-level feature (Experiment 2), five-level feature (Experiment 3), f_0 profile (Experiment 4) and qTA and qTA + f_0 onset (Experiments 5). SVM: Support Vector Machine.

The fairly high tone recognition rates from the best performances in all the five experiments are rather remarkable, given that the f_0 contours used were extracted from fluent speech [51] in multiple tonal contexts and two different syllable positions, yet no contextual or positional information was provided during either training or testing, contrary to the common practice of including tonal context as an input feature in speech technology [101,115,116]. This means that, despite the extensive variability, tones produced in contexts by multiple speakers of both genders are still sufficiently distinct to allow a pattern recognition model (SVM) to accurately identify the tonal categories based on syllable-sized f_0 contours alone. In other words, once implemented as trainable systems, most theory-motivated schemes may be able to perform phonetic recognition to some extent, though with varying levels of success. Thus, the acoustic variability that has prompted much of the early theoretical debates [30,117] does not seem to pose an impenetrable barrier. Instead, there seems to be plenty of consistency underneath the apparent variability for any recognition scheme to capture.

On the other hand, it is still the case that the tone recognition rates achieved by most of the feature extraction schemes in Experiments 2–5 were lower than that of the full f_0 contour baseline in Experiment 1. Only the qTA + initial f_0 scheme nearly matched the full f_0 contour performance. Therefore, for both the qTA + initial f_0 condition and the other feature extraction schemes, a further question is whether the extra processing required by the extraction of the intermediate features is cost-effective when compared to direct processing of full f_0 contours. One way to compare the cost-effectiveness of different tone recognition schemes is to calculate their time complexity [118] in addition to their recognition rates. Time complexity is the amount of time needed to run an algorithm on a computer, as a function of the size of the input. It measures the time taken to execute each statement of the code in an algorithm and gives information about the variation in execution time when the number of operations changes in an algorithm. It is difficult to compute this function exactly, so it is commonly defined in terms of an asymptotic behaviour of the complexity. Time complexity is expressed with the big O notation, $O[n]$, where n is the size of the input data and O is the order of the relation between n and the number of operations performed. Taking the SVM model as an example, the time complexity of the SVM model at testing phase is a function that involves a loop within a loop, which is $O(d) \times O(m) = O(d \times m)$, where d is the number of dimensions of input data and m is the number of categories. When two models, A and B, are compared, if A is better than or equal to B in performance, and has a lower time complexity, A can be said to be a better model than B. If A and B differ clearly in performance, but has a lower time complexity, its performance needs to be balanced against time complexity when deciding which model is better. Table 8 shows the time complexity of all the tone recognition schemes tested in Experiments 1–5.

Table 8. Time complexity of different tone recognition schemes.

Scheme	Method	Size of Input	No. Steps	Time Complexity at Testing Phase
Full f_0 contour	SVM	30 points	1	O(30 × 4)
2 f_0 levels	SVM × 2	15 points	3	O(15 × 2) × 2 + 1
	Matching	2 features		
5 f_0 levels	SVM × 2	15 points	3	O(15 × 5) × 2
	Matching	2 features		
f_0 profile features	Parabola/Broken Line	30 points	2	O(30^3) + O(2 × 4)
	SVM	2 features		
qTA features	qTA Extraction	30 points	2	O(30^3) + O(3 × 4)
	SVM	3 features		

SVM: Support Vector Machine. Qta: quantitative target approximation.

As can be seen, most feature extraction schemes have greater time complexity than the baseline full f_0 contour scheme. The only feature extraction scheme with lower time complexity is the two-level condition. However, its tone recognition accuracy is 3.7% lower than the full f_0 contour condition, as shown in Figure 5. Therefore, its reduced time complexity was not beneficial. In addition, as found in Chen and Xu [119], the time complexity of full f_0 contour scheme does not need to be as high as in Experiment 1, because the temporal resolution of f_0 contours could be greatly reduced without lowering tone recognition accuracy. When the number of f_0 points were reduced to as few as 3, the tone recognition rate was still 95.7%, only a 1.7% drop from the 30-point condition. Therefore, compared to 3 points per contour, the two-level feature extraction would even lose its advantage in time complexity. Overall, therefore, there is no advantage in cost-effectiveness in any of the features extraction schemes over the full f_0 contour scheme.

Worth particular mentioning is the qTA scheme tested in Experiment 5, as it served as a test case for the motor theory of speech perception [31]. The theory assumes that speech perception is a process of recovering articulatory gestures, and the recovery is done by listeners using their own articulatory system to perform analysis-by-synthesis. However, analysis-by-synthesis is a time-consuming process of testing numerous candidate model parameters until an optimal fit is found. As shown in Table 8, the qTA scheme has the greatest time complexity of all the feature extraction schemes. Although its tone recognition accuracy was nearly as high as that of full f_0 contours benchmark when initial f_0 was included as the fourth parameter, one may have to wonder why speech perception would develop such an effortful strategy when direct processing of raw f_0 contours can already achieve top performance at a much lower computational cost.

An implication of the experiments in the present study is that listeners' sensitivity to certain feature-like properties, such as f_0 slope [5], height [4] or alignment of turning point [120,121] does not necessarily mean that those properties are separately extracted during perception. Rather, the sensitivity patterns are what can be observed when various acoustic dimensions are independently manipulated under laboratory conditions. They do not necessarily tell us how speech perception operates. The step-by-step modelling simulations conducted in the current study demonstrate that there may be no need to focus on any specific features. Instead, the process of recognition training allows the perception system to learn how to make use of all the relevant phonetic properties, both major and minor, to achieve optimal phonetic recognition. The dynamic learning operation may in fact reflect how phonetic categories are developed in the first place. That is, speech production and perception probably form a reciprocal feedback loop that guarantees that articulation generates sufficiently distinct cues that perception can make use of during decoding. As a result, those articulatory gestures that can produce the greatest number of effective phonetic cues would tend to be retained in a language. At the same time, perceptual strategies would tend to be those that can make the fullest, and the most economical, use of all the available acoustic cues from detailed acoustic signals.

Finally, a few caveats and clarifications are in order. First, the tone recognition rates obtained in the present study may not directly reflect perception efficacy in real life. On the one hand, they could be too high because the f_0 contours tested here did not contain some of the known adverse effects like consonantal perturbation of f_0 [122,123], intonational confounds [99,124], etc. On the other hand, they could also be too low because not all tonal information is carried by f_0. Listeners are also known to make use of other cues such as duration, intensity, voice quality, etc. [100,101]. Second, there is a need to make a distinction between data transformation and feature extraction. Conversion of f_0 from Hz to semitones and from waveform to MFCC are examples of data transformation. Both have been shown to be beneficial [125,126], and are likely equivalent to certain signal processing performed by the human auditory system. They are therefore different from the feature extraction schemes tested in the present study. In addition, there is another kind of data processing that has been highly recommended [106,127], namely, speaker/data normalization schemes in the frequency dimension such as Z-score transformation (rather than in the temporal dimension). The justification is based on the need to handle variability within and especially across speakers. The difficulty from an operational perspective is that Z-score transformation is based on the total pitch range of multiple speakers in previously processed data. However, Z-score would be hard to compute when processing data from a new speaker, which happens frequently in real life. Furthermore, the present results have shown that, once an operational model is developed, explicit speaker normalization is not really needed, as the training process is already one of learning to handle variability, and the results showed that all models were capable of resolving this problem to various extends. Finally, the present findings do not suggest that a data representation of up to 30 points per syllable is necessary for tone recognition from continuous speech. As mentioned earlier, in a separate study [119] we found that just three f_0 points (taken from the beginning, middle and end of a syllable) are enough for a model equivalent to the full contour model in Experiment 1 to achieve a tone recognition rate close to that of 30 f_0 points, so long as the data points are in the original f_0 values rather than discretized pitch height bands. The study also found that discretization of continuous acoustic signal into categorical values (equivalent to reduction of frequency resolution), which is prototypical of featural representations, is the most likely to adversely affect tone recognition. In other words, a temporal resolution of up to 30 samples per syllable as tested in the present study is not always needed, and may in fact be excessive when time complexity is taken into consideration, whereas high precision of data representation, which is exactly the opposite of featural representation, may be the most important guarantee of effective speech perception.

5. Conclusions

We have used tone recognition as a test case for a re-examination of the widely assumed feature-to-percept assumption about speech perception. Through computational simulation of tone recognition that applied step-by-step modelling procedures, we put various theoretical accounts of speech perception to test by making all of them process continuous acoustic signals. The results show that syllable-sized f_0 contours can be used to directly train pattern recognition models to achieve high tone recognition rates, without extracting intermediate features. In comparison, extracting discrete pitch levels or continuous profile features from the original f_0 contours resulted in reduced rates of tone recognition. Furthermore, when articulatory-based qTA parameters were extracted through analysis-by-synthesis, an operation reminiscent of the motor theory of perception, the recognition rate approached that of original f_0 contours only when syllable-initial f_0 was used as an additional parameter. Finally, we showed through calculation of time complexity relative to model performance that all the feature extraction schemes are less cost effective than the full f_0 contour condition. Based on these findings, we conclude that raw acoustic signal, after certain transformations such as semitone (or MFCC for segments) conversion, can be processed directly in speech perception to recognize phonetic categories. Therefore, feature detection, while useful for analysis and observational purposes, is unlikely to be the core mechanism of speech perception. While the present findings still cannot tell us how exactly speech perception works, they have at least presented a computational reason why real-life speech perception is unlikely a two-phase process, something that is hard to observe through behavioral or state-of-the-art neural investigations alone.

Author Contributions: Conceptualization, Y.C. and Y.X.; methodology, Y.C.; formal analysis, Y.C. and Y.G.; data curation, Y.X. and Y.C.; writing—original draft preparation, Y.C. and Y.X.; writing—review and editing, Y.C., Y.X. and Y.G.; project administration, Y.C. All authors have read and agreed to the published version of the manuscript.

Funding: This research received no external funding.

Institutional Review Board Statement: Not applicable.

Informed Consent Statement: Not applicable.

Data Availability Statement: The original corpus is available upon request from the third author. The data from the simulations are available upon request from the first author.

Conflicts of Interest: The authors declare no conflict of interest.

References

1. Ladefoged, P. What Are Linguistic Sounds Made of? *Language* **1980**, *56*, 485–502. [CrossRef]
2. Wright, R. A Review of Perceptual Cues and Cue Robustness. *Phon. Based Phonol.* **2004**, *34*, 57.
3. Abramson, A.S.; Whalen, D.H. Voice Onset Time (VOT) at 50: Theoretical and Practical Issues in Measuring Voicing Distinctions. *J. Phon.* **2017**, *63*, 75–86. [CrossRef]
4. Abramson, A.S. Static and Dynamic Acoustic Cues in Distinctive Tones. *Lang. Speech* **1978**, *21*, 319–325. [CrossRef]
5. Gandour, J. Perceptual Dimensions of Tone: Evidence from Cantonese. *J. Chin. Linguist.* **1981**, *9*, 20–36.
6. Ladefoged, P.; Johnson, K. *A Course in Phonetics*; Cengage Learning: Boston, MA, USA, 2014.
7. Jakobson, R.; Fant, C.G.; Halle, M. Preliminaries to Speech Analysis: The Distinctive Features and Their Correlates. *Language* **1951**, *29*, 472–481.
8. Jones, D. The History and Meaning of the Term "Phoneme". *Maître Phonétique* **1957**, *35*, 1–20.
9. Trubetzkoy, N.S. *Principles of Phonology*; University of California Press: Berkeley, CA, USA, 1939.
10. Jakobson, R. The Concept of Phoneme. In *On Language*; 1942, reprint; Waugh, L.R., Monique, M.-B., Eds.; Harvard University Press: Cambridge, MA, USA, 1995; pp. 217–241.
11. Chomsky, N.; Halle, M. *The Sound Pattern of English*; Harper & Row: Manhattan, NY, USA, 1968.
12. Clements, G.N. The Geometry of Phonological Features. *Phonology* **1985**, *2*, 225–252. [CrossRef]
13. Jakobson, R.; Halle, M. *Phonology in Relation to Phonteics*; North-Holland Publishing Company: Amsterdam, The Netherlands, 1968.
14. Slifka, J.; Stevens, K.N.; Manuel, S.; Shattuck-Hufnagel, S. A Landmark-Based Model of Speech Perception: History and Recent Developments. *Sound Sense* **2004**, 85–90.
15. Stevens, K.N. Toward a Model for Lexical Access Based on Acoustic Landmarks and Distinctive Features. *J. Acoust. Soc. Am.* **2002**, *111*, 1872–1891. [CrossRef]
16. Flemming, E.S. *Auditory Representations in Phonology*; Routledge: Oxfordshire, UK, 2013.
17. Kingston, J.; Diehl, R.L. Intermediate Properties in the Perception of Distinctive Feature Values. *Pap. Lab. Phonol.* **1995**, *4*, 7–27.
18. Diehl, R.L.; Kluender, K.R. On the Objects of Speech Perception. *Ecol. Psychol.* **1989**, *1*, 121–144. [CrossRef]
19. Kingston, J. The Phonetics and Phonology of Perceptually Motivated Articulatory Covariation. *Lang. Speech* **1992**, *35*, 99–113. [CrossRef] [PubMed]
20. Lotto, A.J.; Kluender, K.R. General Contrast Effects in Speech Perception: Effect of Preceding Liquid on Stop Consonant Identification. *Percept. Psychophys.* **1998**, *60*, 602–619. [CrossRef]
21. Diehl, R.L.; Lotto, A.J.; Holt, L.L. Speech Perception. *Annu. Rev. Psychol.* **2004**, *55*, 149–179. [CrossRef]
22. Stevens, K.N.; Blumstein, S.E. Invariant Cues for Place of Articulation in Stop Consonants. *J. Acoust. Soc. Am.* **1978**, *64*, 1358–1368. [CrossRef]
23. Stevens, K.N.; Keyser, S.J. Quantal Theory, Enhancement and Overlap. *J. Phon.* **2010**, *38*, 10–19. [CrossRef]
24. Stevens, K.N. On the Quantal Nature of Speech. *J. Phon.* **1989**, *17*, 3–45. [CrossRef]
25. Stevens, K.N. The Acoustic/Articulatory Interface. *Acoust. Sci. Technol.* **2005**, *26*, 410–417. [CrossRef]
26. Stevens, K.N.; Keyser, S.J. Primary Features and Their Enhancement in Consonants. *Language* **1989**, *65*, 81–106. [CrossRef]
27. Diehl, R.L.; Kluender, K.R.; Walsh, M.A.; Parker, E.M. Auditory Enhancement in Speech Perception and Phonology. In *Cognition and the Symbolic Processes: Applied and Ecological Perspectives*; Psychology Press: Hove, UK, 1991; pp. 59–76.
28. Lotto, A.J.; Hickok, G.S.; Holt, L.L. Reflections on Mirror Neurons and Speech Perception. *Trends Cogn. Sci.* **2009**, *13*, 110–114. [CrossRef] [PubMed]
29. Galantucci, B.; Fowler, C.A.; Turvey, M.T. The Motor Theory of Speech Perception Reviewed. *Psychon. Bull. Rev.* **2006**, *13*, 361–377. [CrossRef] [PubMed]
30. Liberman, A.M.; Cooper, F.S.; Shankweiler, D.P.; Studdert-Kennedy, M. Perception of the Speech Code. *Psychol. Rev.* **1967**, *74*, 431. [CrossRef] [PubMed]
31. Liberman, A.M.; Mattingly, I.G. The Motor Theory of Speech Perception Revised. *Cognition* **1985**, *21*, 1–36. [CrossRef]
32. Cooper, F.S.; Delattre, P.C.; Liberman, A.M.; Borst, J.M.; Gerstman, L.J. Some Experiments on the Perception of Synthetic Speech Sounds. *J. Acoust. Soc. Am.* **1952**, *24*, 597–606. [CrossRef]
33. Liberman, A.M.; Harris, K.S.; Hoffman, H.S.; Griffith, B.C. The Discrimination of Speech Sounds within and across Phoneme Boundaries. *J. Exp. Psychol.* **1957**, *54*, 358. [CrossRef]
34. Eimas, P.D.; Siqueland, E.R.; Jusczyk, P.; Vigorito, J. Speech Perception in Infants. *Science* **1971**, *171*, 303–306. [CrossRef]
35. Kuhl, P.K.; Miller, J.D. Speech Perception by the Chinchilla: Voiced-Voiceless Distinction in Alveolar PLoSive Consonants. *Science* **1975**, *190*, 69–72. [CrossRef]
36. Damasio, A.R. Aphasia. *N. Engl. J. Med.* **1992**, *326*, 531–539. [CrossRef]
37. Goodglass, H. *Understanding Aphasia*; Academic Press: Cambridge, MA, USA, 1993.
38. Hickok, G.; Okada, K.; Barr, W.; Pa, J.; Rogalsky, C.; Donnelly, K.; Barde, L.; Grant, A. Bilateral Capacity for Speech Sound Processing in Auditory Comprehension: Evidence from Wada Procedures. *Brain Lang.* **2008**, *107*, 179–184. [CrossRef]
39. Fadiga, L.; Craighero, L.; Buccino, G.; Rizzolatti, G. Speech Listening Specifically Modulates the Excitability of Tongue Muscles: A TMS Study. *Eur. J. Neurosci.* **2002**, *15*, 399–402. [CrossRef] [PubMed]
40. Watkins, K.E.; Strafella, A.P.; Paus, T. Seeing and Hearing Speech Excites the Motor System Involved in Speech Production. *Neuropsychologia* **2003**, *41*, 989–994. [CrossRef]

41. Fischer, M.H.; Zwaan, R.A. Embodied Language: A Review of the Role of the Motor System in Language Comprehension. *Q. J. Exp. Psychol.* **2008**, *61*, 825–850. [CrossRef] [PubMed]
42. Hickok, G.; Poeppel, D. The Cortical Organization of Speech Processing. *Nat. Rev. Neurosci.* **2007**, *8*, 393–402. [CrossRef] [PubMed]
43. Pickering, M.J.; Garrod, S. Do People Use Language Production to Make Predictions during Comprehension? *Trends Cogn. Sci.* **2007**, *11*, 105–110. [CrossRef]
44. Pulvermüller, F.; Fadiga, L. Active Perception: Sensorimotor Circuits as a Cortical Basis for Language. *Nat. Rev. Neurosci.* **2010**, *11*, 351–360. [CrossRef]
45. Bartoli, E.; D'Ausilio, A.; Berry, J.; Badino, L.; Bever, T.; Fadiga, L. Listener–Speaker Perceived Distance Predicts the Degree of Motor Contribution to Speech Perception. *Cereb. Cortex* **2015**, *25*, 281–288. [CrossRef]
46. D'Ausilio, A.; Pulvermüller, F.; Salmas, P.; Bufalari, I.; Begliomini, C.; Fadiga, L. The Motor Somatotopy of Speech Perception. *Curr. Biol.* **2009**, *19*, 381–385. [CrossRef]
47. Meister, I.G.; Wilson, S.M.; Deblieck, C.; Wu, A.D.; Iacoboni, M. The Essential Role of Premotor Cortex in Speech Perception. *Curr. Biol.* **2007**, *17*, 1692–1696. [CrossRef]
48. Sato, M.; Tremblay, P.; Gracco, V.L. A Mediating Role of the Premotor Cortex in Phoneme Segmentation. *Brain Lang.* **2009**, *111*, 1–7. [CrossRef]
49. Schmitz, J.; Bartoli, E.; Maffongelli, L.; Fadiga, L.; Sebastian-Galles, N.; D'Ausilio, A. Motor Cortex Compensates for Lack of Sensory and Motor Experience during Auditory Speech Perception. *Neuropsychologia* **2019**, *128*, 290–296. [CrossRef] [PubMed]
50. Birkholz, P.; Kröger, B.J.; Neuschaefer-Rube, C. Synthesis of Breathy, Normal, and Pressed Phonation Using a Two-Mass Model with a Triangular Glottis. In Proceedings of the Twelfth Annual Conference of the International Speech Communication Association, Lorence, Italy, 28–31 August 2011.
51. Xu, Y. Speech as Articulatory Encoding of Communicative Functions. In Proceedings of the 16th International Congress of Phonetic Sciences, Saarbrucken, Germany, 6–10 August 2007; pp. 25–30.
52. Xu, Y.; Wang, Q.E. Pitch Targets and Their Realization: Evidence from Mandarin Chinese. *Speech Commun.* **2001**, *33*, 319–337. [CrossRef]
53. Fant, G. Auditory Patterns of Speech. *Models Percept. Speech Vis.* **1967**, *5*, 111–125.
54. Lisker, L. "Voicing" in English: A Catalogue of Acoustic Features Signaling/b/versus/p/in Trochees. *Lang. Speech* **1986**, *29*, 3–11. [CrossRef] [PubMed]
55. Browman, C.P.; Goldstein, L.M. Towards an Articulatory Phonology. *Phonology* **1986**, *3*, 219–252.
56. Carré, R. Dynamic Properties of an Acoustic Tube: Prediction of Vowel Systems. *Speech Commun.* **2009**, *51*, 26–41. [CrossRef]
57. Fowler, C.A. Coarticulation and Theories of Extrinsic Timing. *J. Phon.* **1980**, *8*, 113–133. [CrossRef]
58. Öhman, S.E.G. Coarticulation in VCV Utterances: Spectrographic Measurements. *J. Acoust. Soc. Am.* **1966**, *39*, 151–168. [CrossRef]
59. Prom-On, S.; Xu, Y.; Thipakorn, B. Modeling Tone and Intonation in Mandarin and English as a Process of Target Approximation. *J. Acoust. Soc. Am.* **2009**, *125*, 405–424. [CrossRef]
60. Xu, Y.; Liu, F. Tonal Alignment, Syllable Structure and Coarticulation: Toward an Integrated Model. *Ital. J. Linguist.* **2006**, *18*, 125.
61. Xu, Y.; Prom-On, S. Economy of Effort or Maximum Rate of Information? Exploring Basic Principles of Articulatory Dynamics. *Front. Psychol.* **2019**, *10*, 2469. [CrossRef] [PubMed]
62. Nam, H.; Goldstein, L.; Saltzman, E. Self-Organization of Syllable Structure: A Coupled Oscillator Model. In *Approaches to Phonological Complexity*; De Gruyter Mouton: Berlin, Germany, 2009; pp. 297–328.
63. Cover, T.M. *Elements of Information Theory*; John Wiley & Sons: Hoboken, NJ, USA, 1999.
64. Fowler, C.A. An Event Approach to the Study of Speech Perception from a Direct–Realist Perspective. *J. Phon.* **1986**, *14*, 3–28. [CrossRef]
65. Hay, J.; Nolan, A.; Drager, K. From *Fush* to *Feesh*: Exemplar Priming in Speech Perception. *Linguist. Rev.* **2006**, *23*, 351–379. [CrossRef]
66. Johnson, K. Resonance in an Exemplar-Based Lexicon: The Emergence of Social Identity and Phonology. *J. Phon.* **2006**, *34*, 485–499. [CrossRef]
67. Pierrehumbert, J.B. Exemplar Dynamics: Word Frequency, Lenition and Contrast. *Typol. Stud. Lang.* **2001**, *45*, 137–158.
68. Lacerda, F. Phonology: An Emergent Consequence of Memory Constraints and Sensory Input. *Read. Writ.* **2003**, *16*, 41–59. [CrossRef]
69. Lindblom, B. Emergent Phonology. In Proceedings of the 25th Annual Meeting of the Berkeley Linguistics Society, Berkeley, CA, USA, 12–15 February 1999; Volume 25, pp. 195–209.
70. Hinton, G.; Deng, L.; Yu, D.; Dahl, G.E.; Mohamed, A.; Jaitly, N.; Senior, A.; Vanhoucke, V.; Nguyen, P.; Sainath, T.N. Deep Neural Networks for Acoustic Modeling in Speech Recognition: The Shared Views of Four Research Groups. *IEEE Signal Process. Mag.* **2012**, *29*, 82–97. [CrossRef]
71. Seide, F.; Li, G.; Yu, D. Conversational Speech Transcription Using Context-Dependent Deep Neural Networks. In Proceedings of the Twelfth Annual Conference of the International Speech Communication Association, Florence, Italy, 27–31 August 2011.
72. Zhang, Y.; Qin, J.; Park, D.S.; Han, W.; Chiu, C.-C.; Pang, R.; Le, Q.V.; Wu, Y. Pushing the Limits of Semi-Supervised Learning for Automatic Speech Recognition. *arXiv* **2020**, arXiv:2010.10504.
73. Benzeghiba, M.; de Mori, R.; Deroo, O.; Dupont, S.; Erbes, T.; Jouvet, D.; Fissore, L.; Laface, P.; Mertins, A.; Ris, C. Automatic Speech Recognition and Speech Variability: A Review. *Speech Commun.* **2007**, *49*, 763–786. [CrossRef]

74. Lee, K.-F. Context-Independent Phonetic Hidden Markov Models for Speaker-Independent Continuous Speech Recognition. *IEEE Trans. Acoust. Speech Signal Process.* **1990**, *38*, 599–609. [CrossRef]
75. Agrawal, P.; Ganapathy, S. Robust Raw Waveform Speech Recognition Using Relevance Weighted Representations. *arXiv* **2020**, arXiv:2011.00721.
76. Sainath, T.; Weiss, R.J.; Wilson, K.; Senior, A.W.; Vinyals, O. Learning the Speech Front-End with Raw Waveform CLDNNs. In Proceedings of the Interspeech 2015, Dresden, Germany, 6–10 September 2015.
77. Zeghidour, N.; Usunier, N.; Synnaeve, G.; Collobert, R.; Dupoux, E. End-to-End Speech Recognition from the Raw Waveform. *arXiv* **2018**, arXiv:1806.07098.
78. Deng, L.; Sun, D. Speech Recognition Using the Atomic Speech Units Constructed from Overlapping Articulatory Features. In Proceedings of the Third European Conference on Speech Communication and Technology, Berlin, Germany, 19–23 September 1993.
79. Liu, S.A. Landmark Detection for Distinctive Feature-based Speech Recognition. *J. Acoust. Soc. Am.* **1996**, *100*, 3417–3430. [CrossRef]
80. Stevens, K.N.; Manuel, S.Y.; Shattuck-Hufnagel, S.; Liu, S. Implementation of a Model for Lexical Access Based on Features. In Proceedings of the Second International Conference on Spoken Language Processing, Banff, AB, Canada, 13–16 October 1992.
81. Eide, E. Distinctive Features for Use in an Automatic Speech Recognition System. In Proceedings of the Seventh European Conference on Speech Communication and Technology, Aalborg, Denmark, 3–7 September 2001.
82. Erler, K.; Freeman, G.H. An HMM-based Speech Recognizer Using Overlapping Articulatory Features. *J. Acoust. Soc. Am.* **1996**, *100*, 2500–2513. [CrossRef]
83. Espy-Wilson, C.Y.; Pruthi, T.; Juneja, A.; Deshmukh, O. Landmark-Based Approach to Speech Recognition: An Alternative to HMMs. In Proceedings of the Eighth Annual Conference of the International Speech Communication Association, Antwerp, Belgium, 27–31 August 2007; Citeseer: Princeton, NJ, USA, 2007; pp. 886–889.
84. Hasegawa-Johnson, M.; Baker, J.; Borys, S.; Chen, K.; Coogan, E.; Greenberg, S.; Juneja, A.; Kirchhoff, K.; Livescu, K.; Mohan, S. Landmark-Based Speech Recognition: Report of the 2004 Johns Hopkins Summer Workshop. In Proceedings of the IEEE International Conference on Acoustics, Speech, and Signal Processing (ICASSP'05), Philadelphia, PA, USA, 23 March 2005; IEEE: Piscataway, NJ, USA, 2005; Volume 1, p. I-213.
85. Xie, Y.; Hasegawa-Johnson, M.; Qu, L.; Zhang, J. Landmark of Mandarin Nasal Codas and Its Application in Pronunciation Error Detection. In Proceedings of the 2016 IEEE International Conference on Acoustics, Speech and Signal Processing (ICASSP), Shanghai, China, 20–25 March 2016; IEEE: Piscataway, NJ, USA, 2016; pp. 5370–5374.
86. Yang, X.; Kong, X.; Hasegawa-Johnson, M.; Xie, Y. Landmark-Based Pronunciation Error Identification on Chinese Learning. In Proceedings of the Speech Prosody, Boston, MA, USA, 31 May–3 June 2016; pp. 247–251.
87. Lin, H.B.; Repp, B.H. Cues to the perception of Taiwanese tones. *Lang. Speech* **1989**, *32*, 25–44. [CrossRef]
88. Wang, S.W. Phonological Features of Tone. *Int. J. Am. Linguist.* **1967**, *33*, 93–105.
89. Chao, Y.R. *Language and Symbolic Systems*; Cambridge University Press: Cambridge, UK, 1968; Volume 260.
90. Clements, G.N.; Michaud, A.; Patin, C. Do We Need Tone Features? In *Tones and Features*; De Gruyter Mouton: Berlin, Germany, 2011; pp. 3–24.
91. Hyman, L.M. Do Tones Have Features? In *Tones and Features*; De Gruyter Mouton: Berlin, Germany, 2011; pp. 50–80.
92. Laniran, Y.O. Intonation in Tone Languages: The Phonetic Implementation of Tones in Yoruba. Ph.D. Thesis, Cornell University, Ithaca, NY, USA, 1992.
93. Morén, B.; Zsiga, E. The Lexical and Post-Lexical Phonology of Thai Tones. *Nat. Lang. Linguist. Theory* **2006**, *24*, 113–178. [CrossRef]
94. Zsiga, E.; Nitisaroj, R. Tone Features, Tone Perception, and Peak Alignment in Thai. *Lang. Speech* **2007**, *50*, 343–383. [CrossRef]
95. Shi, F.; Liao, R. *Essays on Phonetics*; Beijing Language and Culture Press: Beijing, China, 1994.
96. Zhu, X. *Records of Shanghai Tonal Experiments*; Shanghai Education Press: Shanghai, China, 2005.
97. Zhu, X. *Phonetics*; Commercial Press: Beijing, China, 2010.
98. Xu, Y. Contextual Tonal Variations in Mandarin. *J. Phon.* **1997**, *25*, 61–83. [CrossRef]
99. Xu, Y. Effects of Tone and Focus on the Formation and Alignment of F0 contours. *J. Phon.* **1999**, *27*, 55–105. [CrossRef]
100. Yuan, J.; Ryant, N.; Cai, X.; Church, K.; Liberman, M. Automatic Recognition of Suprasegmentals in Speech. *arXiv* **2021**, arXiv:2108.01122.
101. Lin, J.; Li, W.; Gao, Y.; Xie, Y.; Chen, N.F.; Siniscalchi, S.M.; Zhang, J.; Lee, C.-H. Improving Mandarin Tone Recognition Based on DNN by Combining Acoustic and Articulatory Features Using Extended Recognition Networks. *J. Signal Process. Syst.* **2018**, *90*, 1077–1087. [CrossRef]
102. Gauthier, B.; Shi, R.; Xu, Y. Learning Phonetic Categories by Tracking Movements. *Cognition* **2007**, *103*, 80–106. [CrossRef]
103. Chang, C.-C.; Lin, C.-J. LIBSVM: A Library for Support Vector Machines. *ACM Trans. Intell. Syst. Technol.* **2011**, *2*, 1–27. [CrossRef]
104. Krebel, U.-G. Pairwise Classification and Support Vector Machines. In *Advances in Kernel Methods: Support Vector Learning*; The MIT Press: Cambridge, MA, USA, 1999; pp. 255–268.
105. Wehrens, R.; Kruisselbrink, J. Flexible Self-Organizing Maps in Kohonen 3.0. *J. Stat. Softw.* **2018**, *87*, 1–18. [CrossRef]
106. Rose, P. Considerations in the Normalisation of the Fundamental Frequency of Linguistic Tone. *Speech Commun.* **1987**, *6*, 343–352. [CrossRef]

107. Xu, Y.; Prom-On, S. Toward Invariant Functional Representations of Variable Surface Fundamental Frequency Contours: Synthesizing Speech Melody via Model-Based Stochastic Learning. *Speech Commun.* **2014**, *57*, 181–208. [CrossRef]
108. McLoughlin, I.V.; Xu, Y.; Song, Y. Tone Confusion in Spoken and Whispered Mandarin Chinese. In Proceedings of the 9th International Symposium on Chinese Spoken Language Processing, Singapore, 12–14 September 2014; IEEE: Piscataway, NJ, USA, 2014; pp. 313–316.
109. Tupper, P.; Leung, K.; Wang, Y.; Jongman, A.; Sereno, J.A. Characterizing the Distinctive Acoustic Cues of Mandarin Tones. *J. Acoust. Soc. Am.* **2020**, *147*, 2570–2580. [CrossRef]
110. Fujisaki, H.; Hirose, K. Analysis of Voice Fundamental Frequency Contours for Declarative Sentences of Japanese. *J. Acoust. Soc. Jpn.* **1984**, *5*, 233–242. [CrossRef]
111. Saltzman, E.L.; Munhall, K.G. A Dynamical Approach to Gestural Patterning in Speech Production. *Ecol. Psychol.* **1989**, *1*, 333–382. [CrossRef]
112. Halle, M.; Stevens, K. Mechanism of Glottal Vibration for Vowels and Consonants. *J. Acoust. Soc. Am.* **1967**, *41*, 1613. [CrossRef]
113. Liu, F.; Xu, Y.; Prom-on, S.; Yu, A.C.L. Morpheme-like Prosodic Functions: Evidence from Acoustic Analysis and Computational Modeling. *J. Speech Sci.* **2013**, *3*, 85–140.
114. Birkholz, P.; Schmaser, P.; Xu, Y. Estimation of Pitch Targets from Speech Signals by Joint Regularized Optimization. In Proceedings of the 2018 26th European Signal Processing Conference (EUSIPCO), Rome, Italy, 3–7 September 2018; IEEE: Piscataway, NJ, USA, 2018; pp. 2075–2079.
115. Chen, S.-H.; Wang, Y.-R. Tone Recognition of Continuous Mandarin Speech Based on Neural Networks. *IEEE Trans. Speech Audio Process.* **1995**, *3*, 146–150. [CrossRef]
116. Peng, G.; Wang, W.S.-Y. Tone Recognition of Continuous Cantonese Speech Based on Support Vector Machines. *Speech Commun.* **2005**, *45*, 49–62. [CrossRef]
117. Perkell, J.S.; Klatt, D.H. *Invariance and Variability in Speech Processes*; Lawrence Erlbaum Associates, Inc.: Mahwah, NJ, USA, 1986.
118. Sipser, M. Introduction to the Theory of Computation. *ACM Sigact News* **1996**, *27*, 27–29. [CrossRef]
119. Chen, Y.; Xu, Y. Intermediate Features Are Not Useful for Tone Perception. In Proceedings of the International Conference on Speech Prosody, Tokyo, Japan, 25–28 May 2020; ISCA: Singapore, 2020; pp. 513–517.
120. DiCanio, C.; Nam, H.; Whalen, D.H.; Timothy Bunnell, H.; Amith, J.D.; García, R.C. Using Automatic Alignment to Analyze Endangered Language Data: Testing the Viability of Untrained Alignment. *J. Acoust. Soc. Am.* **2013**, *134*, 2235–2246. [CrossRef] [PubMed]
121. Remijsen, B.; Ayoker, O.G. Contrastive Tonal Alignment in Falling Contours in Shilluk. *Phonology* **2014**, *31*, 435–462. [CrossRef]
122. Hombert, J.-M. Consonant Types, Vowel Quality, and Tone. In *Tone*; Elsevier: Amsterdam, The Netherlands, 1978; pp. 77–111.
123. Xu, Y.; Xu, A. Consonantal F0 Perturbation in American English Involves Multiple Mechanisms. *J. Acoust. Soc. Am.* **2021**, *149*, 2877–2895. [CrossRef]
124. Lin, M.; Li, Z. Focus and Boundary in Chinese Intonation. In Proceedings of the ICPhS, Hong Kong, China, 17–21 August 2011; Volume 17, pp. 1246–1249.
125. Ittichaichareon, C.; Suksri, S.; Yingthawornsuk, T. Speech Recognition Using MFCC. In Proceedings of the International Conference on Computer Graphics, Simulation and Modeling, Pattaya, Thailand, 28–29 July 2012; pp. 135–138.
126. Nolan, F. A Recent Voice Parade. *Int. J. Speech Lang. Law* **2003**, *10*, 277–291. [CrossRef]
127. Barras, C.; Gauvain, J.-L. Feature and Score Normalization for Speaker Verification of Cellular Data. In Proceedings of the 2003 IEEE International Conference on Acoustics, Speech, and Signal Processing (ICASSP'03), Hong Kong, China, 6–10 April 2003; IEEE: Piscataway, NJ, USA, 2003; Volume 2, pp. 49–52.

brain sciences

MDPI

Article

Perceived Anger in Clear and Conversational Speech: Contributions of Age and Hearing Loss

Shae D. Morgan [1,*], Sarah Hargus Ferguson [2], Ashton D. Crain [2] and Skyler G. Jennings [2]

1 Department of Otolaryngology—Head and Neck Surgery and Communicative Disorders, University of Louisville, Louisville, KY 40241, USA

2 Department of Communication Sciences and Disorders, University of Utah, Salt Lake City, UT 84111, USA; sarah.ferguson@hsc.utah.edu (S.H.F.); a.crain1991@gmail.com (A.D.C.); skyler.jennings@hsc.utah.edu (S.G.J.)

* Correspondence: shae.morgan@louisville.edu

Abstract: A previous investigation demonstrated differences between younger adult normal-hearing listeners and older adult hearing-impaired listeners in the perceived emotion of clear and conversational speech. Specifically, clear speech sounded angry more often than conversational speech for both groups, but the effect was smaller for the older listeners. These listener groups differed by two confounding factors, age (younger vs. older adults) and hearing status (normal vs. impaired). The objective of the present study was to evaluate the contributions of aging and hearing loss to the reduced perception of anger in older adults with hearing loss. We investigated perceived anger in clear and conversational speech in younger adults with and without a simulated age-related hearing loss, and in older adults with normal hearing. Younger adults with simulated hearing loss performed similarly to normal-hearing peers, while normal-hearing older adults performed similarly to hearing-impaired peers, suggesting that aging was the primary contributor to the decreased anger perception seen in previous work. These findings confirm reduced anger perception for older adults compared to younger adults, though the significant speaking style effect—regardless of age and hearing status—highlights the need to identify methods of producing clear speech that is emotionally neutral or positive.

Keywords: clear speech; conversational speech; perceived emotion; aging; hearing loss

Citation: Morgan, S.D.; Ferguson, S.H.; Crain, A.D.; Jennings, S.G. Perceived Anger in Clear and Conversational Speech: Contributions of Age and Hearing Loss. *Brain Sci.* **2022**, *12*, 210. https://doi.org/10.3390/brainsci12020210

Academic Editors: Yang Zhang and Paul E. Engelhardt

Received: 13 December 2021
Accepted: 27 January 2022
Published: 2 February 2022

1. Introduction

Talkers may adopt a more clear speaking style to overcome perceived barriers to communication (e.g., when the listener has a hearing loss or is a non-native speaker of the talker's language). This clear speaking style serves to improve the communicative experience by affording the listener a speech-understanding benefit compared to when a more conversational style is used (e.g., [1,2]). Clear speech has been found to be associated with numerous acoustic changes from habitual/conversational speech, which are thought to result in this clear speech benefit [3]. At the suprasegmental level, these changes include raised and more variable voice fundamental frequency, decreased speaking rate, increased energy in the 1000–3000 Hz range of the long-term speech spectrum, and larger fluctuations in the temporal envelope [4,5]. At the segmental level, stop consonants are released more often [5,6], voice onset time of word-initial voiceless stop consonants increases [5], and short-term vowel spectra change in a manner that expands the acoustic vowel space [7,8].

While adopting these segmental and suprasegmental modifications makes speech easier to understand for many different listener groups, there has been some speculation recently about other, unintended consequences of clear speech, such as the increased perception of anger [9]. Indeed, many of the suprasegmental acoustic features found in clear speech, including increased high-frequency energy and greater pitch variability, have also been reported in angry speech [10,11]. Thus, the pattern of acoustic modifications that talkers use to make themselves easier to understand may also promote the perception of anger and other negative emotions.

Directly investigating perceived anger in clear and conversational speech, Morgan and Ferguson [9] found that both younger adults with normal hearing and older adults with hearing loss rated clear speech as sounding angry more often than conversational speech. The older adult group with hearing loss, however, showed generally reduced emotion perception, opting to choose "neutral" more often than younger adults with normal hearing. As these listener groups differed primarily on two variables, age and hearing status, it was impossible for the authors to infer the source of this group difference from their data. The following sections will review auditory emotion recognition, and propose how aging and hearing loss may result in the reduced perception of anger observed in clear and conversational speech.

1.1. Emotion Recognition

In psychological sciences, emotion is generally described using one of three models: discrete [12], dimensional [13], or a combination of the two. For the purposes of this research, we will focus on a discrete model, where emotions are thought to have distinct categories [12,14]. Discrete models are considered the simplest for describing emotions, as they use basic labels to categorize emotions, such as happiness, sadness, fear, anger, disgust, contempt, and surprise. According to Ekman and Cordaro [14], these seven emotions are general enough to be cross-cultural, and therefore are appropriate for use across generations.

1.2. Effect of Age on Emotion Recognition

Aging differences have been observed in the emotional perception of both visual and acoustic stimuli [15,16], suggesting that reduced perception of negative emotions is associated with aging. Some groups have even named this observation the "positivity effect", though the robustness and universality of that effect is in question [17]. These differences are thought to be caused by sociocognitive and/or neuropsychological factors. Sociocognitive explanations suggest that aging is accompanied by an increased ability to understand and regulate emotions. This is attributed to extensive life experience in analyzing emotional cues [18]. Neuropsychological explanations, in contrast, posit that the aging process contributes to decreased activity in the brain regions responsible for emotion processing and regulation. Compared to visual perception, these neuropsychological accounts are less well-documented in auditory perception of emotion (e.g., [19]).

A recent study on auditory emotion recognition across the lifespan suggests that emotion recognition performance improves and approaches adult-like levels in early adolescence [20]. Early adulthood marks a stable period of auditory emotion recognition ability, followed by a gradual decline into later adulthood. Older adults (>60 years old) showed a sharp reduction in the accuracy of emotional identification. This may be explained by socioemotional selectivity theory, which suggests that older adults tend to neutralize (or even positively valence) their perceptions of the emotions of others. Another explanation is that older adults may require more dramatic examples of emotions to warrant classification as such compared to younger adults, who are more attentive to the subtle emotionality needed to advance through adulthood (e.g., to attract a partner, secure advancement at work, etc.). It is also possible, however, that declines in hearing contribute to emotion perception in older adults. These declines include reduced access to and processing of high-frequency acoustic cues [21], diminished temporal processing [22], and age-related loss of auditory nerve fibers [23].

1.3. Effect of Hearing Loss on Emotion Recognition

The effect of hearing loss on auditory emotion recognition is an expanding area of research interest. Recent studies have shown conflicting reports, with some suggesting no relationship between hearing loss and emotion recognition ability (e.g., [24]), while others show clear associations [25–29]. Investigations involving participants with mild-to-moderate hearing loss usually find minimal or no reduction of emotion recognition when compared with typically hearing peers. In contrast, studies centering on individuals with more severe hearing loss, and especially individuals with cochlear implants, show clear deficits. There seems to be a critical threshold of hearing loss required to significantly impact the ability to access emotional content from auditory stimuli. Older adults naturally tend to have more significant and severe hearing losses than younger adults, and it may be that this hearing loss is significant enough to reduce older adults' access to the acoustic cues associated with different emotion categories. For example, one acoustic feature of angry speech is increased high-frequency energy compared to emotionally neutral speech (e.g., [10]). An inability to discern this increased energy (due to hearing loss at high frequencies) may result in reduced anger perception for individuals with hearing loss compared to peers with normal hearing.

1.4. Previous Research

In this manuscript we present two experiments that, when compared with previously published data from our research group [9], provide additional insight into the independent contributions of aging and hearing status for the perception of anger in clear and conversational speech. These experiments include (1) assessing the effects of a simulated hearing loss on emotion perception in younger adults with normal hearing, and (2) comparing emotion perception among older adults with normal hearing and hearing loss. We hypothesized that both aging (through sociocognitive and neuropsychological changes) and hearing loss (through reduced sensitivity to acoustic cues that typify emotional productions) would contribute to the overall reduction in the perception of anger for older adults with hearing loss when listening to clear and conversational speech. Understanding which of these contributions is dominant will help determine when using clear speech carries the greatest risk of invoking a negative sentiment. If aging better explains reduced anger perception than hearing loss, then hearing health professionals should take extra care when counseling younger patients with hearing loss (a primary population to whom clear speech is directed) to be wary of unintended negativity when communication partners use this speaking style. On a more foundational level, the present study also provides critical insight regarding the primary mechanisms surrounding emotional perception (i.e., cognitive mechanisms versus peripheral sensitivity) and their independent or combined contributions.

2. Materials and Methods

2.1. Experiment I: Young Adults with a Simulated Hearing Loss

The objective of this experiment was to determine the extent to which the reduced audibility of high frequencies, independent of age, affects the perceived emotion of clear and conversational speech by simulating hearing loss in young adults with normal hearing.

2.1.1. Stimuli

Speech stimuli used in this experiment came from the Ferguson Clear Speech Database (described in [30]). The database consists of 41 talkers, from whom the same list of 188 sentences were recorded in clear and conversational speaking styles. To reduce semantic priming of specific emotions, 14 emotionally neutral sentences were chosen for the current study (e.g., "Use the word bead in a sentence"). In total, 8 talkers producing these 14 sentences in each speaking style were selected from the database for the current study (2 speaking styles × 8 talkers × 14 sentences = 224 total stimuli). The talkers were combined into two groups based on their perceived clarity (see [9] for additional details): Group A and Group B. Talkers in Group A were rated to have the largest clear speech effect and were considered "good" clear speakers, while the talkers in Group B had the smallest clear speech effect and were considered "poor" clear speakers.

We processed the stimuli used in Morgan and Ferguson [9] to simulate aspects of age-related hearing loss following methods and procedures described by Moore and Glasberg [31]. Their approach accounts for the effects of reduced sensitivity and abnormal loudness growth across a bank of simulated auditory filters. Specifically, the following processing steps were completed: (1) split the stimulus into frequency bands; (2) isolate the temporal envelope and fine structure within each band; (3) apply an exponent to the envelope to simulate envelope expansion and loudness recruitment; (4) smooth the expanded envelope of each band; (5) calculate and apply the attenuation level for each band; and (6) sum across bands after recombining and temporally aligning the processed envelope and fine structure. Figure 1 shows the attenuation applied to each frequency band (based on the average audiogram of the older adults with hearing loss from Morgan and Ferguson [9]), as well as the realization of that attenuation (following procedures by Moore and Glasberg [31]) in the long-term average spectra for the original and unprocessed stimuli (created using a 30 s sample of concatenated stimuli from each condition and calculating the fast-Fourier transform via the *fft* function in MATLAB). Similarly, according to Moore and Glasberg, the degree of envelope expansion was directly determined by the attenuation level of each band. The simulation of envelope expansion accounts for loss of cochlear compression; however, such expansion may also occur within the central auditory system due to increased central gain [32]. Our simulations are limited to accounting for aspects of age-related hearing loss that are attributed to changes in the auditory periphery (i.e., reduced audibility and loss of cochlear compression). Thus, any effects of hearing loss attributed to central auditory function (e.g., [33]) are not accounted for by the model. Despite this, our simulations address our hypothesis that age-related changes in emotion perception may be due to the reduced audibility of the acoustic cues that mark a talker's emotion.

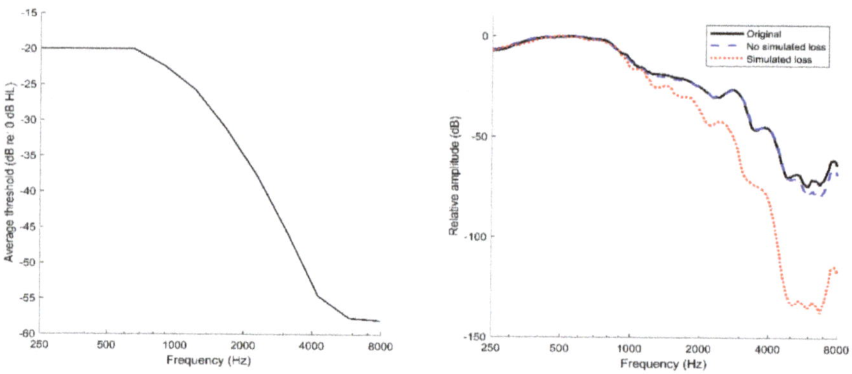

Figure 1. Average simulated hearing loss based on the older adult data from Morgan and Ferguson [9] (left) and the long-term average speech spectrum of 30 s samples of concatenated stimuli from the original unprocessed stimuli, the stimuli processed with no simulated loss, and the stimuli processed with the simulated hearing loss.

In addition to processing the stimuli to simulate hearing loss, we included a control set of stimuli, which underwent the same processing (i.e., band-pass filtering, envelope expansion, etc.) as the experimental stimuli, except no attenuation was applied to the frequency bands (effectively simulating hearing thresholds at 0 dB HL). The stimuli were scaled to the same average root-mean-square amplitude in Cool Edit 2000 to limit the use of amplitude differences as an emotional cue between the speaking styles; this procedure is common in clear speech perception research [34].

2.1.2. Listeners

A total of 44 younger adults (18 to 32 years old; M = 21, SD = 3.2) with normal hearing were recruited for this experiment. Normal hearing was established via a pure tone screening at the time of testing, with all participants demonstrating thresholds better than or equal to 20 dB HL at octave intervals from 250 to 8000 Hz. These participants were divided into two groups (22 listeners per group) that were similar in their average age and hearing thresholds. One group rated the processed control stimuli (i.e., without high-frequency attenuation) and the other group rated the processed experimental stimuli (i.e., with high-frequency attenuation; YSIM). All listeners in this experiment passed the Montreal Cognitive Assessment [35]. All listeners denied any history of speech or language disorders or therapy and were native speakers of American English.

2.1.3. Procedures

All subject recruitment and other procedures were in accordance with a protocol approved by the University of Utah Institutional Review Board. Participants sat in a sound-treated room in front of a computer monitor, mouse, and keyboard. Test methods were similar to those used in Morgan and Ferguson [9]. A custom MATLAB graphical user interface (GUI) presented the instructions and guided the listeners through the experiment. Instructions were provided verbally and on screen. Participants were told, "Judge the emotion you think you hear when listening to each sentence." The emotional category options were "anger," "fear," "disgust," "sadness," "happiness," and "neutral," presented in a six-alternative, forced-choice task paradigm. Participants were instructed to choose "neutral" if they heard no specific emotion in the speech. Ekman and Cordaro [14] recommend these emotions as "basic" emotion categories. Contempt and surprise are also considered "basic" emotions, but they were excluded from this study. Contempt was excluded because it is often difficult to conceptualize due to its lack of common use [36], and surprise was excluded because it is a brief emotion, quickly shifting to another emotion immediately after its expression [14].

Stimuli were presented through a Tucker-Davis Technologies (TDT) RP2.1 real-time processor and then attenuated by a TDT programmable attenuator (PA-5) to a comfortable listening level (70 dB SPL using a 1 kHz calibration tone). The stimuli were then routed via a headphone buffer (TDT HB7) to Sennheiser HD 515 headphones for monaural presentation to the participant's test ear, which was selected by the participant as their perceived better ear. Monaural presentation was chosen to mimic procedures used for the older adult group with hearing loss (OHL) in Morgan and Ferguson [9]. Task familiarization consisted of 20 items that were presented in a practice block prior to testing. All participants confirmed that they felt confident with the task after completing the practice block. For the experiment, listeners heard 224 sentences in a random order without replacement. Participants responded by clicking on the emotion category they felt best corresponded to the sentence and then pressed "Enter" on the keyboard to advance to the next presentation. Listeners were given the opportunity to listen to each stimulus one additional time by clicking a button in the GUI labeled "Listen again". Listeners were offered breaks at regular intervals during each test block.

2.1.4. Statistical Analysis

The responses were combined across the listeners in each group to create a dependent variable of the percentage of listeners who judged a given stimulus to be angry. These data were then analyzed using a linear mixed-effects models via the *lme4* [37] and *lmerTest* [38] packages in *R* (version 3.5.1; [39]). In each model, the fixed effects of speaking style (clear and conversational), listener group (YNH, OHL, and YSIM), and talker group (A and B) were analyzed, along with all two- and three-way interactions among them, and talker was included as a random factor. Main effects were assessed using the *anova* function in *lmerTest*, which implements the Satterthwaite approximation for degrees of freedom to calculate *p*-values [40]. In contrast to using likelihood ratio tests (another popular method for assessing significance of variables in linear mixed-effects models), this approach has been shown to yield more conservative estimates that are less prone to Type I errors [41].

A data integrity check comparing data obtained from YNH listeners in Morgan and Ferguson [9] and the processed control data (with no applied attenuation) from this study showed no significant fixed or interaction effects involving listener group (all $p > 0.26$). This confirmed that any effects related to the simulated hearing loss are a result of the attenuation and recruitment simulation and not the other aspects of the processing.

2.1.5. Results

Figure 2 shows the perceived anger in clear and conversational speech for the two talker groups and the three listener groups (panels a, b, and d). Statistical analyses showed a significant main effect of talker ($F_{(1, 658)} = 17.4, p < 0.001$), a significant interaction between talker and speaking style ($F_{(1, 658)} = 129.5, p < 0.001$), and a significant three-way interaction between speaking style, talker group, and listener group ($F_{(1, 658)} = 6.9, p < 0.001$). In general, clear speech was rated as sounding angry more often than conversational speech, and that effect was more pronounced for "good" clear speakers than for the "poor" clear speakers in this study. The nature of the three-way interaction can be seen in Figure 2, which shows a greater interaction between speaking style and talker group for the YNH and YSIM groups (which appear to be more similar) than the ONH group (which has a smaller interaction relative to the other two groups).

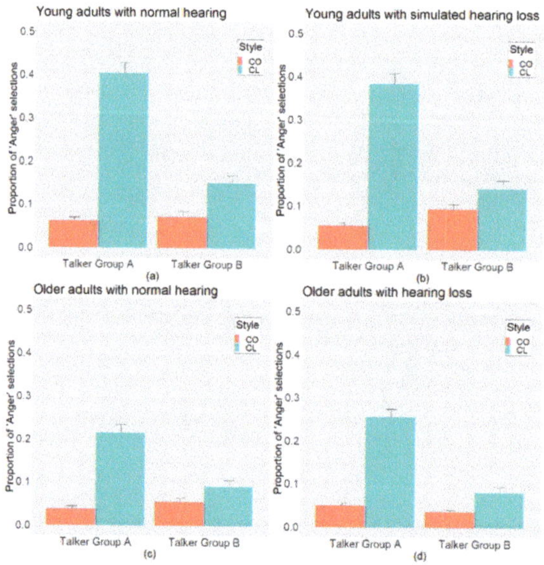

Figure 2. Proportion of "Anger" selections for clear (CL) and conversational (CO) speech by (**a**) young adults with normal hearing; YNH, (**b**) young adults with a simulated hearing loss; YSIM, (**c**) older adults with normal hearing; ONH, and (**d**) older adults with hearing loss; OHL. Error bars represent one standard error.

2.2. Experiment II: Older Adults with Essentially Normal Hearing

Experiment I showed no significant effect of simulated hearing loss on the perception of anger in clear and conversational speech in a group of young adult listeners. This finding suggests that the reduced perception of anger in clear and conversational speech in older adults with hearing loss is driven by the effects of age, rather than the effects of hearing sensitivity. To confirm this, in Experiment II we repeated the study/task of auditory emotion perception for unprocessed clear and conversational speech materials with a sample of older adults with normal or borderline normal hearing (ONH; described below).

2.2.1. Stimuli

Stimuli for Experiment II were the same as Experiment I, except the stimuli were not processed to simulate hearing loss or produce control stimuli. These unprocessed stimuli were the same as those used in Morgan and Ferguson [9].

2.2.2. Listeners

A total of 17 older adult listeners (66 to 83 years old; M = 72.77, SD 4.47) were recruited from the Utah Senior Ears database. This database recruits community members who are interested in participating in hearing-related research in exchange for periodic hearing tests at no charge. The listeners in this study had pure tone thresholds less than or equal to 25 dB HL for 250–3000 Hz and no greater than 40 dB HL for 4000 Hz.

In addition to these hearing status criteria, all listeners in this experiment were administered the Montreal Cognitive Assessment [35]. Two participants in this study did not pass the cognitive screening due to deficits in delayed recall ($n = 1$) or visuospatial processing ($n = 1$). However, data from these participants were included in the data analysis, as there was no significant difference found when their data were excluded. All listeners denied any history of speech or language disorders or therapy and were native speakers of American English. Participants were either compensated for their time ($n = 6$) or opted to participate as volunteers ($n = 11$).

2.2.3. Procedures

The experimental methods and procedures for Experiment II were the same as those used for Experiment I, except the stimuli used in this experiment were not processed to simulate hearing loss.

2.2.4. Statistical Analysis

In total, 17 listeners judged 224 sentences, resulting in 3808 data points for ONH listeners. A total of 121 missing data points occurred due to an experiment software error encountered by some participants which resulted in the premature termination of the experiment ($n = 6$). The data were analyzed in the same manner detailed for Experiment I, but comparing YNH and OHL data with ONH participants rather than YSIM participants from Experiment I.

2.2.5. Results

In addition to the three listener groups from Experiment I, Figure 2 also shows the perceived anger in clear and conversational speech for the two talker groups by ONH listeners (panel c) Statistical analyses revealed a significant main effect of talker ($F_{(1, 658)} = 17.4$, $p < 0.001$), a significant interaction between talker and speaking style ($F_{(1, 658)} = 98.1$, $p < 0.001$), and a significant three-way interaction between speaking style, talker group, and listener group ($F_{(1, 658)} = 10.4$, $p < 0.01$). The nature of this three-way interaction can be seen in Figure 2. The interaction between speaking style and talker group was greater for the YNH group than for the ONH and OHL groups, which appear to be more similar, but with reduced overall perception of anger compared to the YNH group.

Finally, a confirmatory analysis showed a significant listener group effect between YSIM participants in Experiment I and ONH participants in Experiment II ($F_{(1, 434)} = 4.2$, $p < 0.05$), where YSIM participants (Figure 2b) reported hearing anger more often than the ONH group (Figure 2c). In combination with the significant three-way interactions involving the listener group, this analysis confirms that any reduced perception of anger between participant groups is primarily driven by age and not by the hearing status of the participants. Thus, older adults, regardless of hearing status, reported speech as "sounding angry" less often than younger adults, even when younger adults listened to stimuli filtered to simulate the high-frequency hearing loss. The results of this study support the conclusions of Experiment I, which suggested that the differences between the YNH and

OHL listeners in Morgan and Ferguson (2017) were due to age effects and not to differences in signal audibility.

3. General Discussion

The compiled results from the two experiments presented here and the two experiments presented in Morgan and Ferguson [9] reveal that younger adults, regardless of hearing status (YNH and YSIM listeners) were similar in how often they judged clear speech to sound angry, but that older adults (ONH and OHL listeners), whose ratings were similar regardless of hearing status, rated clear speech as sounding angry less often than either of the younger adult groups. Thus, it appears that the presence or simulation of mild to moderate hearing loss did not significantly affect listeners' perceived anger in clear and conversational speech, confirming that hearing loss is not the primary contributing factor underlying age-related changes to the perception of anger in clear and conversational speech. Similarities and differences among the younger and older listener groups were found to be larger for clear speech, but these patterns were also observed in ratings of conversational speech, suggesting that even habitual speech is differentially perceived across a person's lifespan.

Ruffman et al. [15] found that older adults had poorer performance on emotion recognition tasks than younger adults. Older adults in this study perceived anger in vocal expressions less often when compared to younger adults—that is, older adults perceived stimuli to be neutral more often than younger adults. Anatomically, there has been evidence that these findings may be at least partially explained by a consistent age-related decline of the orbitofrontal cortex (one of the crucial areas of the brain related to recognition of anger, sadness, and fearful expressions; [19]). This research provides some general support for the idea that some of the neurological regions involved in emotion recognition are also involved in recognizing facial, auditory, and bodily expressions, and that these regions experience an age-related decline [42].

The discussion thus far has characterized the group differences in terms of age-related decline in emotion recognition. However, sociocognitive theories propose that with aging, one's ability to understand and regulate emotions actually improves [43]. This suggests that the decrease in perceived anger observed in the current experiments may actually be a result of more accurate perception of the stimuli, which were originally recorded as neutral speech with no emotional intention. Older adults have had a lifetime of analyzing emotional cues in interpersonal communication, and it has been assumed that this skill might be preserved or even improve with aging [44].

Clinical Implications

The results from this study, as well as supporting evidence from previous studies, confirm that listeners generally judge clear speech as angry more often than conversational speech, regardless of age and hearing status. Clinical audiologists and other health professionals often counsel communication partners of their hearing-impaired patients to adopt a clear speaking style to improve their partner's speech understanding. The findings of these studies indicate that talkers using this style of speaking may give the impression that the talker is upset, resulting in negative emotions accompanying the improved clarity of speech. This negative perception of clear speech may cause excess stress on relationships for individuals with hearing loss and their communication partners, relationships which are already negatively affected by hearing loss (e.g., [45]). This added stress may have negative social effects on both the hearing-impaired individual and their communication partners. Thus, communication professionals should make these recommendations with caution, particularly for younger adult patients who may be more likely to rate clear speech negatively. While clear speech will likely improve understanding, it may come with unwanted emotional undertones and perceived anger.

4. Conclusions

The two experiments detailed in this manuscript confirm that the differences in perceived anger in clear and conversational speech between YNH and OHL listeners are primarily an effect of aging rather than decreased peripheral sensitivity to emotional acoustic cues. Listeners continued to perceive clear speech as angry more often than conversational speech. This finding is consistent with previous research in the emotional perception of clear speech, as well as clinical patient reports. In addition, anger was perceived more often in the clear speaking style with talkers with good clear speech, compared to talkers who had poor clear speech. YNH and YSIM listeners had a larger perception of anger in clear speech when compared to OHL and ONH listeners. These findings support aging as the primary mechanism in reduced perception of anger in clear speech.

Future research could target other listener populations, such as adolescents and middle-aged adults. This research would provide information to help support or refute the sociocognitive theories surrounding emotional perception of speech. Another direction to be explored could examine the acoustic features of clear speech. By examining the Group A talkers, we could potentially identify speaking strategies that make speech clearer without perceived anger. When these acoustic strategies are found, informational materials could be created that educate audiologists and aural rehabilitation specialists to modify their counseling and training of frequent communication partners.

Author Contributions: Conceptualization, S.D.M., S.G.J. and S.H.F.; methodology, S.D.M., S.G.J. and S.H.F.; software, S.D.M. and S.G.J.; validation, S.D.M.; formal analysis, S.D.M.; investigation, S.D.M. and A.D.C.; resources, S.G.J. and S.H.F.; data curation, S.D.M.; writing—original draft preparation, S.D.M., S.G.J., A.D.C. and S.H.F.; writing—review and editing, S.D.M, S.G.J., A.D.C. and S.H.F.; visualization, S.D.M.; supervision, S.D.M.; project administration, S.D.M.; funding acquisition, S.H.F. and S.G.J. All authors have read and agreed to the published version of the manuscript.

Funding: This work was supported, in part, by grant K23 DC014752 (PI: Jennings) and by grant R01 DC012315 (PI: Hunter) from NIH/NIDCD.

Institutional Review Board Statement: The study was conducted according to the guidelines of the Declaration of Helsinki, and approved by the Institutional Review Board of the University of Utah (protocol code 0051529, original approval 10/26/2011).

Informed Consent Statement: Informed consent was obtained from all subjects involved in the study.

Data Availability Statement: The statistical analyses with accompanying figures for visualization are available via an online Rmarkdown file accessible at: https://rpubs.com/shaedmorgan/agingaffectsangerperception. Other data and materials used in this study are available upon request from the corresponding author.

Conflicts of Interest: The authors declare no conflict of interest.

References

1. Ferguson, S.H. Talker Differences in Clear and Conversational Speech: Vowel Intelligibility for Older Adults with Hearing Loss. *J. Speech Lang. Hear. Res.* **2012**, *55*, 779–790. [CrossRef]
2. Rodman, C.; Moberly, A.C.; Janse, E.; Başkent, D.; Tamati, T.N. The impact of speaking style on speech recognition in quiet and multi-talker babble in adult cochlear implant users. *J. Acoust. Soc. Am.* **2020**, *147*, 101–107. [CrossRef]
3. Smiljanić, R.; Bradlow, A.R. Speaking and Hearing Clearly: Talker and Listener Factors in Speaking Style Changes. *Lang. Linguist. Compass* **2009**, *3*, 236–264. [CrossRef] [PubMed]
4. Krause, J.C.; Braida, L.D. Evaluating the role of spectral and envelope characteristics in the intelligibility advantage of clear speech. *J. Acoust. Soc. Am.* **2009**, *125*, 3346–3357. [CrossRef] [PubMed]
5. Picheny, M.A.; Durlach, N.I.; Braida, L.D. Speaking clearly for the hard of hearing II: Acoustic characteristics of clear and conversational speech. *J. Speech Lang. Hear. Res.* **1986**, *29*, 434–446. [CrossRef] [PubMed]
6. Ferguson, S.H.; Morgana, S.D. Acoustic correlates of reported clear speech strategies. *J. Acad. Rehabil. Audiol.* **2010**, *43*, 45–64.
7. Ferguson, S.H.; Kewley-Port, D. Talker Differences in Clear and Conversational Speech: Acoustic Characteristics of Vowels. *J. Speech Lang. Hear. Res.* **2007**, *50*, 1241–1255. [CrossRef]
8. Whitfield, J.A.; Goberman, A.M. Articulatory-acoustic vowel space: Associations between acoustic and perceptual measures of clear speech. *Int. J. Speech-Lang. Pathol.* **2017**, *19*, 184–194. [CrossRef]

9. Morgan, S.D.; Ferguson, S.H. Judgments of Emotion in Clear and Conversational Speech by Young Adults with Normal Hearing and Older Adults with Hearing Impairment. *J. Speech Lang. Hear. Res.* **2017**, *60*, 2271–2280. [CrossRef]
10. Banse, R.; Scherer, K.R. Acoustic profiles in vocal emotion expression. *J. Personal. Soc. Psychol.* **1996**, *70*, 614–636. [CrossRef]
11. Whiteside, S.P. Acoustic characteristics of vocal emotions simulated by actors. *Percept. Mot. Ski.* **1999**, *89*, 1195–1208. [CrossRef]
12. Ekman, P. An argument for basic emotions. *Cogn. Emot.* **1992**, *6*, 169–200. [CrossRef]
13. Russell, J.A. A Circumplex Model of Affect. *J. Personal. Soc. Psychol.* **1980**, *39*, 1161–1178. [CrossRef]
14. Ekman, P.; Cordaro, D. What is Meant by Calling Emotions Basic. *Emot. Rev.* **2011**, *3*, 364–370. [CrossRef]
15. Ruffman, T.; Halberstadt, J.; Murray, J. Recognition of Facial, Auditory, and Bodily Emotions in Older Adults. *J. Gerontol. Ser. B Psychol. Sci. Soc. Sci.* **2009**, *64B*, 696–703. [CrossRef]
16. Ruffman, T.; Henry, J.; Livingstone, V.; Phillips, L.H. A meta-analytic review of emotion recognition and aging: Implications for neuropsychological models of aging. *Neurosci. Biobehav. Rev.* **2008**, *32*, 863–881. [CrossRef]
17. Carstensen, L.L.; DeLiema, M. The positivity effect: A negativity bias in youth fades with age. *Curr. Opin. Behav. Sci.* **2018**, *19*, 7–12. [CrossRef] [PubMed]
18. Carstensen, L.L.; Isaacowitz, D.M.; Charles, S.T. Taking time seriously: A theory of socioemotional selectivity. *Am. Psychol.* **1999**, *54*, 165. [CrossRef]
19. Sander, D.; Grandjean, D.; Pourtois, G.; Schwartz, S.; Seghier, M.; Scherer, K.R.; Vuilleumier, P. Emotion and attention interactions in social cognition: Brain regions involved in processing anger prosody. *NeuroImage* **2005**, *28*, 848–858. [CrossRef]
20. Amorim, M.; Anikin, A.; Mendes, A.J.; Lima, C.F.; Kotz, S.A.; Pinheiro, A.P. Changes in vocal emotion recognition across the life span. *Emotion* **2021**, *21*, 315–325. [CrossRef] [PubMed]
21. Humes, L.E. The Contributions of Audibility and Cognitive Factors to the Benefit Provided by Amplified Speech to Older Adults. *J. Am. Acad. Audiol.* **2007**, *18*, 590–603. [CrossRef]
22. Clinard, C.G.; Tremblay, K.L. Aging Degrades the Neural Encoding of Simple and Complex Sounds in the Human Brainstem. *J. Am. Acad. Audiol.* **2013**, *24*, 590–599. [CrossRef]
23. Makary, C.A.; Shin, J.; Kujawa, S.G.; Liberman, M.C.; Merchant, S.N. Age-Related Primary Cochlear Neuronal Degeneration in Human Temporal Bones. *J. Assoc. Res. Otolaryngol.* **2011**, *12*, 711–717. [CrossRef] [PubMed]
24. Dupuis, K.; Pichora-Fuller, M.K. Aging Affects Identification of Vocal Emotions in Semantically Neutral Sentences. *J. Speech Lang. Hear. Res.* **2015**, *58*, 1061–1076. [CrossRef] [PubMed]
25. Chatterjee, M.; Zion, D.J.; Deroche, M.L.; Burianek, B.A.; Limb, C.J.; Goren, A.P.; Kulkarni, A.M.; Christensen, J.A. Voice emotion recognition by cochlear-implanted children and their normally-hearing peers. *Hear. Res.* **2015**, *322*, 151–162. [CrossRef]
26. Christensen, J.A.; Sis, J.; Kulkarni, A.M.; Chatterjee, M. Effects of Age and Hearing Loss on the Recognition of Emotions in Speech. *Ear Hear.* **2019**, *40*, 1069–1083. [CrossRef]
27. Tinnemore, A.R.; Zion, D.J.; Kulkarni, A.M.; Chatterjee, M. Children's Recognition of Emotional Prosody in Spectrally Degraded Speech Is Predicted by Their Age and Cognitive Status. *Ear Hear.* **2018**, *39*, 874–880. [CrossRef]
28. Luo, X.; Fu, Q.-J.; Galvin, J.J. Cochlear Implants Special Issue Article: Vocal Emotion Recognition by Normal-Hearing Listeners and Cochlear Implant Users. *Trends Amplif.* **2007**, *11*, 301–315. [CrossRef] [PubMed]
29. Luo, X.; Kern, A.; Pulling, K.R. Vocal emotion recognition performance predicts the quality of life in adult cochlear implant users. *J. Acoust. Soc. Am.* **2018**, *144*, EL429–EL435. [CrossRef] [PubMed]
30. Ferguson, S.H. Talker differences in clear and conversational speech: Vowel intelligibility for normal-hearing listeners. *J. Acoust. Soc. Am.* **2004**, *116*, 2365–2373. [CrossRef]
31. Moore, B.C.J.; Glasberg, B.R. Simulation of the effects of loudness recruitment and threshold elevation on the intelligibility of speech in quiet and in a background of speech. *J. Acoust. Soc. Am.* **1993**, *94*, 2050–2062. [CrossRef]
32. Auerbach, B.D.; Radziwon, K.; Salvi, R. Testing the Central Gain Model: Loudness Growth Correlates with Central Auditory Gain Enhancement in a Rodent Model of Hyperacusis. *Neuroscience* **2019**, *407*, 93–107. [CrossRef]
33. Sardone, R.; Battista, P.; Panza, F.; Lozupone, M.; Griseta, C.; Castellana, F.; Capozzo, R.; Ruccia, M.; Resta, E.; Seripa, D.; et al. The Age-Related Central Auditory Processing Disorder: Silent Impairment of the Cognitive Ear. *Front. Neurosci.* **2019**, *13*, 619. [CrossRef] [PubMed]
34. Uchanski, R.M. Clear speech. In *The Handbook of Speech Perception*; Pisoni, D.B., Remez, R.E., Eds.; Blackwell: Malden, MA, USA/Oxford, UK, 2005.
35. Julayanont, P.; Nasreddine, Z.S. Montreal Cognitive Assessment (MoCA): Concept and clinical review. In *Cognitive Screening Instruments*; Springer: Berlin/Heidelberg, Germany, 2017; pp. 139–195.
36. Matsumoto, D.; Ekman, P. The relationship among expressions, labels, and descriptions of contempt. *J. Personal. Soc. Psychol.* **2004**, *87*, 529–540. [CrossRef]
37. Bates, D.; Mächler, M.; Bolker, B.; Walker, S. Fitting Linear Mixed-Effects Models Using lme4. *J. Stat. Softw.* **2015**, *67*, 1–48. [CrossRef]
38. Kuznetsova, A.; Brockhoff, P.B.; Christensen, R.H.B. lmerTest Package: Tests in linear mixed effects models. *J. Stat. Softw.* **2017**, *82*, 1–38. [CrossRef]
39. R Development Core Team. *R: A Language and Environment for Statistical Computing*; R Foundation for Statistical Computing: Vienna, Austria, 2018.
40. Satterthwaite, F.E. Synthesis of variance. *Psychometrika* **1941**, *6*, 309–316. [CrossRef]

41. Luke, S.G. Evaluating significance in linear mixed-effects models in R. *Behav. Res. Methods* **2017**, *49*, 1494–1502. [CrossRef]
42. Resnick, S.M.; Lamar, M.; Driscoll, I. Vulnerability of the Orbitofrontal Cortex to Age-Associated Structural and Functional Brain Changes. *Ann. NY Acad. Sci.* **2007**, *1121*, 562–575. [CrossRef] [PubMed]
43. Phillips, L.H.; MacLean, R.; Allen, R. Age and the Understanding of Emotions: Neuropsychological and Sociocognitive Perspectives. *J. Gerontol. Ser. B Psychol. Sci. Soc. Sci.* **2002**, *57*, P526–P530. [CrossRef]
44. Dougherty, L.M.; Abe, J.A.; Izard, C.E. Differential Emotions Theory and Emotional Development in Adulthood and Later Life. In *Handbook of Emotion, Adult Development, and Aging*; Elsevier: Amsterdam, The Netherlands, 1996; pp. 27–41.
45. Hétu, R.; Jones, L.; Getty, L. The Impact of Acquired Hearing Impairment on Intimate Relationships: Implications for Rehabilitation. *Int. J. Audiol.* **1993**, *32*, 363–380. [CrossRef] [PubMed]

brain sciences

MDPI

Article

What Do Cognitive Networks Do? Simulations of Spoken Word Recognition Using the Cognitive Network Science Approach

Michael S. Vitevitch * and Gavin J. D. Mullin

Department of Psychology, University of Kansas, Lawrence, KS 66045, USA; gavin.mullin@ku.edu
* Correspondence: mvitevit@ku.edu

Abstract: Cognitive network science is an emerging approach that uses the mathematical tools of network science to map the relationships among representations stored in memory to examine how that structure might influence processing. In the present study, we used computer simulations to compare the ability of a well-known model of spoken word recognition, TRACE, to the ability of a cognitive network model with a spreading activation-like process to account for the findings from several previously published behavioral studies of language processing. In all four simulations, the TRACE model failed to retrieve a sufficient number of words to assess if it could replicate the behavioral findings. The cognitive network model successfully replicated the behavioral findings in Simulations 1 and 2. However, in Simulation 3a, the cognitive network did not replicate the behavioral findings, perhaps because an additional mechanism was not implemented in the model. However, in Simulation 3b, when the decay parameter in *spreadr* was manipulated to model this mechanism the cognitive network model successfully replicated the behavioral findings. The results suggest that models of cognition need to take into account the multi-scale structure that exists among representations in memory, and how that structure can influence processing.

Keywords: phonology; network science; one-phoneme metric; phonological neighbors; spoken word recognition; computer simulation; TRACE; cognitive network

Citation: Vitevitch, M.S.; Mullin, G.J.D. What Do Cognitive Networks Do? Simulations of Spoken Word Recognition Using the Cognitive Network Science Approach. *Brain Sci.* **2021**, *11*, 1628. https://doi.org/10.3390/brainsci11121628

Academic Editors: Richard Wright and Benjamin V. Tucker

Received: 15 November 2021
Accepted: 9 December 2021
Published: 10 December 2021

1. Introduction

Various metaphors have been used to increase our understanding of the mind, with the computer perhaps being the most well-known and fundamental metaphor in Cognitive Psychology [1]. Another metaphor that has been used repeatedly by Cognitive Psychologists to examine representations and processing of various kinds is a "network" of some sort. An early use of the network metaphor in Cognitive Psychology is exemplified in the spreading activation theory of semantic memory proposed by [2]. They suggested that information stored in semantic memory—such as perceptual features (e.g., colors) and common nouns (e.g., fire engine)—could be represented as nodes, and relationships among nodes could be represented by labeled connections between nodes (e.g., "IS-A" and "HAS" links to indicate that a fire engine IS-A type of vehicle and HAS the color red). The spreading of activation across the semantic network proposed by Collins and Loftus has been used to understand numerous memory and language phenomena.

Another use of the network metaphor in Cognitive Psychology is the "artificial neural network" approach exemplified in (localist) connectionist models and in parallel distributed processing (PDP) models. Both types of artificial neural network saw a rise in popularity in the late 1980s and early 1990s [3,4].

In localist connectionist models, nodes represent specific pieces of information, such as a phoneme, a syllable, or a word, and connections link together those pieces of information, often in a hierarchical manner. For example, nodes representing the phonemes /k/, /æ/, and /t/ would be connected to nodes representing words such as *at*, *cat*, *tack*, etc. Those word nodes, in turn, might be connected to another layer of nodes that contain semantic information.

In contrast, in the PDP approach, " … active representations in the mind are thought to correspond to the *patterns of activation* generated over a set of units" [4] (pg. 1038; emphasis added). In this approach, knowledge in memory " … does not exist as a set of dormant data structures in a separate store but is encoded directly in the network architecture, in the values of the connection weights that allow the system to generate useful internal representations and outputs" [4] (pg. 1039). In other words, representations are not symbols stored in a separate memory store (as in the localist connectionist approach), but instead are ephemeral and emerge from the processing that occurs over the many distributed processing units in this type of artificial neural network. The artificial neural network approach was (and remains) a significant driver of research on memory, speech production [5,6], and spoken word recognition [7–9].

A more recent use of the network metaphor in Psychology can be found in what is becoming known as Cognitive Network Science [10,11]. The Cognitive Network Science approach applies the quantitative tools of network science [12]—used to understand a wide range of complex systems—to address questions about human cognition. In this approach, networks are used to map the relationships that exist among representations stored in memory. In this case, the term network is not referring to an artificial neural network as described above. In the Cognitive Network Science approach, the network consists of *nodes* that represent entities in a system, and *edges* that connect nodes that are related in some way. Cognitive Networks have been used to represent words in the mental lexicon that are semantically related [13], but this approach differs from the earlier semantic network in [2], because the links in cognitive networks are not labeled in the same way as they are in [2] (e.g., "IS-A" and "HAS" links to indicate that a fire engine IS-A type of vehicle and HAS the color red). Rather, the connections between nodes in cognitive networks typically represent a single type of relationship, such as the words being semantic associates of each other [13]. Cognitive networks such as this can also be used to represent other types of information and relationships among words, such as words that are phonologically related, as in Figure 1 [14].

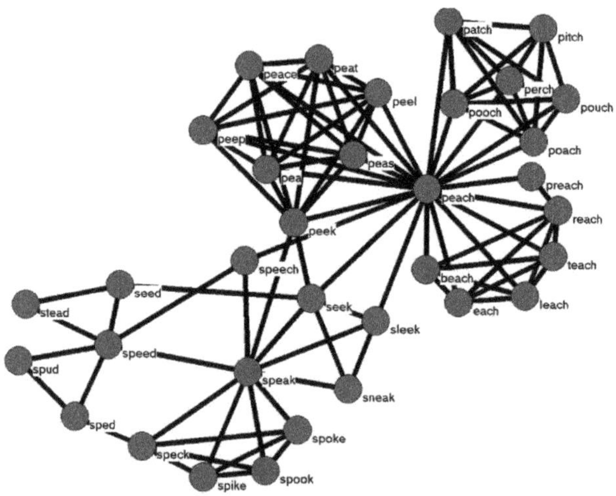

Figure 1. An example of a phonological network where nodes represent words in the mental lexicon, and edges connect words that are phonologically similar to each other (based on the addition, deletion, or substitution of a phoneme in one word to form another word). Phonological similarity can be defined in other ways as well. This network represents the 2-hop neighborhood of the word speech; a 2-hop network contains the neighbors of a word, and the neighbors of the neighbors.

Important to the Cognitive Network Science approach is the fact that the way in which these representations are organized or structured in memory influences how effectively and efficiently processes operate in the system [15,16]. That is, two networks with the same number of nodes and the same number of connections that are just connected in different ways in the two networks will have drastically different outcomes for a simple search algorithm [17]. This central tenet of the Cognitive Network Science approach contrasts with the semantic network of [2] and the artificial neural network approaches, which do not make this assumption, nor measure the structure of their respective types of "networks" in the manner described below.

The structure of a cognitive network can be measured at multiple scales: micro, macro, and meso. The micro-scale refers to measures of individual nodes in the network. Macro-scale measures assess the whole network. At the meso-scale, measures are made of subsets of nodes in the network. Because the structure of a network can influence processing, it is important to measure a cognitive network at all three scales, and to examine how the structure at each scale might influence cognitive processing.

The results of a number of behavioral experiments using conventional psycholinguistic tasks in laboratory settings have shown that certain network structures at various scales of the phonological network influence the production, recognition, and learning of spoken words in English. For example, the experiments in [18] considered a micro-scale measure, the (local) clustering coefficient, and how it influenced spoken word recognition (see also [19–22]). At the macro-scale, experiments by [23] examined how the location of words in the giant component (i.e., the largest group of connected nodes in a network) or in "lexical islands" (i.e., smaller groups of words that are connected to each other, but not to words in the giant component) of the phonological network influenced spoken word recognition. Finally, at the meso-scale, [24] found that a set of words in key positions, whose removal would disconnect the network, tended to be recognized more quickly than foil words that were similar to the keywords in a variety of lexical characteristics.

Given that the structure of the network influences processing, behavioral studies as well as a computer simulation with an artificial neural network—namely the TRACE model [7]—further demonstrated the importance of considering how nodes in a network are organized [18]. TRACE has been described as "... arguably the most successful model of spoken word recognition (SWR) to date" [25] (pg. 19). Indeed, as of 13-NOV-2021, the paper by [7] was cited over 3650 times (as per Google Scholar).

TRACE is a localist artificial neural network that contains processing units organized into three layers: (1) units representing acoustic–phonetic-like features, (2) units representing phonemes, and (3) units representing words. The units in each layer are excited or inhibited based on how well they match the speech input that is presented to the model. For more details about the TRACE model, we refer the reader to the original work [7], and to the more recent implementation of TRACE, dubbed jTRACE [25], which is used in the simulations reported below.

Twenty-eight monosyllabic words with three phonemes with higher clustering coefficients and 28 monosyllabic words with three phonemes with lower clustering coefficients were selected by [18] from the *initial_lexicon* that was used in the original simulations of TRACE. Using the default parameters, the model ran for 180 time-cycles [18]. At the end of the 180 time-cycles, the difference in the maximum activation levels for words with higher (*mean* = 0.55, *SD* = 0.010) compared to lower clustering coefficient (*mean* = 0.55, *SD* = 0.004), was not statistically significant (F (1,54) = 2.012, p = 0.16; as reported in [18]).

When the maximum activation levels were reached was also examined [18], and the difference in the number of time-cycles required to reach maximum activation also was not statistically significant (F (1,54) = 1.294, p = 0.26). As reported in [18], words with high clustering coefficient reached maximum activation on average in the 105th cycle (*SD* = 16.28), and words with a low clustering coefficient reached maximum activation on average in the 99th cycle (*SD* = 17.98). In combination with the behavioral data that they obtained, [18] viewed the inability of TRACE to simulate the results of their behavioral experiments as an indication that the structure of the phonological lexicon is in fact important to consider in models of spoken word recognition. Specifically, as suggested by the Cognitive Network Science approach, the structure of the phonological lexicon influences lexical processing.

Despite the success of the cognitive network approach in accounting for certain aspects of spoken word recognition (and other language-related and memory processes) this approach has been criticized because " … these networks do not 'do' anything; they have no function" [26] (pg. 16). One could argue that what cognitive networks "do" is capture in their structure certain regularities and relationships among entities in the world. By adding a simple process such as a random walk or the diffusion of activation across the network, one can examine how the structure of the network at multiple scales might influence cognitive processing.

The three behavioral experiments described above, which demonstrated that human performance in language-related tasks is influenced by structural characteristics at various scales in the phonological network, were simulated in the leading model of spoken word recognition—TRACE [7] (more recently implemented in Java as jTRACE by [25])—and on a cognitive network model based on the phonological network of [14]. If the structure at various scales of the phonological network influences processing, as claimed in the Cognitive Network Science approach, then we expect the phonological network model to qualitatively replicate the results of the three behavioral experiments described above. Further, given that the way in which words are connected to each other (i.e., how the lexicon is structured) is not considered in TRACE, we expect TRACE to fail to qualitatively replicate the results of the three behavioral experiments described above, as it did in [18].

In the Cognitive Network Science approach, edges are used to capture some sort of relationship between nodes resulting in a network that maps the structural organization of information in memory. Processing in these structural models can be modeled by either a random walk [27] or the diffusion of activation—akin to spreading activation—across the network [28]. An *R* package called *spreadr* has recently been created that can diffuse activation across a network provided by the user over a range of timesteps, initial activation levels, etc. [29]. We used *spreadr* in the simulations that follow to diffuse activation across the network from [14], allowing us to examine if a cognitive network can account for the results from the three behavioral experiments that previously examined the influence of the structure of the phonological network at various scales on human language processing.

2. Simulation 1: Clustering Coefficient

The first study to demonstrate that one of the network structures observed in [14] influenced the performance of humans in a conventional psycholinguistic task was reported in [18]. In that study it was observed that the micro-scale measure known as the (local) clustering coefficient influenced the accurate identification of words presented in noise in a perceptual identification task. Informally and in the context of a phonological network, the local clustering coefficient refers to the extent to which phonological neighbors of a given word are also neighbors of each other (see [18] and others for a more formal definition of clustering coefficient). As seen in Figure 2, the words *badge* and *log* in the middle of the two networks represent the target words, with the same number of phonological neighbors encircling each target word. The word *badge* has a higher clustering coefficient than the word *log*, because the phonological neighbors of *badge* tend to be neighbors of each other to a greater extent than the phonological neighbors of *log*.

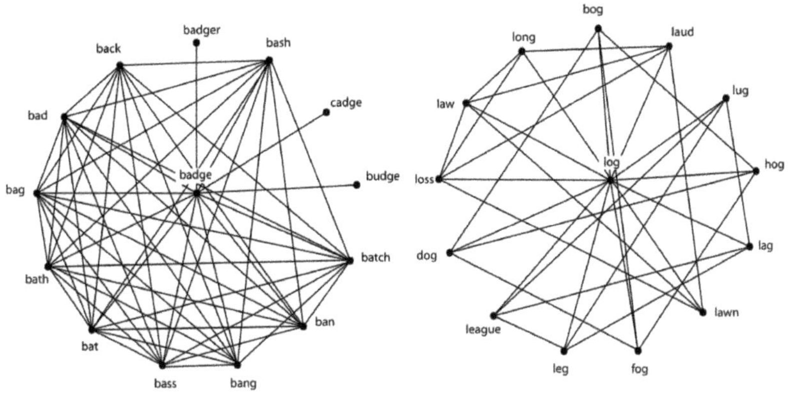

Figure 2. Although both words have the same number of phonological neighbors, the left panel represents a word with a higher clustering coefficient (*badge*), whereas the right panel represents a word with a lower clustering coefficient (*log*). That is, there is a difference in the extent to which the neighbors of each word are also neighbors of each other.

In Experiment 1 of [18] it was found that participants correctly identified words, such as *log*, with lower clustering coefficients more often (72%) than words such as *badge*, with a higher clustering coefficient (58%). Better performance for words with a low clustering coefficient was also obtained in Experiment 2 using the auditory lexical decision task, another conventional and widely used task in psycholinguistics (see also [19–21]).

To account for the results in [18], it was suggested that activation would initially spread from the target word to the phonologically related words, and from those words to other words that were phonologically related, and so on. In the case of words with a lower clustering coefficient, the activation would tend to disperse to the rest of the network, allowing the target word to "stand out" from the background of partially activated phonological neighbors, resulting in rapid and accurate retrieval from the lexicon. In the case of words with a higher clustering coefficient, the spreading activation would recirculate among the highly interconnected phonological neighbors, resulting in the target word being "buried" in the background of partially activated phonological neighbors, and therefore slow and less accurate retrieval from the lexicon. In other words, the micro-structure of the phonological lexicon influenced lexical processing.

Although the mechanism proposed in [18] to account for their results was based on computational work performed in other domains of network science, the model they put forward at the time was a verbal model, which have well-known shortcomings compared to computer simulations [30]. Subsequent computer simulations [28,29], however, showed that activation diffusing across 2-hop networks (such as the network displayed in Figure 1) of a different set of stimulus words successfully simulated the behavioral results originally observed in [18], substantiating the original verbal model and demonstrating further that the structure of the phonological lexicon influences lexical processing.

The *initial_lexicon* often used in the TRACE model has 211 words that contain sounds from a restricted set of phonemes, and that have a frequency of occurrence of 20 or more per million in [31]. One could arguably call *initial_lexicon* a "toy" lexicon rather than a lexicon representative of a typical speaker. Note that simulations on jTRACE have used a larger lexicon (*biglex*) of 907 words [32]. Although slightly larger, this lexicon is also not representative of the lexicon of a typical speaker. Rather than using the "toy" lexicon in TRACE or a subnetwork to model the lexicon as in previous simulations examining the influence of clustering coefficient on processing [18], in the present simulations both TRACE and the network model had as their lexicon the 19,340 words that were examined in an initial network analysis of the phonological lexicon [14]. Using the same lexicon not only makes comparison between the two different models equivalent, but the use of a large lexicon also tests if the two approaches can successfully scale up to a lexicon that is more realistic in size and composition to a human lexicon. Granted, estimates of the number of words known by the average person vary widely, but the number of words in the lexicon used in the present simulations is several orders of magnitude larger than the size of the lexicons used in previous computer simulations, arguably making for a more realistic and computationally challenging test of the two types of models.

2.1. Materials and Methods

This study was not preregistered. The stimuli used in all of the simulations are listed in Appendix A. The data from the simulations are available upon request from the first author.

jTRACE: The lexicon in the present simulation and the simulations that follow consisted of the 19,340 words in the lexicon examined in [14]. We modified the phonemes and phonetic features in the model to accommodate all of the phonemes that were found in the words in the larger lexicon. Aside from the new lexicon, phonemes, and phonetic features, the default parameters and settings were used for all of the simulations reported here (except Simulation 3b).

Appendix A shows the 76 stimulus words from Experiment 1 of [18] that were presented to jTRACE, which was allowed to process each word for 100 timesteps (N.B., the default setting in jTRACE is 99 timesteps). Although 180 timesteps were used in [18] maximum activation was achieved at approximately 100 timesteps, so we used this smaller number of timesteps in the present simulations to reduce computational burden, thereby accelerating data collection. After 100 timesteps had elapsed we examined the 10 most-activated competitors to obtain the activation level for each of the stimulus words. Although the word with the highest activation value is typically considered to be the word that has been retrieved, we documented the activation level of the stimulus word, even if it was not the most active word in the competitor set. If the stimulus word was not among the 10 most activated competitors, then an activation value of 0 was assigned.

spreadR: The lexicon in the present simulation and the simulations that follow consisted of the 19,340 words in the lexicon examined in [14]. This is the same lexicon used in the jTRACE simulations as well. As reported in [33], the network formed from the lexicon contained 19,340 nodes with 31,267 connections placed between nodes if the words differed by the addition, deletion, or substitution of a single phoneme. The giant component of the resulting network contained 6508 (34%) nodes; 10,265 (53%) of the nodes were isolates (i.e., lexical hermits with degree = 0), and the remaining 2567 (13%) of the nodes were found in smaller components (i.e., lexical islands).

The 76 stimulus words from Experiment 1 in [18] were presented to *spreadr* [29] with the following settings for the various parameters in the model. An initial activation value of 20 units was used for each stimulus word in the present simulation. Although *activation* = 100 units in the simulations reported in [28], this value is arbitrary. A smaller value was selected in the present simulations to reduce computational burden, thereby accelerating data collection.

Decay (d) refers to the proportion of activation lost at each time step. This parameter ranges from 0 to 1, and was set to 0 in the simulations reported here (except Simulation 3b) to be consistent with the parameter settings used in [28].

Retention (r) refers to the proportion of activation retained in a given node when it diffused activation to other nodes connected to it. This value ranges from 0 to 1, and was set to 0.5 in the simulations reported here. In [28] values ranged from 0.1 to 0.9 in increments of 0.1. Because the various retention values in [28] produced comparable results across retention values, we selected in the present simulations a single, mid-range value (0.5) for the retention parameter in order to reduce the computational burden, thereby accelerating data collection.

The suppress (s) parameter in *spreadr* will force nodes with activation values lower than the selected value to activation = 0. It was suggested that when this parameter is used a very small value (e.g., $s < 0.001$) should be used [29]. In the present simulations suppress = 0 in order to be consistent with the parameter settings used in [28].

Time (t) refers to the number of time steps that activation diffuses or spreads across the network. In [28] $t = 10$; however, in the present simulations $t = 5$. A smaller value was selected in the present case because as shown in Figure 3 of [29], activation values reach asymptote at approximately 5 timesteps, making additional timesteps uninformative. Further, as shown in the hop-plot depicted in Figure 2 in [34] approximately 50% of the network has been reached by traversing on average 5 connections (i.e., hops) in every direction from a given node, suggesting that the network has been sufficiently saturated. We selected in the present simulations a smaller value ($t = 5$) for the time parameter in order to reduce the computational burden, thereby accelerating data collection. At the end of 5 timesteps we documented the activation level of each of the stimulus words.

2.2. Results

Given the variety of dependent measures used in the various behavioral experiments that we attempted to simulate in the present study, and the different activation levels in TRACE and the cognitive network model, we attempted in the simulations reported here to replicate only qualitatively the findings from each of the behavioral experiments. For both jTRACE and *spreadr*, larger activation values correspond to better performance in the behavioral tasks (e.g., faster reaction times, more accurate responses, etc.).

In Experiment 1 in [18], words with a lower clustering coefficient were identified more accurately than words with a higher clustering coefficient when presented in noise in a perceptual identification task. In the cognitive network model implemented on *spreadr*, we found that words with a lower clustering coefficient had higher activation levels (*mean* = 1.28 units; *sd* = 0.26) indicating they were identified more accurately than words with a higher clustering coefficient (*mean* = 1.13 units; *sd* = 0.09). An independent samples t-test shows that this difference is statistically significant (t (74) = 3.29, $p = 0.0015$).

For jTRACE, activation levels could only be obtained for 2 of the 38 words with higher clustering coefficient (*bath* and *wire*), and no activation levels could be obtained for the 38 words with lower clustering coefficient. For the two words from the higher clustering coefficient condition, both words were the most active items in the candidate set, indicating that they had been correctly retrieved from the lexicon. For the remaining 74 words, the stimulus word was not among the 10 most-active candidates that emerged after 100 timesteps, and was therefore assigned an activation value of zero.

2.3. Discussion

The results of the simulation of Experiment 1 in [18] show that the cognitive network model implemented in *spreadr* was able to qualitatively replicate the results obtained in [18]. Specifically, words with lower clustering coefficient were identified more accurately (as indicated by higher activation levels in *spreadr*) than words with higher clustering coefficient. This result not only replicates the behavioral study reported in [18], but also replicates the simulations performed by [28,29] on 2-hop networks using slightly different parameter settings. Given that the cognitive network model successfully replicated the results in [18], one could argue that the structure among the words in the lexicon is important, and may indeed influence processing (lexical retrieval in this case).

Replicating results in a simulation with different parameter settings is one way of qualitatively assessing the global performance of a model [35], making the present simulation with *spreadr* more than a simple replication of previous simulations or behavioral studies. Rather, even though the cognitive network model appears simple on the surface, the present results using different parameter settings in *spreadr* help us better understand the complex behaviors of the model.

One of the major differences between the previous and the present simulation is the significantly larger size of the lexicon/phonological network used in the present simulations. Replicating previous results with a large lexicon suggests that the principles found in the cognitive network approach scale up to a more realistically sized vocabulary.

In contrast, using a more realistically sized lexicon with jTRACE in the present simulation led to performance that was significantly worse than the previous simulation of jTRACE reported in [18] using the toy lexicon often used in TRACE simulations (i.e., *initial_lexicon*). In the previous simulation of jTRACE in [18], the model was able to successfully retrieve from the toy lexicon all 56 stimulus words that varied in clustering coefficient (as measured in the network created from the words in *initial_lexicon*). However, TRACE was not able to differentially retrieve the stimulus words based on their clustering coefficient, thus failing to simulate the behavioral results reported in [18].

In the present simulation, in which jTRACE had a more realistic (i.e., not just high-frequency words) and realistically sized lexicon and was given the task of retrieving the 76 stimulus words used in Experiment 1 in [18], jTRACE was only able to retrieve 2 of the 76 stimulus words, making it difficult to assess whether the model could replicate the results reported in [18]. Is the failure of jTRACE indicative that models of spoken word recognition that "ignore" the structure among words in the lexicon do so at their own peril? Is the performance of TRACE in this case indicative that the representations and processes implemented in the model are problematic, incorrect, or simply do not scale-up to a more realistically sized lexicon?

Perhaps the poor performance of TRACE is just a computational/engineering limitation? Indeed, a new model of spoken word recognition called TISK has been proposed that uses computationally more efficient time invariant string kernels to represent incoming speech input [36]. String kernels are commonly used in machine learning applications to represent sequences of symbols. As noted in [36] (page 4): "To our knowledge, however, there have been no published investigations of string kernels in the domain of spoken word recognition." Although a model such as TISK may indeed be more computationally efficient than TRACE, it is not clear what such engineering approaches say about human performance or cognitive processing (see also [37,38]).

3. Simulation 2: Giant Component/Islands

To examine how the organization of representations in memory at the macro-scale influence cognitive processing we simulated the findings in [23], where words in lexical islands were retrieved more quickly in a naming and a lexical decision task than words located in the giant component. The giant component refers to the largest group of connected nodes in a network. Lexical islands refer to smaller groups of words that are connected to each other, but not to words in the giant component. ("Lexical islands" are referred to simply as "components" in the field of network science.)

3.1. Materials and Methods

The same methods and parameter settings used in Simulation 1 for jTRACE and *spreadr* were used in the present simulation. In the present simulation the 96 words used in Experiments 1 and 2 in [23] were presented to jTRACE and *spreadr* (see Appendix A for the words). Forty-eight of the words were found in the giant component, and the remaining words were found in other components/lexical islands in the phonological network.

3.2. Results

It was reported in [23] that words located in lexical islands were retrieved more quickly in a naming and a lexical decision task than words located in the giant component of the phonological network. For the cognitive network model implemented in *spreadr*, we found that words located in lexical islands had higher activation levels (*mean* = 5.98 units; *sd* = 2.09) indicating that they were retrieved more quickly than words located in the giant component (*mean* = 3.89 units; *sd* = 1.56). An independent samples t-test shows that this difference is statistically significant (t (94) = 5.56, p = 0.0001).

For jTRACE, activation levels could only be obtained for 2 of the 48 words located in the lexical islands (*beckon* and *lizard*), and no activation levels could be obtained for the 48 words located in the giant component. For the two words located in the lexical islands, one word was the most active item in the candidate set (indicating that it had been correctly retrieved from the lexicon), and the other word was simply among the 10 most-active candidates, but was not the most active candidate. For the remaining 94 words, the stimulus word was not among the 10 most-active candidates that emerged after 100 timesteps, and was therefore assigned an activation value of zero.

3.3. Discussion

The results of the present simulation show that the cognitive network model was able to qualitatively replicate the results obtained in Experiments 1 and 2 in [23]. Specifically, words located in lexical islands were retrieved more quickly in a naming and a lexical decision task (as indicated by higher activation levels in *spreadr*) than words located in the giant component of the phonological network (see also [39]). As in Simulation 1, TRACE did not recognize most of the stimulus words, making it difficult to assess if TRACE can replicate the results of [23].

Given the success of the cognitive network model, the present result may again suggest that the structure of the lexicon has an important influence on processing. Indeed, the structure of the phonological network is responsible for the higher activation levels obtained for the words in the present simulation compared to the activation levels obtained for the words in Simulation 1. Recall that lexical islands are groups of words that are connected to each other, but not connected to the giant component. In the giant component there are many more words for activation to spread to [34], resulting in less activation remaining in the target words in the giant component. In the lexical islands, however, which are smaller than the giant component, the activation will spread among the words in the island, but because there is nowhere else to spread to, activation will remain trapped in the island, resulting in relatively higher activation levels for the words in the present simulation compared to the activation levels obtained in Simulation 1.

4. Simulation 3a: Key Players

To examine structure at the meso-scale of the phonological network we simulated the results of [24], who examined how a set of words in "key" positions in the network might influence lexical processing. When asked to identify a node in a "key" position in the network in Figure 3, many people select node 1, because it is connected to many other nodes in the network. In network science terms, node 1 has high degree centrality. In contrast, the Keyplayer algorithm developed by [40] would identify node 8 as being in a "key" position in this network, because when node 8 is removed from the network the network becomes disconnected, forming two smaller components.

It was reported in [24] that a set of words in key positions (such as node 8 in Figure 3), whose removal would disconnect the network, tended to be responded to in the lexical decision task used in Experiment 3 more quickly than foil words. Foil words were similar to the set of keywords in word frequency, neighborhood density, word length, and a variety of other lexical characteristics; they just were not located in those key positions in the network.

Because of their strategic position in the network, it was suggested that words in those key positions would be indirectly and partially activated more often than words not in key positions when nearby words were retrieved [24]. Over time that indirect and partial activation from nearby words might, for example, lower the activation threshold or raise the resting activation level of keywords more than foils, making the keywords easier to retrieve than comparable words that were not in those key positions.

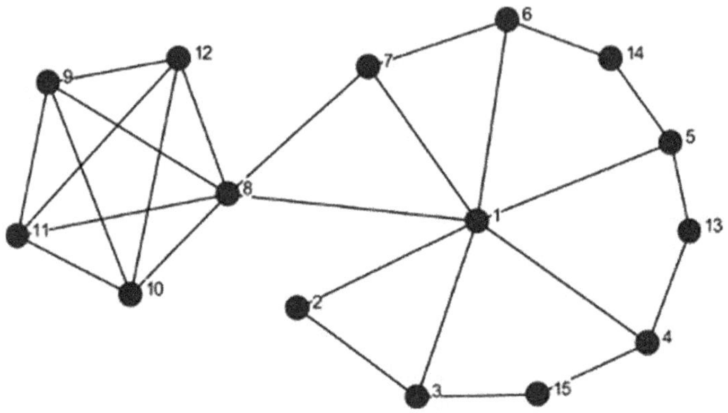

Figure 3. Node 8 is a key player in this network because the removal of that node results in the disconnection of the network (i.e., instead of there being the single component depicted above, two smaller components are formed).

4.1. Materials and Methods

The same methods and parameter settings used in the previous simulations for jTRACE and *spreadr* were used in the present simulation. In the present simulation, the 50 words used in the three experiments reported in [24] (see Appendix A) were presented to jTRACE and *spreadr*. Twenty-five of the words were in key positions, and the remaining words were referred to as foil words. As reported in [24], the foil words were comparable to the key words in word length, subjective familiarity, word frequency, neighborhood density, neighborhood frequency, phonotactic probability, the duration of the stimulus sound files, clustering coefficient, and closeness centrality. All of the words were found in the giant component of the phonological network.

4.2. Results

It was reported in [24] that a set of words in key positions (such as node 8 in Figure 3), whose removal would disconnect the network, tended to be responded to more quickly and accurately than foil words. For the cognitive network model (implemented on *spreadr*) we found that key words had activation levels (*mean* = 2.65 units; *sd* = 0.82) that were statistically indistinguishable from the foil words (*mean* = 2.63 units; *sd* = 1.46; t (48) = 0.05, $p = 0.9544$).

For jTRACE, activation levels could only be obtained for 2 of the 25 key words (*amend* and *auricle*), and for 4 of the 25 foil words (*album, aloft, attest*, and *party*). For the two key words both stimulus words were the most active item in the candidate set (indicating that they had been correctly retrieved from the lexicon). For the four foil words, two were the most active item in the candidate set, and two were simply among the 10 most-active candidates (but were not the most active candidate). For the remaining 44 words, the stimulus word was not among the 10 most-active candidates that emerged after 100 timesteps, and was therefore assigned an activation value of zero.

4.3. Discussion

As in the previous simulations, TRACE did not recognize most of the stimulus words, making it difficult to assess if TRACE can replicate the results of [24]. We note, however, that TRACE performed better in this simulation than in the other simulations, successfully retrieving six words compared to two words in Simulation 1 and two words in Simulation 2.

Although it was found in [24] that words in key positions were responded to more quickly than foil words, the cognitive network model with the same parameters as used in the previous simulations was not able to simulate that finding. Recall that a verbal model was proposed in [24] that suggested that words in key positions would be indirectly and partially activated more often than words not in key positions when nearby words were retrieved. Over time that indirect and partial activation might, for example, lower the activation threshold or raise the resting activation level of key words, making them easier to retrieve than comparable words that were not in those key positions. Several examples in the literature that demonstrated that partial activation of competitors can affect subsequent processing were discussed in [24], but the observed effects were not computationally modeled.

Our attempt in the present simulation to model the effects observed in [24] overlooked the crucial mechanism of partial and indirect activation modifying subsequent ease of lexical access. Because *spreadr* simply performs a single retrieval event and does not have a mechanism in it to allow for previous retrievals to affect subsequent retrievals, it should not be surprising that the cognitive network model implemented with the current set of parameters in *spreadr* was not able to reproduce the results observed in [24] with human listeners. Further, given all of the lexical variables that were comparable in the key and foil words, it should not be surprising that *spreadr* retrieved both sets of words with comparable (and statistically indistinguishable) ease. In the next simulation, we manipulated one of the other parameters in *spreadr* to try to mimic the changes that occur to keywords over time that were proposed in [24].

5. Simulation 3b: Key Players Manipulating Decay in *spreadr*

Recall that it was suggested in [24] that words in key positions in the phonological network would be indirectly and partially activated more often than words not in key positions when nearby words were retrieved. Over time indirect and partial activation from nearby words might, for example, lower the activation threshold or raise the resting activation level of keywords more than foils, making the keywords easier to retrieve than comparable words that were not in those key positions.

Another alternative not discussed in [24] is that previous or partial activation of a node might also influence subsequent activations of that node not by directly "strengthening" the node (i.e., lowering the threshold, or raising the resting activation level), but by "strengthening" the connections to the node. Indeed, such a mechanism is described in Node Structure Theory (NST), a model of language processing proposed in [41]. In NST the connections between nodes become stronger or more efficient with use, enabling the rate and amount of priming transmitted across the connections to increase over time. (Note that "priming" in NST is akin to spreading activation in other types of models.) It is this mechanism that allows NST to account for the well-known effects of the frequency of occurrence of a word in language processing.

To alter the efficiency of the connections in the phonological network, thereby affecting the rate and amount of activation that diffuses through the network, we decided to manipulate the *decay* (*d*) parameter in *spreadr*. The *d* parameter determines the proportion of activation that is lost at each time step. More efficient (or stronger) connections should lose a small amount of activation at each time step, whereas less efficient (or weaker) connections should lose a larger amount of activation at each time step. This parameter ranges from 0 to 1, and was set to 0 in the previous simulations to be consistent with the parameter settings used in [28]. In the present simulation we manipulated d in an attempt to vary the efficiency of the connections for the foil and key words, similar to the mechanism in NST put forward in [41]. Based on the argument in [24] that keywords become "stronger" than foils over time, we set in the present simulation $d = 0.1$ for the keywords and $d = 0.3$ for the foils, but the rest of the parameters remained as they were in the previous simulations.

Given that we changed a parameter in *spreadr*, we decided to try a different parameter setting for TRACE as well. The performance of TRACE across the three previous simulations was best in Simulation 3a, with six words being successfully retrieved from the lexicon. That level of performance will provide us with a reasonable baseline to allow us to determine if different parameters in TRACE would increase or decrease the number of words it successfully retrieved from the lexicon (and perhaps even allow us to evaluate whether TRACE can account for the behavioral finding being simulated). As noted in endnote 2 in [25] (page 30), there is some risk involved in changing parameters in TRACE:

> *As Frauenfelder once put it in a conference presentation [42], the large number of parameters in TRACE are in "delicate equilibrium." Caution must be exercised when changing any parameters, since a small change in one parameter may result in large changes in the model's behavior, and one cannot be sure that the model will successfully perform simulations conducted with other parameter settings.*

Heeding this warning, we therefore decided to change just one parameter in jTRACE; namely, we turned off lexical feedback. This parameter was also turned off in the simulations reported in [18] to test if a different model of spoken word recognition—Shortlist [8], which eschews feedback—could account for the behavioral results that they found (and which were replicated in Simulation 1 reported here). Like the TRACE simulation reported in [18], the Shortlist simulation successfully retrieved all of the words from the toy lexicon that was used, but did not have differential activation values for the words that varied in clustering coefficient.

We recognize that there is debate about the utility of lexical feedback in TRACE. For example, it was reported in [43] that reducing feedback from 0.030 to 0.025 improved performance with a larger lexicon of 977 words (referred to as *Biglex*), and that turning off lexical feedback sped recognition time for about half of the small set of words ($n = 21$) they examined. In contrast, it was reported in [32] that when a larger set of words ($n = 900$) was examined (without noise), 27% of the words were recognized more quickly without feedback, 57% were recognized more quickly with feedback, and 16% had equivalent retrieval times with and without feedback. It was also observed in [32] that feedback increased accuracy when increasing levels of noise were added to the input. Given that we are not adding noise to the input in the present simulation, and given the partial success of turning off lexical feedback reported in [18], we decided to examine if turning off lexical feedback might improve performance when TRACE has the much larger lexicon being used in the present simulations.

5.1. Materials and Methods

The same methods and parameter settings used in the previous simulations for jTRACE and *spreadr* were used in the present simulation, with the exception of *decay* (d) being manipulated in *spreadr*, and lexical feedback was now turned off in jTRACE. The 50 words used in the three experiments reported by [24] (see Appendix A) and in Simulation 3a were presented to jTRACE and spreadr in the present simulation. Twenty-five of the words were in key positions (and had the *decay* parameter, d, set to 0.1 in *spreadr*), and the remaining words were referred to as foil words (and had the decay parameter, d, set to 0.3 in *spreadr*) to mimic the change in processing efficiency that occurs over time proposed by [24].

5.2. Results

It was found in [24] that a set of words in key positions (such as node 8 in Figure 3), whose removal would disconnect the network, tended to be responded to more quickly and accurately than foil words. For the cognitive network model implemented in *spreadr*, we found that key words (with the *decay* parameter $d = 0.1$) had higher activation levels (*mean* = 1.57 units; *sd* = 0.49) indicating that they were retrieved more quickly than the foil words (with the *decay* parameter $d = 0.3$; *mean* = 0.44 units; *sd* = 0.25). An independent samples t-test shows that this difference is statistically significant (t (48) = 10.29, $p < 0.0001$).

For jTRACE with no lexical feedback, activation levels could be obtained for 3 of the 25 key words (*amend, auricle* (the same words retrieved in Simulation 3a), with the addition of *pallet*), and for 4 of the 25 foil words (*album, aloft, attest*, and *party*; the same words retrieved in Simulation 3a). For both the foil and key words, all of the words were the most active item in the candidate set. Instead of assigning zero activation to the remaining items, in this simulation we simply compared the mean activation values for the three key words (*mean* = 0.6585; *sd* = 0.006) to the mean activation values for the four foil words (*mean* = 0.6700; *sd* = 0.009) that jTRACE successfully retrieved. The difference in activation levels was not statistically significant (t (5) = 1.76, p = 0.14). Further, the direction of the difference was the opposite of what was predicted based on the behavioral results reported in [24].

5.3. Discussion

It was found in [24] that participants responded to words in key positions more quickly than foil words. They accounted for that result by suggesting that words in key positions would be indirectly and partially activated more often than words not in key positions when nearby words were retrieved. Over time that indirect and partial activation might, for example, lower the activation threshold of key words, raise the resting activation level of key words or, as suggested in NST [41], might strengthen or increase the efficiency of the connections between nodes for keywords, making keywords easier to retrieve than words that are not in those key positions.

To mimic the differences in connection efficiency as suggested in NST [41] we manipulated in this simulation the decay (d) parameter in *spreadr*. With the manipulation of d in the present simulation (compared to Simulation 3a) we now observed that words in key positions with more efficient/stronger connections were responded to more quickly than foil words with less efficient/weaker connections (as indicated by higher activation levels for keywords compared to foils). The result of Simulation 3b qualitatively replicates the behavioral result observed in [24].

We also manipulated a parameter in jTRACE in an attempt to improve the performance of the model. In this case, we turned off lexical feedback, which did improve performance. In the present simulation seven words were retrieved, compared to six words in Simulation 3a. Further, all of the words that were retrieved in the present simulation were actually the most active item in the candidate set (compared to only four of the six words being the most active item in the candidate set in Simulation 3a). Although there is some debate about whether feedback improves performance in TRACE (cf., [32,43]), in the present case turning off lexical feedback did improve the overall performance of the model with a larger set of phonetic features and phonemes, and a much larger lexicon. Although the overall performance of TRACE was improved by the manipulation of this parameter, the difference in the activation values of the words that were retrieved was not statistically different, and trended in the opposite direction to what was predicted based on the behavioral results reported in [24].

6. Conclusions

In the present study we simulated in TRACE [7,25] and a phonological network [14] using the R package *spreadr* [29] the results of three psycholinguistic experiments that examined how the structure of a phonological network at the micro-, macro- and meso-scale might influence lexical retrieval. At the micro-scale (Simulation 1), measuring the characteristics of individual nodes, we simulated the results in [18] examining the measure known as the (local) clustering coefficient, which measures the extent to which neighbors of a word are also neighbors of each other. At the macro-scale (Simulation 2), measuring the characteristics of the entire network, we simulated the results of [23] who looked at whether a word being located in the giant component or in a lexical island influenced lexical retrieval. At the meso-scale (Simulations 3a,b), which considers groups or subsets of nodes rather than individual nodes or the whole network, we simulated the results of [24] who looked at how key players in the network might influence lexical processing. Key players refer to a set of nodes whose removal from the network results in maximal disconnection of the network.

In Simulations 1 and 2 the cognitive network model qualitatively replicated the results observed in the psycholinguistic experiments, but TRACE was not able to successfully retrieve a sufficient number of words to assess the ability of this model to simulate the behavioral results. In Simulation 3a the cognitive network model was not able to successfully replicate the results observed in the psycholinguistic experiment. In this simulation, TRACE was able to successfully retrieve a larger number of words compared to the previous simulations, but still not enough words to assess statistically the ability of this model to simulate the behavioral results.

The failure of the cognitive network model in Simulation 3a led us to reconsider the mechanism proposed in a verbal model in [24]: previous activation and retrievals of nearby words can influence subsequent retrievals of the target word. Although verbal models are useful in the initial stages of a theory, many have written about the value of using formal, computational models to more precisely examine the representational and processing aspects of cognition [30]. Therefore, in Simulation 3b we manipulated another parameter in *spreadr*, namely the *decay* (*d*) parameter to mimic the changes that occur over time to the key words. When the keywords and foils had different values for the *d* parameter to model differences in the strength/efficiency of the connections to those words, we now observed a qualitative replication of [24] in the cognitive network model.

Given that we manipulated a parameter in *spreadr* in Simulation 3b, we decided to also manipulate a parameter in jTRACE to see if performance could be improved enough to assess the ability of the model to qualitatively replicate the results of [24]. In Simulation 3b, we turned off feedback from the word level to the phoneme level as had been carried out in [18] and in [43] (cf., [32]). Here we found that overall performance did improve enough to statistically analyze the activation values of the key and foil words. However, the difference in the activation values was not statistically different.

The poor performance of TRACE in the present set of simulations is troubling, especially given that [7] (p. 22) reported that the behavior of TRACE was qualitatively robust over a wide range of parameter values (with minor changes in the magnitude or timing of various effects when using different parameter settings). We grant that the default parameters in TRACE established with the original, very small lexicon and a limited set of phonetic features and phonemes may be optimal only under those conditions. We further grant that in the present situation with different and larger sets of phonetic features and phonemes, and a much larger lexicon, that the default parameters may be suboptimal. However, even shutting off lexical feedback as we did in Simulation 3b did little to change the performance of the model.

Others have discussed the importance of testing model performance across a range of parameter settings [35], and of assessing the scope of a model [44]. Here we simply note that the cognitive network model in Simulation 1 of the present study performed accurately with different parameter settings than used previously [28,29], suggesting that the performance of cognitive network models may not be as sensitive to a unique set of parameter settings as other types of models.

We believe there is much to learn about cognition by using formal, computational models [45] as long as realistic contexts rather than idealized or over-simplified settings are used in those models [46]. Recall that in the TRACE simulation reported in [18] the model successfully retrieved from the toy lexicon all of the stimulus words that varied in clustering coefficient (as determined by measuring the clustering coefficient of the 211 words in the *initial_lexicon*), but no difference as a function of clustering coefficient was observed. In the present simulations where TRACE and *spreadr* were given a more realistically sized lexicon of 19,340 words, TRACE retrieved such a small fraction of the stimulus words that statistical analyses could not be performed in most cases. In contrast, the cognitive network model was able to scale-up from smaller subnetworks (i.e., the 2-hop networks used in [28,29]) to a lexicon that was several orders of magnitude larger, demonstrating the robustness of the cognitive network approach in simplified and in more complex/realistic settings.

In addition to highlighting the importance of using formal, computational models to increase our understanding of cognition, the results of the present study suggest that future models of cognitive processing should consider how representations are organized in memory, and how that structure influences processing. The psycholinguistic experiments simulated in the present study demonstrated that the structure of a phonological network at multiple scales (micro, meso, and macro) influences various language processes. We believe the influence of structure on processing also applies to other areas of Cognitive Psychology [21]. Indeed, experiments from cognitive science, neuroscience, and linguistics demonstrate that humans are able to learn about the meso- and macro-scale of network representations in memory, and that those structures influence processing in a variety of cognitive tasks [47].

The present results suggest that what cognitive networks "do" is capture the regularities and relationships that exist among representations in memory. With the "smarts" of the system captured in the structure of the network (i.e., how the nodes are connected to each other), a much simpler processing algorithm—such as a random walk, the diffusion of activation across the network [27], or a mixture of random and directed walks [48]—may be sufficient to reproduce the behavior exhibited by humans in various tasks [49].

Cognitive networks may not be the only way to model the regularities and relationships that exist among representations in memory, and how that structure influences processing. Indeed, just as there are limits to what a network can model in other domains, there may be some limits to what cognitive networks can model [50]. Similarly, the richness and variety of cognition may be too complex to be captured by simple processes such as a random walk or diffusion of activation across the network.

In addition, the present simulations used a cognitive network that captured a "snap shot" of only the phonological lexicon of the "average" language user at one point in time. Advances in network science and in the application of networks to psychology are rapidly being made to address some of these limitations inherent in the present simulations. Work reported in [51] demonstrated that networks that grow over time can be used to provide insight into how typically developing and "late talking" children learn the meanings of new words. A similar approach has been used in [52] to capture changes/declines in semantic information in older adults. These studies also demonstrate that cognitive networks may hold much promise for increasing our understanding of various speech, language, and hearing disorders as well [39].

Cognitive networks need not be limited—such as the phonological network examined in [14]—to one type of representation or information. Work on multilevel networks, which enable researchers to look at, for example, a network of words with phonological relationships overlayed on a network of words with semantic relationships have increased our understanding of word-learning in children [53], and of acquired language disorders in adults [54].

Further, increasingly sophisticated network analyses are providing tools to track changes in behavior over time in an individual, rather than the average behavior of a group [55]. Such analysis techniques have significant implications for individualized- or personalized-treatment in a number of domains (e.g., psychopathology, speech-language-hearing disorders, etc.).

Although the TRACE I and TRACE II models accounted for a wide range of phenomena in speech perception and spoken word recognition [7], we find it troubling that the most successful model of spoken word recognition did not scale up to a more realistic lexicon. Further, as described above, we believe that the cognitive network approach has much potential to account for an increasing number of phenomena in speech perception and spoken word recognition, as well as other areas of Cognition. Furthermore, the cognitive network approach may not only account for the same phenomena in speech perception and spoken word recognition that the TRACE models can [56], but may also account for phenomena that TRACE and other contemporary models of spoken word recognition cannot account for, such as the influence that the structure of the lexicon at various scales has on processing.

Author Contributions: Conceptualization, M.S.V.; methodology, M.S.V.; software, G.J.D.M.; formal analysis, M.S.V.; writing—original draft preparation, M.S.V. and G.J.D.M.; writing—review and editing, M.S.V. and G.J.D.M.; supervision, M.S.V.; project administration, M.S.V. All authors have read and agreed to the published version of the manuscript.

Funding: This research received no external funding.

Institutional Review Board Statement: Not applicable.

Informed Consent Statement: Not applicable.

Data Availability Statement: The data from the simulations are available upon request from the first author.

Conflicts of Interest: The authors declare no conflict of interest.

Appendix A

Table A1. Stimulus words from Chan and Vitevitch (2009) that were used in Simulation 1.

High Clustering Coefficient	Low Clustering Coefficient
bash	beach
bath	bead
bib	beat
bull	bush
bug	boot
dot	dog
dig	dead
dish	deck
dug	debt
feel	fat
full	fell
foul	fate
gang	gas
gain	goat
gum	gull
call	cough
case	couch
lag	lock
leaf	log
leap	lose
lease	ledge
leave	lick
look	lip
lose	live
lull	lime
love	luck
math	miss
mall	merge
meal	mood
mouse	mile
perk	pass
pearl	purse
ring	rhyme
ripe	rise
seal	sauce
size	save
weak	word
wire	wide

Table A2. Stimulus words from Siew and Vitevitch (2016) that were used in Simulation 2.

Giant Component	Lexical Islands
brittle	banish
cartridge	beckon
ceiling	central
century	coffin
chapter	concede
colleague	concern
collect	confine
comic	consign
coroner	cunning
cumber	deafen
danger	domain
defend	felon
device	furnish
drench	gallop
dribble	happen
driven	lizard
facet	locus
filing	manage
grunt	margin
hamper	marriage
hardly	memory
knowledge	mission
languor	nervous
limber	nominee
magnet	notice
mention	partition
minute	peasant
mountain	permission
mustard	petition
panther	plaza
parable	portion
parcel	position
receive	radio
remind	regain
remit	remain
repeat	report
reverse	retail
rollick	retain
salvage	revolve
scant	service
scepter	siphon
spiral	soften
squid	solemn
straighten	taken
stutter	treasure
supposed	trophy
temple	village
temporal	warrant

Table A3. Stimulus words from Vitevitch and Goldstein (2014) that were used in Simulation 3a,b.

Keywords	Foils
Amend	Album
Auricle	Aloft
Bring	Attest
Colic	Brief
Defy	Cockney
Filing	Downy
Fish	Espy
Inurn	Firm
Leva	Feudal
Ling	Lave
Lion	Lighten
Milling	Manna
Misty	Mystic
Opine	Osprey
Over	Party
Packet	Pasty
Pallet	Pilot
Pocket	Poster
Polite	Rent
Scrawl	Rupee
Spring	Squirt
Tenet	Stilt
Tense	Test
Void	Torrid
Wrist	Vest

References

1. Casey, G.; Moran, A. The computational metaphor and Cognitive Psychology. *Ir. J. Psychol.* **2012**, *10*, 143–161. [CrossRef]
2. Collins, A.M.; Loftus, E.F. A spreading-activation theory of semantic processing. *Psychol. Rev.* **1975**, *82*, 407–428. [CrossRef]
3. Rumelhart, D.E.; McClelland, J.L. The PDP Research Group. Parallel Distributed Processing: Explorations in the Microstructure of Cognition. In *Volume I: Foundations & Volume II: Psychological and Biological Models*; MIT Press: Cambridge, MA, USA, 1986.
4. Rogers, T.T.; McClelland, J.L. Parallel Distributed Processing at 25: Further explorations in the microstructure of cognition. *Cogn. Sci.* **2014**, *38*, 1024–1077. [CrossRef]
5. Dell, G.S. A spreading-activation theory of retrieval in sentence production. *Psychol. Rev.* **1986**, *93*, 283–321. [CrossRef]
6. Levelt, W.J.M.; Roelofs, A.; Meyer, A.S. A theory of lexical access in speech production. *Behav. Brain Sci.* **1999**, *22*, 1–38. [CrossRef] [PubMed]
7. McClelland, J.L.; Elman, J.L. The TRACE model of speech perception. *Cogn. Psychol.* **1986**, *18*, 1–86. [CrossRef]
8. Norris, D. Shortlist: A connectionist model of continuous speech recognition. *Cognition* **1994**, *52*, 189–234. [CrossRef]
9. Norris, D.; McQueen, J.M.; Cutler, A. Merging information in speech recognition: Feedback is never necessary. *Behav. Brain Sci.* **2000**, *23*, 299–370. [CrossRef]
10. Siew, C.S.Q.; Wulff, D.U.; Beckage, N.M.; Kenett, Y.N. Cognitive Network Science: A review of research on cognition through the lens of representations, processes, and dynamics. *Complexity* **2019**, *2019*, 1–24. [CrossRef]
11. Vitevitch, M.S. *Network Science in Cognitive Psychology*; Routledge: London, UK, 2019.
12. Barabási, A.-L. *Network Science*; Cambridge University Press: Cambridge, UK, 2016.
13. Steyvers, M.; Tenenbaum, J.B. The large-scale structure of semantic networks: Statistical analyses and a model of semantic growth. *Cogn. Sci.* **2005**, *29*, 41–78. [CrossRef] [PubMed]
14. Vitevitch, M.S. What can graph theory tell us about word learning and lexical retrieval? *J. Speech Lang. Hear. Res.* **2008**, *51*, 408–422. [CrossRef]
15. Kleinberg, J.M. Navigation in a small world. *Nature* **2000**, *406*, 845. [CrossRef]
16. Latora, V.; Marchiori, M. Efficient behavior of small-world networks. *Phys. Rev. Lett.* **2001**, *87*, 198701. [CrossRef]
17. Karuza, E.A.; Kahn, A.E.; Thompson-Schill, S.L.; Bassett, D.S. Process reveals structure: How a network is traversed mediates expectations about its architecture. *Sci. Rep. UK* **2017**, *7*, 12733. [CrossRef]
18. Chan, K.Y.; Vitevitch, M.S. The Influence of the Phonological Neighborhood Clustering-Coefficient on Spoken Word Recognition. *J. Exp. Psychol. Hum.* **2009**, *35*, 1934–1949. [CrossRef]
19. Chan, K.Y.; Vitevitch, M.S. Network structure influences speech production. *Cogn. Sci.* **2010**, *34*, 685–697. [CrossRef]
20. Goldstein, R.; Vitevitch, M.S. The influence of clustering coefficient on word-learning: How groups of similar sounding words facilitate acquisition. *Front. Lang Sci.* **2014**, *5*, 01307. [CrossRef]

21. Vitevitch, M.S.; Chan, K.Y.; Roodenrys, S. Complex network structure influences processing in long-term and short-term memory. *J. Mem. Lang.* **2012**, *67*, 30–44. [CrossRef]
22. Vitevitch, M.S.; Ng, J.W.; Hatley, E.; Castro, N. Phonological but not semantic influences on the speech-to-song illusion. *Q. J. Exp. Psychol.* **2021**, *74*, 585–597. [CrossRef]
23. Siew, C.S.Q.; Vitevitch, M.S. Spoken word recognition and serial recall of words from components in the phonological network. *J. Exp. Psychol. Learn.* **2016**, *42*, 394–410. [CrossRef]
24. Vitevitch, M.S.; Goldstein, R. Keywords in the mental lexicon. *J. Mem. Lang.* **2014**, *73*, 131–147. [CrossRef]
25. Strauss, T.; Harris, H.; Magnuson, J. jTRACE: A reimplementation and extension of the TRACE model of speech perception and spoken word recognition. *Behav. Res. Methods* **2007**, *39*, 19–30. [CrossRef]
26. Brown, K.S.; Allopenna, P.D.; Hunt, W.R.; Steiner, R.; Saltzman, E.; McRae, K.; Magnuson, J.S. Universal features in phonological neighbor networks. *Entropy* **2018**, *20*, 526. [CrossRef] [PubMed]
27. Abbott, J.T.; Austerweil, J.L.; Griffiths, T.L. Random walks on semantic networks can resemble optimal foraging. *Psychol. Rev.* **2015**, *122*, 558–569. [CrossRef] [PubMed]
28. Vitevitch, M.S.; Ercal, G.; Adagarla, B. Simulating retrieval from a highly clustered network: Implications for spoken word recognition. *Front. Lang. Sci.* **2011**, *2*, 369. [CrossRef] [PubMed]
29. Siew, C.S.Q. *Spreadr*: An R package to simulate spreading activation in a network. *Behav. Res. Methods* **2019**, *51*, 910–929. [CrossRef] [PubMed]
30. Lewandowsky, S. The rewards and hazards of computer simulations. *Psychol. Sci.* **1993**, *4*, 236–243. [CrossRef]
31. Kucera, H.; Francis, W. *Computational Analysis of Present-Day American English*; Brown University Press: Providence, RI, USA, 1967.
32. Magnuson, J.S.; Mirman, D.; Luthra, S.; Strauss, T.; Harris, H.D. Interaction in Spoken Word Recognition Models: Feedback Helps. *Front. Psychol.* **2018**, *9*, 369. [CrossRef] [PubMed]
33. Vitevitch, M.S.; Castro, N.; Mullin, G.J.D.; Kulphongpatana, Z. *Using Cognitive Network Science and Computer Simulations to Examine Aphasia*; University of Kansas: Lawrence, KS, USA, 2021; submitted, Unpublished manuscript.
34. Vitevitch, M.S.; Goldstein, R.; Johnson, E. Path-Length and the Misperception of Speech: Insights from Network Science and Psycholinguistics. In *Towards a Theoretical Framework for Analyzing Complex Linguistic Networks*; Mehler, A., Blanchard, P., Job, B., Banish, S., Eds.; Springer: Berlin/Heidelberg, Germany, 2016.
35. Pitt, M.A.; Kim, W.; Navarro, D.J.; Myung, J.I. Global model analysis by parameter space partitioning. *Psychol. Rev.* **2006**, *113*, 57–83. [CrossRef] [PubMed]
36. Hannagan, T.; Magnuson, J.; Grainger, J. Spoken word recognition without a TRACE. *Front. Psychol.* **2013**, *4*, 563. [CrossRef]
37. Nenadić, F.; Tucker, B.V. Computational modelling of an auditory lexical decision experiment using jTRACE and TISK. *Lang. Cogn. Neurosci.* **2020**, *35*, 1326–1354. [CrossRef]
38. Taylor, J.E.T.; Taylor, G.W. Artificial cognition: How experimental psychology can help generate explainable artificial intelligence. *Psychon. B Rev.* **2021**, *28*, 454–475. [CrossRef] [PubMed]
39. Vitevitch, M.S.; Castro, N. Using network science in the language sciences and clinic. *Int. J. Speech-Language Pathol.* **2015**, *17*, 13–25. [CrossRef] [PubMed]
40. Borgatti, S.P. Identifying sets of key players in a network. *Comput. Math. Organ. Theory* **2006**, *12*, 21–34. [CrossRef]
41. MacKay, D.G. *The Organization of Perception and Action: A Theory for Language and Other Cognitive Skills*; Springer: New York, NY, USA, 1987.
42. Frauenfelder, U.H.; Content, A. Activation Flow in Models of Spoken Word Recognition. In Proceedings of the Workshop on Spoken Word Access Processes, Nijmegen, The Netherlands, 29–31 May 2000; pp. 79–82.
43. Frauenfelder, U.H.; Peeters, G. Simulating the Time Course of Spoken Word Recognition: An Analysis of Lexical Competition in TRACE. In *Localist Connectionist Approaches to Human Cognition*; Grainger, J., Jacobs, A.M., Eds.; Erlbaum: Mahwah, NJ, USA, 1998; pp. 101–146.
44. Cutting, J. Accuracy, scope, and flexibility of models. *J. Math. Psychol.* **2000**, *44*, 3–19. [CrossRef]
45. Farrell, S.; Lewandowsky, S. Computational models as aids to better reasoning in psychology. *Curr. Dir. Psychol. Sci.* **2010**, *19*, 329–335. [CrossRef]
46. Pinker, S.; Prince, A. On language and connectionism: Analysis of a parallel distributed processing model of language acquisition. *Cognition* **1988**, *28*, 73–193. [CrossRef]
47. Lynn, C.W.; Bassett, D.S. How humans learn and represent networks. *Proc. Natl. Acad. Sci. USA* **2020**, *117*, 29407–29415. [CrossRef] [PubMed]
48. O'Keeffe, K.P.; Anjomshoaa, A.; Strogatz, S.H.; Santi, P.; Ratti, C. Quantifying the sensing power of crowd-sourced vehicle fleets. *arXiv* **2020**, arXiv:1811.10744.
49. Hills, T.T.; Jones, M.N.; Todd, P.M. Optimal foraging in semantic memory. *Psychol. Rev.* **2012**, *119*, 431–440. [CrossRef] [PubMed]
50. Butts, C.T. Revisiting the Foundations of Network Analysis. *Science* **2009**, *325*, 414–416. [CrossRef] [PubMed]
51. Beckage, N.; Smith, L.; Hills, T. Small worlds and semantic network growth in typical and late talkers. *PLoS ONE* **2011**, *6*, e19348.
52. Dubossarsky, H.; De Deyne, S.; Hills, T.T. Quantifying the structure of free association networks across the lifespan. *Dev. Psychol.* **2017**, *53*, 1560. [CrossRef] [PubMed]
53. Stella, M.; Beckage, N.M.; Brede, M.; De Domenico, M. Multiplex model of mental lexicon reveals explosive learning in humans. *Sci. Rep. UK* **2018**, *8*, 2259. [CrossRef] [PubMed]

54. Castro, N.; Stella, M.; Siew, C.S.Q. Quantifying the interplay of semantics and phonology during failures of word retrieval by people with aphasia using a multiplex lexical network. *Cogn. Sci.* **2020**, *44*, e12881. [CrossRef] [PubMed]
55. Epskamp, S.; van Borkulo, C.D.; van der Veen, D.C.; Servaas, M.N.; Isvoranu, A.M.; Riese, H.; Cramer, A. Personalized Network Modeling in Psychopathology: The Importance of Contemporaneous and Temporal Connections. *Clin. Psychol. Sci.* **2018**, *6*, 416–427. [CrossRef] [PubMed]
56. Vitevitch, M.S.; Niehorster-Cook, L.; Niehorster-Cook, S. Exploring How Phonotactic Knowledge Can Be Represented in Cognitive Networks. *Big Data Cogn. Comput.* **2021**, *5*, 47. [CrossRef]

MDPI
St. Alban-Anlage 66
4052 Basel
Switzerland
Tel. +41 61 683 77 34
Fax +41 61 302 89 18
www.mdpi.com

Brain Sciences Editorial Office
E-mail: brainsci@mdpi.com
www.mdpi.com/journal/brainsci

www.ingramcontent.com/pod-product-compliance
Lightning Source LLC
Chambersburg PA
CBHW061813210926
51850CB00006B/695
9783036574134